Fatigue Damage Mechanism and Fatigue Life Prediction of Metallic Materials

Fatigue Damage Mechanism and Fatigue Life Prediction of Metallic Materials

Editors

Haitao Cui
Qinan Han

MDPI

Basel • Beijing • Wuhan • Barcelona • Belgrade • Novi Sad • Cluj • Manchester

Editors

Haitao Cui
College of Energy and Power Engineering
Nanjing University of Aeronautics and Astronautics
Nanjing
China

Qinan Han
College of Energy and Power Engineering
Nanjing University of Aeronautics and Astronautics
Nanjing
China

Editorial Office
MDPI
St. Alban-Anlage 66
4052 Basel, Switzerland

This is a reprint of articles from the Special Issue published online in the open access journal *Metals* (ISSN 2075-4701) (available at: www.mdpi.com/journal/metals/special_issues/fatigue_damage_fatigue_life_metals).

For citation purposes, cite each article independently as indicated on the article page online and as indicated below:

Lastname, A.A.; Lastname, B.B. Article Title. *Journal Name* **Year**, *Volume Number*, Page Range.

ISBN 978-3-0365-9890-1 (Hbk)
ISBN 978-3-0365-9889-5 (PDF)
doi.org/10.3390/books978-3-0365-9889-5

Contents

Preface

The fatigue of metallic materials stands as a pivotal challenge in engineering applications. Addressing this micro-macro multi-scale issue necessitates interdisciplinary expertise, involving mechanics, materials science, mechanical engineering, numerical computing, and related fields. Exploring the fatigue damage mechanism and fatigue life prediction of metallic materials is of paramount importance. This reprint includes eleven articles that explore relevant topics, covering industrial materials such as steels, superalloys, aluminum alloys, etc. These articles are intended for academic researchers, engineers, and students seeking a comprehensive understanding of recent advancements in this field. We wish to convey our appreciation to all the authors for their valuable contributions, and we also acknowledge the dedicated efforts of the editorial staff for their time and commitment to the development of this reprint.

Haitao Cui and Qinan Han
Editors

metals

MDPI

Editorial

Fatigue Damage Mechanism and Fatigue Life Prediction of Metallic Materials

Haitao Cui * and Qinan Han *

College of Energy and Power Engineering, Nanjing University of Aeronautics and Astronautics, Nanjing 210016, China

* Correspondence: cuiht@nuaa.edu.cn (H.C.); hanqn@nuaa.edu.cn (Q.H.)

1. Introduction and Scope

Metallic materials are crucial in engineering applications and often subjected to complex loads and extreme environments, with fatigue being one of the key problems. At present, the fatigue of metallic materials remains a challenging problem that still needs to be addressed in engineering applications. The behavior of metallic materials under fatigue loading is a micro–macro multi-scale issue encompassing microscopic defect and damage evolution, the formation of small cracks and their coalescence, propagation, and microstructural interaction, ultimately leading to macroscopic component fracture. Therefore, the study of fatigue of metallic materials typically requires interdisciplinary knowledge of advanced mechanics, materials science, mechanical engineering, numerical computing, etc.

In addition, fatigue is affected by many internal and external factors, including material processing and microstructures, geometric size and configuration, load, environment, etc. It is important to study the fatigue mechanism and fatigue behavior considering these factors. Moreover, the research into fatigue life prediction and fatigue performance optimization methods can provide technical understanding of and solutions for engineering applications. This Special Issue aims to report experimental, theoretical, and numerical studies related to the fatigue damage mechanism and fatigue life prediction of metallic materials.

2. Contributions

Eleven articles are included in this Special Issue, including one review article and ten research articles. The research subjects encompass industrial materials, such as superalloys, steels, and aluminum alloys, and span multi-scales, ranging from materials to large-scale components.

Among these articles, four focused on steel, which is one of the most significant metallic materials with extensive industrial applications. Christodoulou et al. [1] studied crack initiation behavior and life using combined experimental and numerical methods. Life prediction was implemented based on the local strain and fracture mechanics parameter. Su et al. [2] developed a digital image correlation (DIC)-based method to measure the relative displacement between contact surfaces in fretting fatigue, and the transition from gross slip to partial slip was identified via the measured relative displacement distribution. Yelemessov et al. [3] assessed the effectiveness of reusing previously used railway rails by applying the fatigue analysis method to the rail steels. Coupled study methods using various microstructural characterization and mechanical tests were adopted to reveal the microstructural and mechanical property changes during operation. Rui et al. [4] studied the evolution of intragranular misorientation parameters KAM and GROD, obtained via 2D-EBSD, and their relationships with 3D plastic deformation were clarified at a crystal level.

Two articles focused on the fatigue behavior and modelling of superalloys under extreme conditions. Wu et al. [5] studied the crack behavior and mechanism of superalloy under the biaxial fatigue loading condition. Biaxial fatigue tests were conducted and fatigue crack formation and propagation processes were captured via a replica method. Hu et al. [6]

Citation: Cui, H.; Han, Q. Fatigue Damage Mechanism and Fatigue Life Prediction of Metallic Materials. *Metals* **2023**, *13*, 1752. https://doi.org/10.3390/met13101752

Received: 30 September 2023
Accepted: 7 October 2023
Published: 16 October 2023

developed constitutive models for superalloy under creep-fatigue loading at different elevated temperatures. The cyclic softening effect was considered in the constitutive models, and the results showed the models' capabilities to describe responses under creep-fatigue conditions.

Two articles focused on the fatigue behavior of aluminum alloys used in various industrial environments. Shi et al. [7] conducted pre-corrosion fatigue tests of an aluminum alloy in a marine atmosphere, and various electron microscopy methods were used to characterize fatigue damage mechanisms and their relationships with the fatigue life. Urrego et al. [8] studied the fatigue behavior of an aluminum alloy used in industrial joining structures. The effects of geometric factors on fatigue behavior were considered, and the relationship between the stress intensity factor and crack length was obtained.

Several articles investigated other important aspects of fatigue, such as fatigue performance optimization design, system fatigue reliability evaluation, and fatigue fracture simulation methods. Cui et al. [9] reviewed the state of the art of phase field methods, including fundamentals, recent progress, and their applications in various conditions across multiple scales. Performance-enhancing strategies for phase field methods were summarized, and the outlook was presented in the aspect of complex loading conditions, fatigue degradation criterion, coupled crystal plasticity, etc. Li et al. [10] conducted gear low-cycle fatigue tests under different stress levels and studied the system reliability estimation method using finite element modelling. The effect of component geometric size was considered, and its effect on the fatigue life of the large aviation planetary system was determined. Wen et al. [11] determined the shape optimization design of the vent hole structure to achieve enhancing fatigue life. Non-parametric- and geometric parameter-based optimization methods were implemented, and the reduction in the stress concentration was analyzed.

3. Conclusions and Outlook

Some of the most recent progress in damage mechanisms, fatigue life prediction, and performance optimization was reported in this Special Issue. The development of experimental, theoretical, and numerical methods and their application in the field of fatigue can serve as good foundations for future research. Finally, as Guest Editors, we would like to express our gratitude to all authors for their contributions, as well as to the reviewers and the *Metals* Editorial Office for their time and efforts.

Conflicts of Interest: The authors declare no conflict of interest.

References

1. Christodoulou, P.I.; Kermanidis, A.T. A Combined Numerical–Analytical Study for Notched Fatigue Crack Initiation Assessment in TRIP Steel: A Local Strain and a Fracture Mechanics Approach. *Metals* **2023**, *13*, 1652. [CrossRef]
2. Su, Y.; Rui, S.-S.; Han, Q.-N.; Shang, Z.-H.; Niu, L.-S.; Li, H.; Ishikawa, H.; Shi, H.-J. Estimation Method of Relative Slip in Fretting Fatigue Contact by Digital Image Correlation. *Metals* **2022**, *12*, 1124. [CrossRef]
3. Yelemessov, K.; Baskanbayeva, D.; Martyushev, N.V.; Skeeba, V.Y.; Gozbenko, V.E.; Karlina, A.I. Change in the Properties of Rail Steels during Operation and Reutilization of Rails. *Metals* **2023**, *13*, 1043. [CrossRef]
4. Rui, S.-S.; Su, Y.; Zhao, J.-M.; Shang, Z.-H.; Shi, H.-J. A 3D Polycrystalline Plasticity Model for Isotropic Linear Evolution of Intragranular Misorientation with Mesoscopic Plastic Strain in Stretched or Cyclically Deformed Metals. *Metals* **2022**, *12*, 2159. [CrossRef]
5. Wu, Z.; Pan, Y.; Lei, H.; Wang, S.; Fang, L. Fatigue Crack Growth Behavior and Failure Mechanism of Nickel-Based Alloy GH4169 under Biaxial Load Based on Fatigue Test of Cruciform Specimen. *Metals* **2023**, *13*, 588. [CrossRef]
6. Hu, X.; Zhuang, S.; Zheng, H.; Zhao, Z.; Jia, X. Non-Unified Constitutive Models for the Simulation of the Asymmetrical Cyclic Behavior of GH4169 at Elevated Temperatures. *Metals* **2022**, *12*, 1868. [CrossRef]
7. Shi, L.; Xiang, L.; Tao, J.; Chen, Q.; Liu, J.; Zhong, Y. Actual Marine Atmospheric Pre-Corrosion Fatigue Performance of 7075-T73 Aluminum Alloy. *Metals* **2022**, *12*, 874. [CrossRef]
8. Urrego, L.F.; García-Beltrán, O.; Arzola, N.; Araque, O. Mechanical Fracture of Aluminium Alloy (AA 2024-T4), Used in the Manufacture of a Bioproducts Plant. *Metals* **2023**, *13*, 1134. [CrossRef]
9. Cui, H.; Du, C.; Zhang, H. Applications of Phase Field Methods in Modeling Fatigue Fracture and Performance Improvement Strategies: A Review. *Metals* **2023**, *13*, 714. [CrossRef]

10. Li, M.; Luo, Y.; Xie, L. Fatigue Reliability Prediction Method of Large Aviation Planetary System Based on Hierarchical Finite Element. *Metals* **2022**, *12*, 1785. [CrossRef]
11. Wen, C.; Zheng, Y.; Mi, D.; Qian, Z.; Zhang, H. Design for the Vent Holes of Gas Turbine Flow Guide Disks Based on the Shape Optimization Method. *Metals* **2023**, *13*, 1151. [CrossRef]

metals

Review

Applications of Phase Field Methods in Modeling Fatigue Fracture and Performance Improvement Strategies: A Review

Haitao Cui [1,2,*], Chenyu Du [1,2] and Hongjian Zhang [1,2,*]

1 College of Energy & Power Engineering, Nanjing University of Aeronautics and Astronautics, Nanjing 210016, China
2 Aero-Engine Thermal Environment and Structure Key Laboratory of Ministry of Industry and Information Technology, Nanjing University of Aeronautics and Astronautics, Nanjing 210016, China
* Correspondence: cuiht@nuaa.edu.cn (H.C.); zhanghongjian@nuaa.edu.cn (H.Z.); Tel.: +86-133-7202-1015 (H.C.); +86-181-6810-1903 (H.Z.)

Abstract: Fatigue fracture simulation based on phase field methods is a promising numerical approach. As a typical continuum approach, phase field methods can naturally simulate complex fatigue fracture behavior. Moreover, the cracking is a natural result of the simulation without additional fracture criterion. This study first introduced the phase field fracture principle, then reviewed some recent advances in phase field methods for fatigue fracture modeling, and gave representative examples in macroscale, microscale, and multiscale structural simulations. In addition, some strategies to improve the performance of phase field models were summarized from different perspectives. The applications of phase field methods to fatigue failure demonstrate the ability to handle complex fracture behaviors under multiple loading forms and their interactions, and the methods have great potential for development. Finally, an outlook was made in four aspects: loading form, fatigue degradation criterion, coupled crystal plasticity, and performance improvement.

Keywords: phase field methods; fatigue fracture; performance improvement strategies

Citation: Cui, H.; Du, C.; Zhang, H. Applications of Phase Field Methods in Modeling Fatigue Fracture and Performance Improvement Strategies: A Review. *Metals* **2023**, *13*, 714. https://doi.org/10.3390/met13040714

Academic Editor: Alain Pasturel

Received: 17 March 2023
Revised: 1 April 2023
Accepted: 3 April 2023
Published: 5 April 2023

1. Introduction

Material failure prediction is of great importance in engineering and materials science, and fatigue fracture as a typical form of failure has been of concern to scholars [1]. The process of generating cracks in components under fatigue loading can be decomposed into stages such as crack nucleation, steady-state propagation, and transient fracture. The fundamental challenge in simulating fatigue fracture is accurately tracing the crack surface under complex load conditions and material constitutive models [2,3]. Therefore, to handle the problem, numerous numerical techniques have been created, which can be classified into discontinuous and continuous methods [4,5]. However, discontinuous approaches frequently call for a way to represent the fracture topology and the related crack-tracking algorithm, which makes numerical integration and convergence challenging [6,7]. As a typical continuous method, the phase field method has recently received a wide range of attention [8,9]. The method can naturally model complex fracture processes including crack nucleation, propagation, coalescence, and branching [10,11]. Furthermore, the crack is a natural result of the simulation, obtained by solving differential equations related to the phase field, without additional fracture criteria. The crack evolution can be modeled on the fixed mesh, which avoids the tedious task of constructing the crack surface [12,13].

In the past few years, researchers have made many efforts to develop phase field methods for fatigue crack growth, and considerable progress has been made [14–17]. Phase field methods are widely applied to reproduce fatigue fracture behaviors, such as brittle [18], ductile [19], hyperelastic [20], and corrosion [21] fracture. Complex loading conditions, such as mechanical [22], thermal [23], and chemical [24] loads, as well as their interactions, are commonly the root cause of these fatigue failure issues. Related phase field modeling has been applied in macroscale [25], microscale [26], and multiscale [27] models, as shown in Figure 1. Most of the phase field models proposed in these studies accurately built the crack morphology and reasonably predicted the fatigue life. Meanwhile, some strategies for improving the performance of phase field models are proposed from various perspectives to further improve efficiency, accuracy, and convergence. The performance of phase field models has been satisfactorily improved due to the application of these techniques. This paper aims to review the advances in these directions.

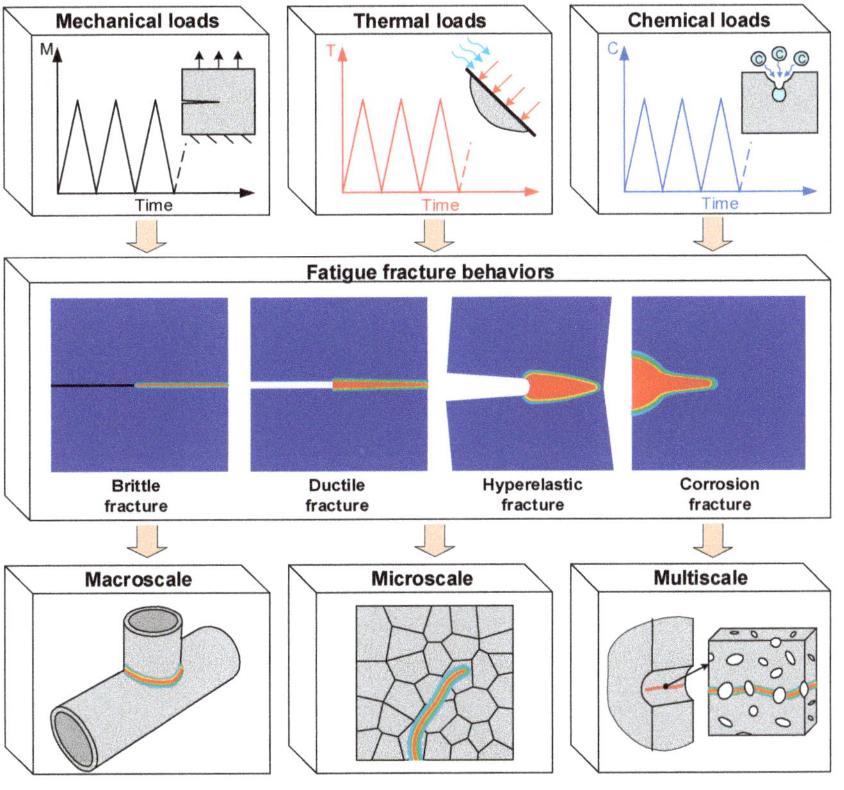

Figure 1. Fatigue fracture behaviors simulated with phase field methods.

The paper is organized as follows: the phase field fracture principle is introduced in Section 2. Brittle, ductile, hyperelastic, and corrosion fracture behaviors in phase field fatigue modeling are given in Section 3, including the macroscale, microscale, and multiscale models. Strategies for improving the performance of phase field models are given in Section 4, including accelerated solution technologies, element processing technologies, and discretization methods. Finally, the concluding remarks and future directions are given in Section 5.

2. Phase Field Fracture Principle

In this section, a linear elastic phase field model based on Griffith's fracture theory is introduced to illustrate the phase field fracture principle.

For a one-dimensional case, the topology of a sharp crack (Figure 2a) can be described by an auxiliary field variable $\overline{\phi} \in [0, 1]$ with:

$$\overline{\phi}(x) = \begin{cases} 1 & \text{if } x = 0 \\ 0 & \text{if } x \neq 0 \end{cases}, \tag{1}$$

which is called the crack phase field order parameter, with ϕ = 0 and ϕ = 1 denoting the intact and fully broken states of the solid, respectively. The non-smooth crack phase field can be approximated by the exponential function, given by [13]:

$$\phi(x) = e^{-\frac{|x|}{l}}, \tag{2}$$

ϕ is phase field variable, representing a diffusive or regularized crack topology (Figure 2b). The regularization parameter l specifies the width of the smearing function, approaching the discrete crack topology as $l \to 0$.

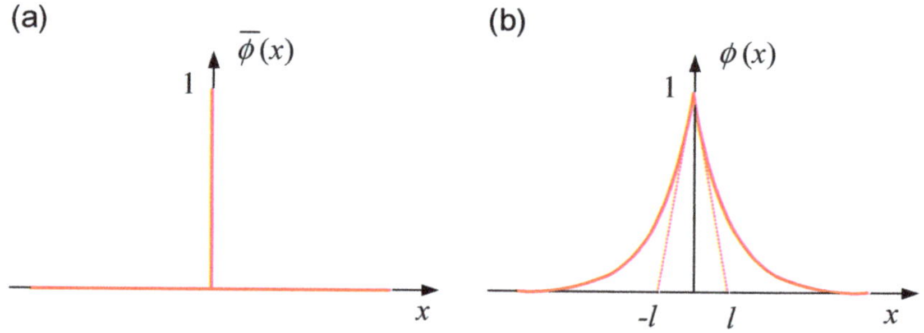

Figure 2. Sharp (**a**) and diffusive (**b**) crack topologies.

Consider a discrete internal discontinuity Γ in a solid body Ω (Figure 3a), a regularized crack function Γ_l can be defined (Figure 3b):

$$\Gamma_l = \int_{\Omega} \Upsilon(\phi, \nabla\phi) \mathrm{d}V, \tag{3}$$

where Υ is the crack surface density function.

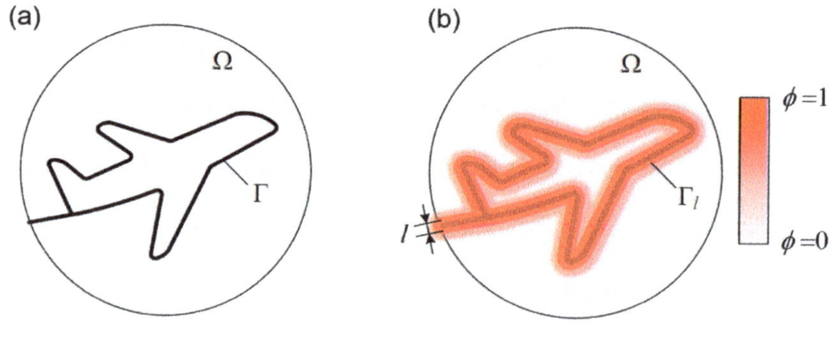

Figure 3. Schematic representation of a solid body. (**a**) Internal discontinuity boundaries, and (**b**) a phase field approximation of the discrete discontinuities.

Phase field variable ϕ is smooth and continuous, and discrete cracks are represented in a diffuse form. Thus, the robust modeling of crack interactions and branches with arbitrary topological complexity is possible. The objective of the diffuse representation is to introduce the following fracture energy approximation over a discrete internal discontinuity Γ:

$$\Phi = \int_{\Gamma} G_c \, \mathrm{d}S \approx \int_{\Omega} G_c \Upsilon(\phi, \nabla\phi) \mathrm{d}V, \tag{4}$$

$$\varphi = G_c \Upsilon(\phi, \nabla\phi), \tag{5}$$

where Φ and φ are the fracture energy and its density, respectively. G_c is fracture toughness. The rate-independent fracture description is extended to include time and history dependent concerns [25]:

$$\Phi = \int_0^t \left[\int_{\Omega} G_c f(\overline{\vartheta}(t)) \dot{\Upsilon}(\phi, \nabla\phi) \mathrm{d}V \right] \mathrm{d}t, \tag{6}$$

$$\varphi = G_c f(\overline{\vartheta}(t)) \dot{\Upsilon}(\phi, \nabla\phi), \tag{7}$$

where $\overline{\vartheta}$ is the cumulative history variable, which fulfils $\dot{\overline{\vartheta}} > 0$, and $f(\overline{\vartheta}(t))$ is a fatigue degradation function. This form achieves fatigue degradation of materials by degrading fracture toughness.

The total potential energy of the solid is defined as the sum of the strain energy and fracture energy, which is expressed as:

$$\prod = \int_{\Omega} (\psi_0(\varepsilon, g(\phi)) + \varphi(\phi, \nabla\phi)) \mathrm{d}V, \tag{8}$$

$$W = \psi_0(\varepsilon, g(\phi)) + \varphi(\phi, \nabla\phi), \tag{9}$$

where $\prod$ and W are total potential energy and its density, respectively. ψ_0 is the undamaged elastic strain energy density. $g(\phi)$ is a phase field degradation function, and one of its general expressions is described as:

$$g(\phi) = \left(1 - \phi^2\right). \tag{10}$$

In order to explain the inconsistency in the contribution of tensile and compressive behavior to material degradation, the elastic strain energy density must be additively divided into active ψ_0^- and inactive ψ_0^- components [13,28]:

$$\psi_0 = \left(1 - \phi^2\right) \psi_0^+(\varepsilon) + \psi_0^-(\varepsilon). \tag{11}$$

Damage is an irreversible process: $\dot{\phi} \geq 0$. To enforce irreversibility, a history field variable $\mathcal{H}$ is introduced, which must satisfy the Karush-Kuhn-Tucker conditions:

$$\psi_0^+ - \mathcal{H} \leq 0, \; \dot{\mathcal{H}} \geq 0, \; \dot{\mathcal{H}}(\psi_0^+ - \mathcal{H}) = 0. \tag{12}$$

For a current time t, over a total time τ, the history field $\mathcal{H}$ is understood as:

$$\mathcal{H} = \max_{t \in [0,\tau]} \psi_0^+(t). \tag{13}$$

The fracture energy density is further defined as:

$$\varphi = f(\overline{\vartheta}) G_c \Upsilon(\phi, \nabla\phi) = f(\overline{\vartheta}) \frac{G_c}{4c_w l} \left(w(\phi) + l^2 \left| \nabla\phi \right|^2 \right). \tag{14}$$

where $w(\phi)$ is the geometric crack function, and c_w is a scaling constant. The dissipation function rules the energy dissipation due to the formation of a new crack. There are two widely used models [18]:

$$\text{AT2 model: } w(\phi) = \phi^2 \text{ and } c_w = \frac{1}{2}, \tag{15}$$

$$\text{AT1 model: } w(\phi) = \phi \text{ and } c_w = \frac{2}{3}. \tag{16}$$

The AT2 model has a vanishing threshold for the onset of damage, resulting in a material behavior without an initial linear elastic branch, while the AT1 model has the ability to reproduce the initial linear elastic constitutive behavior.

The total potential energy and its density of the solid are reformulated as:

$$\Pi = \int_\Omega \left(\left(1-\phi^2\right)\mathcal{H} + f(\overline{\vartheta})\frac{G_c}{4c_w}\left(\frac{w(\phi)}{l} + l|\nabla\phi|^2\right)\right) dV, \tag{17}$$

$$W = \left(1-\phi^2\right)\mathcal{H} + f(\overline{\vartheta})\frac{G_c}{4c_w}\left(\frac{w(\phi)}{l} + l|\nabla\phi|^2\right). \tag{18}$$

Based on the above equation, the weak form of the equilibrium equations for the displacement and phase fields can be derived using the variational principle and applied to the subsequent solutions.

To sum up, phase field models at the mechanical level can be considered an extension of the Griffith fracture theory. The potential energy used for variational analysis includes the fracture energy, which is regularized with the aim to build up a diffuse form of the crack. As a result, phase field methods offer the enormous possibility for handling challenging fracture problems.

3. Typical Fracture Behavior in Phase Field Fatigue Modeling

In this section, four typical fracture behaviors in phase field fatigue modeling will be reviewed, including brittle, ductile, hyperelastic, and corrosion fracture. The scales of the model involve macroscale, microscale, and multiscale. Note that the characteristic length to differentiate between macroscale and microscale is 0.1 mm in this paper.

3.1. Brittle Fracture

3.1.1. Macroscale Brittle Fracture

In the field of research on macroscale brittle fracture, Alessi [29] proposed a new variational fatigue phase field model using a phenomenological approach for the study of the one-dimensional case. The model is capable to describe a typical Wöhler curve and to recover the known trends in the description of the mean-stress effects. Mesgarnejad et al. [30] developed a class of phase field models to describe fatigue fracture phenomenologically through local fracture toughness degradation. The model is the ability to detail the growth of a collection of interacting cracks in complex geometries. A similar approach is also reflected in the work of Grossman-Ponemon et al. [31] and Hasan et al. [14]. Carrara et al. [18] modified the standard regularized free energy function by introducing a sui fatigue history variable and a fatigue degradation function, which effectively reduces the fracture toughness. Numerical results demonstrate that the crack propagation rate curve and the Paris law are naturally recovered, and subsequently demonstrate that the framework is available to handle complex geometries and loading conditions in 2D (Figure 4a) and 3D (Figure 4b).

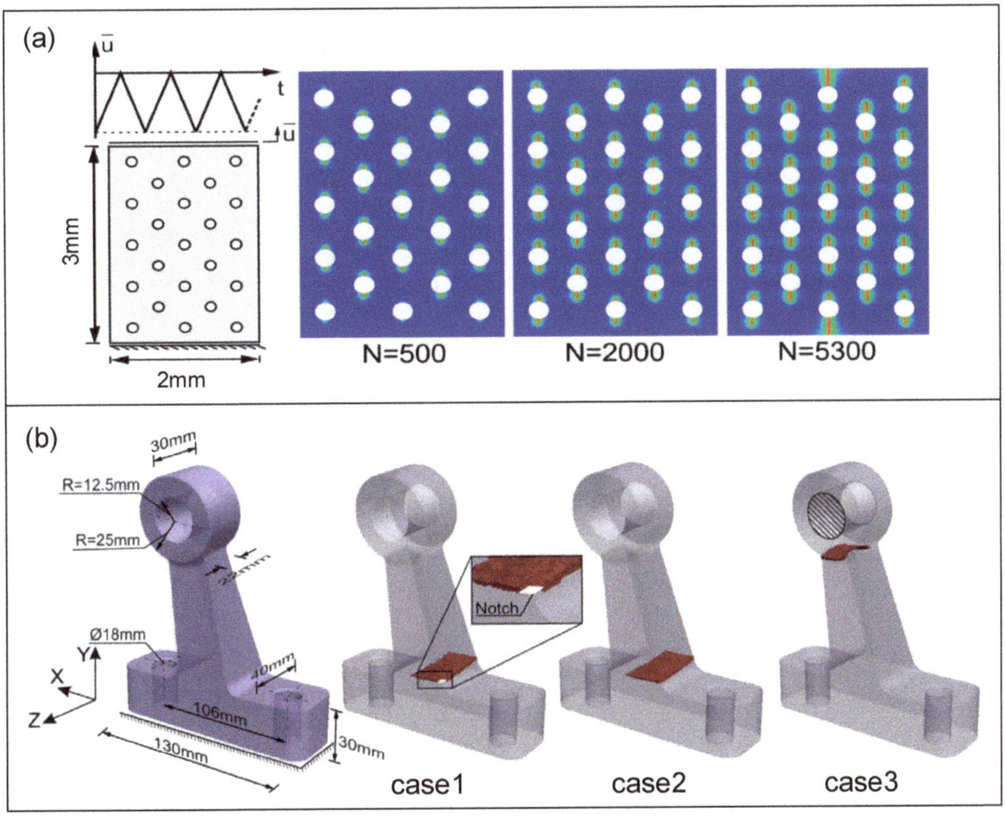

Figure 4. Macroscale brittle fracture. (**a**) A 2D plate with holes model and (**b**) a 3D steering arm model [18].

Lo et al. [32] introduced a crack expansion viscosity parameter into the standard phase field model for the brittle fracture to account for the rate- or period-dependent crack propagation phenomenon. An elastic energy decomposition method is employed to ensure that the compressive stresses do not contribute to the phase field evolution, and a modified J-integral is proposed to reproduce the Paris law using the phase field approach. Finally, the ability of the model to reproduce a 3D crack is illustrated. To effectively simulate fatigue crack growth in thin-walled structures, Liu et al. [33] presented a global-local phase field method for large deformation shells to provide an efficient modeling framework. The global model can be coarsely meshed, while the local model can be meshed more finely to handle the displacement-phase field problem. The capability of the global-local model is proved by simulating crack propagation in a cylindrical thin-walled structure under fatigue cyclic loading.

3.1.2. Microscale Brittle Fracture

The failure mechanism of materials generally depends on their microstructures, such as grain size and orientation, grain boundary characteristics, impurities, and voids. Emdadi and Zaeem [34] proposed a modified phase field model based on Griffith's theory for the study of intergranular and transgranular crack growth in polycrystalline brittle materials. The simulation results demonstrate that the specific combination of grain boundary strength and crack surface energy can promote along-crystal crack growth in, thus affecting the fracture performance of polycrystalline materials.

The micro crack in some components has a significant impact on their performance, such as heat transfer, mass transfer, and energy storage. Relevant failure problems can be handled using the phase field method. For the failure of thermal barrier coatings (TBCs), Xiao et al. [35] presented a growth chemical model for the Thermal growth oxide (TGO) and thermal mechanical interaction phase field model and discussed the delamination mechanism of TBCs under thermal cycling from a microscopic perspective. The result proves that the phase field model is a good reproduction of the complex failure behavior in TBCs. The electrode particles in Lithium-ion batteries are small in size and their cracking drive battery capacity degradation. Ai et al. [36] developed a multi-physics phase field fatigue model to investigate the crack growth in battery electrode particles undergoing hundreds of cycles. In addition, the electro-chemo-mechanical formulation is combined with CT imaging to simulate the fatigue fracture of real particle microstructure. Using the method, the nonlinear crack growth behavior can be accurately predicted, and the crack length is observed to increase exponentially with the number of cycles.

3.2. *Ductile Fracture*

3.2.1. Macroscale Ductile Fracture

Ulloa et al. [37] offered a coupled gradient-enhanced plastic damage model that uniformly captures the features of both low and high circumferential fatigue. The suggested variational model can explain cyclic failure under both force and displacement loading by combining cyclic plasticity and the phase field description of the fatigue fracture. The simulation result of an asymmetric notched plate model is shown in Figure 5. To ensure that the compressive plastic energy does not contribute to crack formation in elastoplastic materials, Shi et al. [19] established a plastic history field and presented an alternate elastic energy decomposition technique. The model can eliminate the fictitious propagation of compression cracks and predict the results closer to the experimental observations. Haveroth et al. [38] proposed a general thermodynamically consistent non-isothermal phase field model for fatigue evolutions in elastoplastic materials under the hypothesis of small strains. The ductile fracture and fatigue processes can be qualitatively and statistically reproduced using the model. Song et al. [39] developed a phase field viscoplasticity coupling method to model the crack growth in a nickel-based superalloy under fatigue. The result shows that the method is very effective in predicting the fatigue crack growth under varied dwell times at peak load, and it is further used to simulate the growth path of the 3D corner crack model. Cyclic plasticity in fatigue behavior often leads to large computational costs. Therefore, Khalil et al. [25] proposed a generalized phase field model for fatigue crack growth in elastoplastic solids with an efficient solver. Both non-linear kinematic and isotropic hardening are investigated, as well as the combination of the two models. By testing typical numerical examples, the framework is found to be efficient in predicting fatigue crack propagation in arbitrary geometries as well as cyclic hardening in elastoplastic materials. Some advanced materials tend to have anisotropic elastoplastic fracture behavior, and phase field methods have been applied to related fields. Li et al. [40] presented a phase field model for anisotropic elastoplastic fracture behavior. The model contains three innovations: an extension to a phase field model of anisotropic elastoplasticity, a transformation from quasi-brittle to elastoplastic fracture behavior, and a novel method to identify the macroscale strain density as a function of microscale interface damage variables. The application of the model for 3D printed polymer materials has been approved through an experimental comparison.

3.2.2. Microscale Ductile Fracture

It is challenging to simulate ductile fracture in microscale models. Seleš et al. [41] simulated complex fracture and fatigue failure processes in the microstructure of nodular cast irons to investigate the transition between brittle and ductile fracture material behavior. It is concluded that under cyclic loading conditions, the transition between low and high cycle fatigue regimes is captured by analyzing the fracture patterns and

Wöhler curve of the specimens. As the load amplitude decreases, the model shows less plasticity and thus transitions to high cyclic regimes. The phase field modeling of microcracks in polycrystalline materials is equally of concern. Tu et al. [42] developed a coupled crystal plasticity-phase field (CP-PF) model for simulating image-based crack growth in polyphase-polycrystalline microstructures of aluminum alloy. The problem is characterized by elastoplasticity anisotropy and cracking features such as nucleation, propagation, coalescence, and branching. Simulations of structural deformation, crack nucleation, and propagation in polycrystalline microstructures were carried out under monotonic and cyclic loading conditions, and the robustness and effectiveness of the model in modeling microcracks was demonstrated.

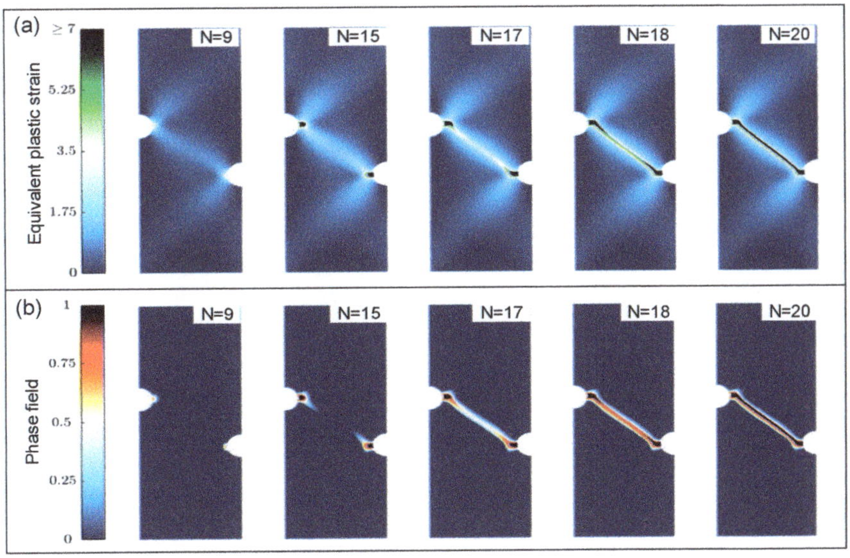

Figure 5. Macroscale ductile fracture. (**a**) Equivalent plastic strain field and (**b**) phase field [37].

3.2.3. Multiscale Ductile Fracture

In terms of multiscale methodological studies, Sadeghirad et al. [27] proposed a method for predicting crack initiation in the microstructure, which uses a phase field model combined with an extended finite element method (XFEM) to achieve an accurate modeling of macroscale crack growth. The macroscopic model assumes linear elastic material behavior while coupling the phase field model and the crystal plasticity model at the microscopic scale to capture the plastic behavior of non-uniform polycrystals. Compared with experimental results, it is found that the proposed multiscale approach utilizes the advantages of each method to provide an accurate prediction with minimal computational cost.

3.3. *Hyperelastic Fracture*

To describe rubber damage caused by cyclic loading, Loew et al. [43–45] extended the previous method to establish a phase field model for rate-dependent rubber fracture in a finite strain condition. A rate-dependent fatigue model is developed based on a load-history dependent fatigue damage source. The Wöhler curve and the rate of the crack growth curve are given by the simulation, and the failure behavior of the single notch under cyclic tensile load is also predicted, which is in good agreement with the experimental findings. The simulation results of two kinds of fatigue tests are shown in Figure 6a,b. Although this model accurately describes fatigue failure, its calculation speed is slow. Therefore, Loew et al. [44] provided an acceleration scheme for the cyclic loading of the phase field damage model, and the simulation speed is improved by using the cyclic skipping technique.

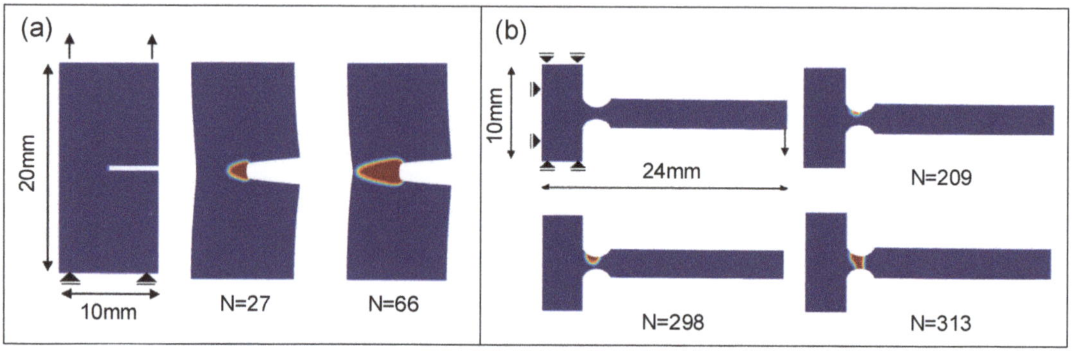

Figure 6. Hyperelastic fracture. (**a**) A single edge notched plate model and (**b**) a bending beam model [44].

Fatigue failure simulation of Shape Memory Alloys (SMAs) has been a difficult problem owing to nonlinear constitutive behavior. Simoes et al. [46,47] developed the first phase field fracture formulation for SMAs. The hyperelasticity and shape memory effects are caught by the constitutive behavior, which includes phase transitions induced by stress and temperature. An implicit time integration scheme with both monolithic and staggered solution strategies is applied for the numerical implementation of the theoretical formulation. Based on this framework, fatigue failure studies are carried out for complex SMAs structures. The result shows that the phase field approach can simulate several typical 2D and 3D problems involving subcritical crack propagation, unstable cracking, crack coalescence, and fatigue crack growth, which demonstrates the potential and robustness of the model. The design of SMAs components can also be optimized using the framework.

3.4. Corrosion Fracture

Predicting the failure of engineering parts in aggressive environments is a long-standing scientific challenge. The interaction of mechanical loading and the corrosive environment promotes the nucleation and propagation of cracks, a process known as stress corrosion cracking (SCC). It occurs in a wide variety of metallic materials and settings and is frequently facilitated by pre-existing flaws [24]. Martínez-Pañeda et al. [48] presented a phase field formulation for the long-term problem of hydrogen assisted cracking. A coupled mechanical-diffusion-phase field finite element framework is used to analyze hydrogen transport to the fracture area and subsequent cracking, and the plate with preexisting defects is considered. The model has the ability to capture complex crack growth caused by inherent defects in the corrosive environment. The strategy has quickly gained in popularity as an approach for combining different hydrogen embrittlement models [49–51], however, the majority of analyses are restricted to 2D boundary value problems. The deformation-diffusion-damage coupling framework developed by Kristensen et al. [52] considers the AT1 and AT2 models, inertial effects, and 3D model implementation for the first time, which can predict the defect evolution under operating conditions until the ultimate failure. Cui et al. [21,24] presented a new framework for modeling mechanically-assisted corrosion in elastoplastic solids and proposed a formulation based on the film fracture-dissolution-repassivation mechanism. The model takes into account the electrochemical driving force of mechanical straining, which speeds up the kinetics of corrosion. The simulated crack of the C-ring reveals a good agreement with the experimental observation, as shown in Figure 7. Golahmar et al. [53] proposed a multi-physics phase field model for hydrogen-assisted fatigue by combining hydrogen assisted failures and mechanical fatigue. To capture how newly formed fracture surfaces are quickly exposed to the environment, a penalty technique was used to implicitly enforce moving chemical boundary requirements. The study shows that the model sufficiently captures the sensitivity of fatigue crack growth rates to hydrogen content, and recovers the Paris law naturally, which can quantify the influence of hydrogen on the Paris law parameters.

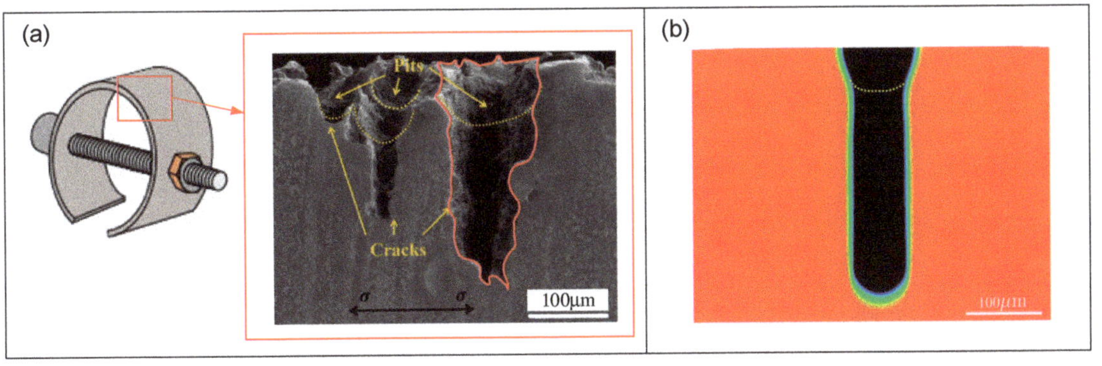

Figure 7. Corrosion fracture. (**a**) Experimental observation and (**b**) simulated crack of the C-ring [21].

Corrosion fracture in the microstructures is equally of attention. To achieve the simulation of the SCC initiated from surface pits, Mai et al. [54,55] implemented the film rupture-dissolution-repassivation model, which relates the interface kinetics parameter to the stress distribution near the crack tip. The evolution of the cracking interface is approximated by coupling the governing equations of electrochemical corrosion with the elastoplastic deformation of metals. It is demonstrated that the model can reliably replicate experimental results after proper calibration and show applications for the material microstructure on SCC.

4. Performance Improvement Strategies for Phase Field Models

The phase field method has advantages for addressing complicated fracture problems, but it has a high computational cost and modeling difficulty. It is necessary to develop studies on the performance improvement of phase field models. In this section, some typical performance improvement strategies for phase field models are reviewed from three aspects, including accelerated solution technologies, element processing technologies, and discretization methods.

4.1. Accelerated Solution Technology

4.1.1. Cycle Skipping Technique

Improving the computational efficiency of phase field models is also a concern for researchers. The computational cost of cyclic loading is quite large, especially for high cycle fatigue. To keep the solution time within acceptable limits, Schreiber et al. [56] collected a similar number of cycles into blocks, and additional fatigue damage processed within one simulation step referred to these blocks. The block size was adaptively selected to control the damage rate. Yan et al. [57] developed an adaptive cyclic incremental adjustment algorithm, which can reduce the computational cost of the simulation without sacrificing accuracy. The whole simulation is divided into three stages: elastic stages, transition phase, and fatigue stages. The damage increment is controlled to obtain a moderate fatigue energy growth, where the number of cycle increments is chosen adaptively according to the damage increment.

When nonlinear material constitutive is involved, Seleš et al. [58] presented a full range phenomenological fatigue fracture model able to reproduce the main features of low and high cycle fatigue which consider cyclic plasticity. A two-part cycle skipping technique is implemented in this work. The technique precisely solves the highly nonlinear time-evolutionary behavior by automatically calculating the lower number of cycles to be skipped or by not skipping cycles at all. Loew et al. [44] presented explicit and implicit cycle skipping schemes to simulate the fatigue crack growth of rubber. It is found that both schemes reduce the calculation time significantly, while the accuracy remains substantially unaffected. The adaptive cycle skipping method for the implicit acceleration framework is dependent on the number of iterations since the size of the cycle step controls how quickly simulations accelerate, but may give rise to incorrect results. The method proposed by Loew can reduce the calculated costs by 99.5%.

4.1.2. Quasi-Newton Monolithic Scheme

One of the main reasons for the high computational cost of phase field models is that the robust but inefficient alternating minimization (AM) or staggered scheme is widely used. Aiming to tackle the difficulty, Wu et al. [59,60] proposed, for the first time, to use the Broyden–Fletcher–Goldfarb–Shanno (BFGS) algorithm to solve the coupled governing equations in a monolithic manner rather than the standard Newton iterative scheme in a staggered manner. Representative example shows that the quasi-Newton monolithic scheme is about 3–7 times faster than the staggered solver, and the larger the model scale is, the more costs can be saved. On this basis, Kristensen and Khalil et al. [25,61] proposed a generalized phase field model for fatigue crack growth in elastic and elastoplastic solids by combining the phase field framework with the quasi-Newton monolithic scheme. In addition, a new adaptive time increment scheme is introduced to further reduce the cost while allowing sudden changes in fracture behavior to be accurately resolved. Liu et al. [33] presented a phase field model for large deformation shells based on the quasi-Newton monolithic scheme, and the excellent convergence performance of the form was proved. To further save computational costs, a specific global-local method using solid shell elements was proposed, and its capability is verified by simulating crack growth in cylindrical structures.

4.2. Element Processing Technology

4.2.1. Adaptive Mesh Refinement Technology

The adaptive mesh refinement technology is an effective means to solving the contradiction between the computational efficiency and accuracy of the phase field model. Freddi et al. [62] investigated the global and global/local refinement strategies. Compared with the global technique, the global/local strategy can reduce the calculation cost by about 20–30%, but it is more sensitive to parameter selection. Xu et al. [63] proposed a multi-level adaptive mesh refinement technique suitable for the phase field model. In addition, an improved backtracking algorithm enables the adaptive mesh to handle complex crack propagation. The calculation efficiency of this method can be increased by around 30 times over global mesh refinement. Aiming at the problem that the existing adaptive mesh refinement technology cannot accurately simulate crack initiation without a prior local refinement, Gupta et al. [64] developed an adaptive mesh refinement algorithm for phase field models using a multi-level mark–unmark strategy. The algorithm utilizes a threshold to mark elements based on effective crack driving energy and reduces the calculation time by 5–50 times as compared to simulations that adopt a priori non-adaptively strategy. Eldahshan et al. [65] combined the framework of ductile fracture phase field modeling with adaptive refinement technology to effectively simulate ductile fracture without the need to predefine the crack initiation site.

4.2.2. Combined Extended Finite Element Method (XFEM)

XFEM is one of the most popular methods to simulate crack propagation. As a discontinuous method, it has the following advantages: (a) Explicit representation is helpful to the modeling of crack morphology; (b) Any wrong interaction between crack surfaces can be avoided; and (c) The rear area of the crack tip may be roughly discretized. Therefore, if the phase field model can be combined with XFEM, the efficiency and performance of the model can be effectively improved. Existing studies have confirmed this conjecture. Giovanardi et al. [66] established a hybrid XFEM-Phase field (Xfield) method. A global solution for the displacement field is calculated with XFEM. Then, using a finer mesh and Dirichlet boundary conditions that are derived from the overall solution, a phase field model in overlapping subdomains including the crack points is solved to determine the growth. Patil et al. [67] proposed a similar strategy called the local moving extended phase field method (LMXPFM). The crack is described by XFEM, and the phase field model is employed in small circular subdomains around the crack tips. These works are limited to 2D problems. Muixí et al. [68] developed a combined XFEM phase field model (PF-XFEM) without remeshing. The same background mesh is used throughout the simulation. The discretization is automatically updated as the crack grows, avoiding remeshing to improve efficiency. The robustness of the model is proved in 2D and 3D instances.

4.3. Discretization Method

4.3.1. Fast Fourier Transform (FFT) Method

The application of phase field methods to simulate fractures in heterogeneous materials usually requires fine meshing and hence a large number of elements in the model, weakening the efficiency of standard finite element solvers. Motivated by this limitation, Chen et al. [69] proposed an FFT solver of a variational phase-field model for brittle fracture. Relying on the staggered scheme, the phase field and elastic problem are resolved by the FFT technique. As an alternative, FFT methods use matrix-free iterative algorithms to solve partial differential equations with periodic boundary conditions. Moreover, segmented digital images can be used as input for the simulation without the meshing step. To further improve the performance of a fixed-point algorithm, an improved convergence acceleration approach is presented. Sadeghirad and Ma et al. [27,70] simulated the crack growth inside the polycrystalline microstructure with the help of the FFT-solver, and the accuracy and efficiency of this method are proved.

4.3.2. Machine Learning: Physics Informed Neural Network (PINN) Algorithm

Goswami et al. [71] proposed a new physics informed neural network (PINN) algorithm for predicting the crack propagation based on the phase field method. The suggested approach offers two key advantages over the conventional residual-based PINN. Firstly, the imposition of the boundary conditions is simpler and more robust. Moreover, training the network is quicker because the derivatives in the functional form of the variational energy are of a lower order than those in the residual form employed in standard PINN. The proposed machine learning method has less code and is easy to implement. After the network is trained, compared with the finite element method, the method can have good calculation savings, but the discrete elements required are significantly smaller. The potential application of this approach is to be a low-fidelity surrogate for the high-fidelity numerical solvers, which is very useful in reliability analysis and design optimization.

4.3.3. Meshfree Method

Amiri et al. [72] developed a fourth order phase field model based on the meshless local maximum entropy (LME) approximants. This meshless method allows the model to directly solve the fourth order phase field equation without splitting the fourth order differential equation into two second order differential equations. Representative examples, such as the mode-I tension specimen and the single edge notched beam, show that both second-order and fourth-order phase field models are capturing complex crack behavior, and the fourth-order model can capture crack surfaces more accurately. In addition, the fourth order model requires fewer nodes. The simulation results are in good agreement with the analytical solutions and experimental data.

4.4. Summary

The seven performance enhancement strategies reviewed above all have achieved effective performance enhancement for phase field models, providing a basis for subsequent related studies. However, these strategies still have certain disadvantages, which are summarized in Table 1.

Table 1. Summary of the main performance improvement strategies.

Strategies	Specific Methods	Disadvantages
Accelerated solution technology	Cycle skipping technique Quasi-Newton monolithic scheme	The strategy improves computational efficiency by appropriately increasing the size of increment step, but larger increment steps may ignore abrupt load changes, which can lead to incorrect results.
Element processing technology	Adaptive mesh refinement technology Combined XFEM	For slightly complex 3D models, the strategy is difficult to generate structured mesh, and only unstructured mesh with low computational efficiency and accuracy can be used.
Discretization method	FFT method PINN algorithm Meshfree method	Compared with the finite element method, the strategy has a greater difficulty of the model development, while there are few related studies, and more samples are urgently needed to verify the accuracy and effectiveness of the method.

5. Conclusions

Phase field methods is a promising tool for numerical simulation. This work first introduces the principle of phase field approximation of cracks and then reviews some recent advances in phase field methods for simulating fatigue fracture. Representative examples from macroscale, microscale, and multiscale structural simulations are given, and some strategies for improving the performance are summarized. The application of phase field methods to fatigue failure fully demonstrates its ability to reproduce complex fracture behaviors under multiple loading forms and their interactions, and the methods have great potential for development. For future work, four aspects are proposed:

1. Phase field models need to be further applied to fatigue failure simulation of advanced materials under complex loading forms, such as thermal fatigue, thermo-mechanical fatigue, and high temperature vibration fatigue.
2. Most of the fatigue degradation criteria in existing phase field models are phenomenological. The establishment of fatigue degradation criteria that correspond to the real physical mechanism can contribute to the improvement of the prediction accuracy of phase field models.
3. The combination of phase field methods and crystal plasticity finite element has great potential to simulate fatigue crack growth in microscale, which can clearly understand the damage mechanism of metal microstructure. More advanced algorithms need to be built to accomplish this coupling process.
4. Some effective strategies for improving the performance of phase field models are proposed. However, these methods have certain disadvantages and continue to have their own limitations when faced with some complex simulations. It is still essential to further improve the performance of phase field models.

Author Contributions: Conceptualization, H.C.; methodology, H.Z.; investigation, C.D.; writing—original draft preparation, H.C. and C.D.; writing—review and editing, H.C., C.D. and H.Z.; supervision, H.C. and H.Z.; project administration, H.C.; funding acquisition, H.C. and H.Z. All authors have read and agreed to the published version of the manuscript.

Funding: This work was funded by the National Natural Science Foundation of China, grant number 91860111.

Data Availability Statement: Not applicable.

Conflicts of Interest: The authors declare no conflict of interest. The funders had no role in the design of the study; in the collection, analyses, or interpretation of data; in the writing of the manuscript, or in the decision to publish the results.

References

1. Qiu, S.; Cui, H.; Zhang, H.; Wen, W.; Guo, J. A dual-threshold modelling approach for fatigue life prediction under combined high and low cycle fatigue. *Int. J. Fatigue* **2022**, *164*, 107110. [CrossRef]
2. Wang, J.; Yang, Y.; Yu, J.; Wang, J.; Du, F.; Zhang, Y. Fatigue Life Evaluation Considering Fatigue Reliability and Fatigue Crack for FV520B-I in VHCF Regime Based on Fracture Mechanics. *Metals* **2020**, *10*, 371. [CrossRef]
3. Sadananda, K.; Nani Babu, M.; Vasudevan, A.K. A review of fatigue crack growth resistance in the short crack growth regime. *Mater. Sci. Eng. A* **2019**, *754*, 674–701. [CrossRef]
4. Egger, A.; Pillai, U.; Agathos, K.; Kakouris, E.; Chatzi, E.; Aschroft, I.A.; Triantafyllou, S.P. Discrete and Phase Field Methods for Linear Elastic Fracture Mechanics: A Comparative Study and State-of-the-Art Review. *Appl. Sci.* **2019**, *9*, 2436. [CrossRef]
5. Fang, J.; Wu, C.; Rabczuk, T.; Wu, C.; Ma, C.; Sun, G.; Li, Q. Phase field fracture in elasto-plastic solids: Abaqus implementation and case studies. *Theor. Appl. Fract. Mec.* **2019**, *103*, 102252. [CrossRef]
6. Brockman, R.A.; Hoffman, R.M.; Golden, P.J.; Musinski, W.D.; Jha, S.K.; John, R. A computational framework for microstructural crack propagation. *Int. J. Fatigue* **2021**, *152*, 106397. [CrossRef]
7. Gairola, S.; Jayaganthan, R. XFEM Simulation of Tensile and Fracture Behavior of Ultrafine-Grained Al 6061 Alloy. *Metals* **2021**, *11*, 1761. [CrossRef]
8. Ambati, M.; Gerasimov, T.; De Lorenzis, L. A review on phase-field models of brittle fracture and a new fast hybrid formulation. *Comput. Mech.* **2015**, *55*, 383–405. [CrossRef]
9. Zhuang, X.; Zhou, S.; Huynh, G.D.; Areias, P.; Rabczuk, T. Phase field modeling and computer implementation: A review. *Eng. Fract. Mech.* **2022**, *262*, 108234. [CrossRef]
10. Wu, J. A geometrically regularized gradient-damage model with energetic equivalence. *Comput. Method. Appl. Mech.* **2018**, *328*, 612–637. [CrossRef]
11. Molnár, G.; Gravouil, A. 2D and 3D Abaqus implementation of a robust staggered phase-field solution for modeling brittle fracture. *Finite Elem. Anal. Des.* **2017**, *130*, 27–38. [CrossRef]
12. Badnava, H.; Etemadi, E.; Msekh, M. A Phase Field Model for Rate-Dependent Ductile Fracture. *Metals* **2017**, *7*, 180. [CrossRef]
13. Miehe, C.; Hofacker, M.; Welschinger, F. A phase field model for rate-independent crack propagation: Robust algorithmic implementation based on operator splits. *Comput. Method. Appl. Mech.* **2010**, *199*, 2765–2778. [CrossRef]
14. Hasan, M.M.; Baxevanis, T. A phase-field model for low-cycle fatigue of brittle materials. *Int. J. Fatigue* **2021**, *150*, 106297. [CrossRef]
15. Živković, J.; Dunić, V.; Milovanović, V.; Pavlović, A.; Živković, M. A Modified Phase-Field Damage Model for Metal Plasticity at Finite Strains: Numerical Development and Experimental Validation. *Metals* **2021**, *11*, 47. [CrossRef]
16. Fang, J.; Wu, C.; Li, J.; Liu, Q.; Wu, C.; Sun, G.; Li, Q. Phase field fracture in elasto-plastic solids: Variational formulation for multi-surface plasticity and effects of plastic yield surfaces and hardening. *Int. J. Mech. Sci.* **2019**, *156*, 382–396. [CrossRef]
17. Molnár, G.; Gravouil, A.; Seghir, R.; Réthoré, J. An open-source Abaqus implementation of the phase-field method to study the effect of plasticity on the instantaneous fracture toughness in dynamic crack propagation. *Comput. Method. Appl. Mech.* **2020**, *365*, 113004. [CrossRef]
18. Carrara, P.; Ambati, M.; Alessi, R.; De Lorenzis, L. A framework to model the fatigue behavior of brittle materials based on a variational phase-field approach. *Comput. Method. Appl. Mech.* **2020**, *361*, 112731. [CrossRef]
19. Shi, Q.; Yu, H.; Guo, L.; Hao, L.; Huang, K. A phase field model with plastic history field for fracture of elasto-plastic materials. *Eng. Fract. Mech.* **2022**, *268*, 108447. [CrossRef]
20. Russ, J.; Slesarenko, V.; Rudykh, S.; Waisman, H. Rupture of 3D-printed hyperelastic composites: Experiments and phase field fracture modeling. *J. Mech. Phys. Solids* **2020**, *140*, 103941. [CrossRef]
21. Cui, C.; Ma, R.; Martínez-Pañeda, E. A phase field formulation for dissolution-driven stress corrosion cracking. *J. Mech. Phys. Solids* **2021**, *147*, 104254. [CrossRef]
22. Schreiber, C.; Kuhn, C.; Müller, R.; Zohdi, T. A phase field modeling approach of cyclic fatigue crack growth. *Int. J. Fract.* **2020**, *225*, 89–100. [CrossRef]
23. Wang, T.; Ye, X.; Liu, Z.; Liu, X.; Chu, D.; Zhuang, Z. A phase-field model of thermo-elastic coupled brittle fracture with explicit time integration. *Comput. Mech.* **2020**, *65*, 1305–1321. [CrossRef]
24. Cui, C.; Ma, R.; Martínez-Pañeda, E. A generalised, multi-phase-field theory for dissolution-driven stress corrosion cracking and hydrogen embrittlement. *J. Mech. Phys. Solids* **2022**, *166*, 104951. [CrossRef]
25. Khalil, Z.; Elghazouli, A.Y.; Martínez-Pañeda, E. A generalised phase field model for fatigue crack growth in elastic–plastic solids with an efficient monolithic solver. *Comput. Method. Appl. Mech.* **2022**, *388*, 114286. [CrossRef]
26. Shanthraj, P.; Sharma, L.; Svendsen, B.; Roters, F.; Raabe, D. A phase field model for damage in elasto-viscoplastic materials. *Comput. Method. Appl. Mech.* **2016**, *312*, 167–185. [CrossRef]
27. Sadeghirad, A.; Momeni, K.; Ji, Y.; Ren, X.; Chen, L.; Lua, J. Multiscale crystal-plasticity phase field and extended finite element methods for fatigue crack initiation and propagation modeling. *Int. J. Fract.* **2019**, *216*, 41–57. [CrossRef]

28. Amor, H.; Marigo, J.; Maurini, C. Regularized formulation of the variational brittle fracture with unilateral contact: Numerical experiments. *J. Mech. Phys. Solids* **2009**, *57*, 1209–1229. [CrossRef]
29. Alessi, R.; Vidoli, S.; De Lorenzis, L. A phenomenological approach to fatigue with a variational phase-field model: The one-dimensional case. *Eng. Fract. Mech.* **2018**, *190*, 53–73. [CrossRef]
30. Mesgarnejad, A.; Imanian, A.; Karma, A. Phase-field models for fatigue crack growth. *Theor. Appl. Fract. Mec.* **2019**, *103*, 102282. [CrossRef]
31. Grossman-Ponemon, B.E.; Mesgarnejad, A.; Karma, A. Phase-field modeling of continuous fatigue via toughness degradation. *Eng. Fract. Mech.* **2022**, *264*, 108255. [CrossRef]
32. Lo, Y.; Borden, M.J.; Ravi-Chandar, K.; Landis, C.M. A phase-field model for fatigue crack growth. *J. Mech. Phys. Solids* **2019**, *132*, 103684. [CrossRef]
33. Liu, Z.; Reinoso, J.; Paggi, M. Phase field modeling of brittle fracture in large-deformation solid shells with the efficient quasi-Newton solution and global–local approach. *Comput. Method. Appl. Mech.* **2022**, *399*, 115410. [CrossRef]
34. Emdadi, A.; Asle Zaeem, M. Phase-field modeling of crack propagation in polycrystalline materials. *Comp. Mater. Sci.* **2021**, *186*, 110057. [CrossRef]
35. Xiao, Y.Q.; Yang, L.; Zhu, W.; Zhou, Y.C.; Pi, Z.P.; Wei, Y.G. Delamination mechanism of thermal barrier coatings induced by thermal cycling and growth stresses. *Eng. Fail. Anal.* **2021**, *121*, 105202. [CrossRef]
36. Ai, W.; Wu, B.; Martínez-Pañeda, E. A coupled phase field formulation for modelling fatigue cracking in lithium-ion battery electrode particles. *J. Power Sources* **2022**, *544*, 231805. [CrossRef]
37. Ulloa, J.; Wambacq, J.; Alessi, R.; Degrande, G.; François, S. Phase-field modeling of fatigue coupled to cyclic plasticity in an energetic formulation. *Comput. Method. Appl. Mech.* **2021**, *373*, 113473. [CrossRef]
38. Haveroth, G.A.; Vale, M.G.; Bittencourt, M.L.; Boldrini, J.L. A non-isothermal thermodynamically consistent phase field model for damage, fracture and fatigue evolutions in elasto-plastic materials. *Comput. Method. Appl. Mech.* **2020**, *364*, 112962. [CrossRef]
39. Song, J.; Zhao, L.G.; Qi, H.; Li, S.; Shi, D.; Huang, J.; Su, Y.; Zhang, K. Coupling of phase field and viscoplasticity for modelling cyclic softening and crack growth under fatigue. *Eur. J. Mech. A/Solids* **2022**, *92*, 104472. [CrossRef]
40. Li, P.; Yvonnet, J.; Combescure, C.; Makich, H.; Nouari, M. Anisotropic elastoplastic phase field fracture modeling of 3D printed materials. *Comput. Method. Appl. Mech.* **2021**, *386*, 114086. [CrossRef]
41. Seleš, K.; Tomić, Z.; Tonković, Z. Microcrack propagation under monotonic and cyclic loading conditions using generalised phase-field formulation. *Eng. Fract. Mech.* **2021**, *255*, 107973. [CrossRef]
42. Tu, X.; Ray, A.; Ghosh, S. A coupled crystal plasticity FEM and phase-field model for crack evolution in microstructures of 7000 series aluminum alloys. *Eng. Fract. Mech.* **2020**, *230*, 106970. [CrossRef]
43. Loew, P.J.; Peters, B.; Beex, L.A.A. Rate-dependent phase-field damage modeling of rubber and its experimental parameter identification. *J. Mech. Phys. Solids* **2019**, *127*, 266–294. [CrossRef]
44. Loew, P.J.; Poh, L.H.; Peters, B.; Beex, L.A.A. Accelerating fatigue simulations of a phase-field damage model for rubber. *Comput. Method. Appl. Mech.* **2020**, *370*, 113247. [CrossRef]
45. Loew, P.J.; Peters, B.; Beex, L.A.A. Fatigue phase-field damage modeling of rubber using viscous dissipation: Crack nucleation and propagation. *Mech. Mater.* **2020**, *142*, 103282. [CrossRef]
46. Simoes, M.; Braithwaite, C.; Makaya, A.; Martínez Pañeda, E. Modelling fatigue crack growth in shape memory alloys. *Fatigue Fract. Eng. Mech.* **2022**, *45*, 1243–1257. [CrossRef]
47. Simoes, M.; Martínez-Pañeda, E. Phase field modelling of fracture and fatigue in Shape Memory Alloys. *Comput. Method. Appl. Mech.* **2021**, *373*, 113504. [CrossRef]
48. Martínez-Pañeda, E.; Golahmar, A.; Niordson, C.F. A phase field formulation for hydrogen assisted cracking. *Comput. Method. Appl. Mech.* **2018**, *342*, 742–761. [CrossRef]
49. Huang, C.; Gao, X. Phase field modeling of hydrogen embrittlement. *Int. J. Hydrogen Energ.* **2020**, *45*, 20053–20068. [CrossRef]
50. Wu, J.; Mandal, T.K.; Nguyen, V.P. A phase-field regularized cohesive zone model for hydrogen assisted cracking. *Comput. Method. Appl. Mech.* **2020**, *358*, 112614. [CrossRef]
51. Duda, F.P.; Ciarbonetti, A.; Toro, S.; Huespe, A.E. A phase-field model for solute-assisted brittle fracture in elastic-plastic solids. *Int. J. Plast.* **2018**, *102*, 16–40. [CrossRef]
52. Kristensen, P.K.; Niordson, C.F.; Martínez-Pañeda, E. Applications of phase field fracture in modelling hydrogen assisted failures. *Theor. Appl. Fract. Mec.* **2020**, *110*, 102837. [CrossRef]
53. Golahmar, A.; Kristensen, P.K.; Niordson, C.F.; Martínez-Pañeda, E. A phase field model for hydrogen-assisted fatigue. *Int. J. Fatigue* **2022**, *154*, 106521. [CrossRef]
54. Mai, W.; Soghrati, S. A phase field model for simulating the stress corrosion cracking initiated from pits. *Corros. Sci.* **2017**, *125*, 87–98. [CrossRef]
55. Mai, W.; Soghrati, S.; Buchheit, R.G. A phase field model for simulating the pitting corrosion. *Corros. Sci.* **2016**, *110*, 157–166. [CrossRef]
56. Schreiber, C.; Müller, R.; Kuhn, C. Phase field simulation of fatigue crack propagation under complex load situations. *Arch. Appl. Mech.* **2021**, *91*, 563–577. [CrossRef]
57. Yan, S.; Schreiber, C.; Müller, R. An efficient implementation of a phase field model for fatigue crack growth. *Int. J. Fract.* **2022**, *237*, 47–60. [CrossRef]

58. Seleš, K.; Aldakheel, F.; Tonković, Z.; Sorić, J.; Wriggers, P. A general phase-field model for fatigue failure in brittle and ductile solids. *Comput. Mech.* **2021**, *67*, 1431–1452. [CrossRef]
59. Wu, J.; Huang, Y.; Nguyen, V.P. On the BFGS monolithic algorithm for the unified phase field damage theory. *Comput. Method. Appl. Mech.* **2020**, *360*, 112704. [CrossRef]
60. Wu, J.; Huang, Y. Comprehensive implementations of phase-field damage models in Abaqus. *Theor. Appl. Fract. Mec.* **2020**, *106*, 102440. [CrossRef]
61. Kristensen, P.K.; Martínez-Pañeda, E. Phase field fracture modelling using quasi-Newton methods and a new adaptive step scheme. *Theor. Appl. Fract. Mec.* **2020**, *107*, 102446. [CrossRef]
62. Freddi, F.; Mingazzi, L. Mesh refinement procedures for the phase field approach to brittle fracture. *Comput. Method. Appl. Mech.* **2022**, *388*, 114214. [CrossRef]
63. Xu, W.; Li, Y.; Li, H.; Qiang, S.; Zhang, C.; Zhang, C. Multi-level adaptive mesh refinement technique for phase-field method. *Eng. Fract. Mech.* **2022**, *276*, 108891. [CrossRef]
64. Gupta, A.; Krishnan, U.M.; Mandal, T.K.; Chowdhury, R.; Nguyen, V.P. An adaptive mesh refinement algorithm for phase-field fracture models: Application to brittle, cohesive, and dynamic fracture. *Comput. Method. Appl. Mech.* **2022**, *399*, 115347. [CrossRef]
65. Eldahshan, H.; Alves, J.; Bouchard, P.; Perchat, E.; Munoz, D.P. CIPFAR: A 3D unified numerical framework for the modeling of ductile fracture based on the phase field model and adaptive remeshing. *Comput. Method. Appl. Mech.* **2021**, *387*, 114171. [CrossRef]
66. Giovanardi, B.; Scotti, A.; Formaggia, L. A hybrid XFEM –Phase field (Xfield) method for crack propagation in brittle elastic materials. *Comput. Method. Appl. Mech.* **2017**, *320*, 396–420. [CrossRef]
67. Patil, R.U.; Mishra, B.K.; Singh, I.V. A local moving extended phase field method (LMXPFM) for failure analysis of brittle materials. *Comput. Method. Appl. Mech.* **2018**, *342*, 674–709. [CrossRef]
68. Muixí, A.; Marco, O.; Rodríguez-Ferran, A.; Fernández-Méndez, S. A combined XFEM phase-field computational model for crack growth without remeshing. *Comput. Mech.* **2021**, *67*, 231–249. [CrossRef]
69. Chen, Y.; Vasiukov, D.; Gélébart, L.; Park, C.H. A FFT solver for variational phase-field modeling of brittle fracture. *Comput. Method. Appl. Mech.* **2019**, *349*, 167–190. [CrossRef]
70. Ma, R.; Sun, W. FFT-based solver for higher-order and multi-phase-field fracture models applied to strongly anisotropic brittle materials. *Comput. Method. Appl. Mech.* **2020**, *362*, 112781. [CrossRef]
71. Goswami, S.; Anitescu, C.; Chakraborty, S.; Rabczuk, T. Transfer learning enhanced physics informed neural network for phase-field modeling of fracture. *Theor. Appl. Fract. Mec.* **2020**, *106*, 102447. [CrossRef]
72. Amiri, F.; Millán, D.; Arroyo, M.; Silani, M.; Rabczuk, T. Fourth order phase-field model for local max-ent approximants applied to crack propagation. *Comput. Method. Appl. Mech.* **2016**, *312*, 254–275. [CrossRef]

metals

MDPI

Article

Estimation Method of Relative Slip in Fretting Fatigue Contact by Digital Image Correlation

Yue Su [1,2], Shao-Shi Rui [1,3], Qi-Nan Han [4,5], Zhi-Hao Shang [1], Li-Sha Niu [1], Hao Li [6], Hiroshi Ishikawa [6] and Hui-Ji Shi [1,*]

1 AML, School of Aerospace Engineering, Tsinghua University, Beijing 100084, China; suy@nwpu.edu.cn (Y.S.); ruishaoshi@imech.ac.cn (S.-S.R.); troygol@126.com (Z.-H.S.); niulsh@tsinghua.edu.cn (L.-S.N.)
2 School of Power and Energy, Northwestern Polytechnical University, Xi'an 710129, China
3 LNM, Institute of Mechanics, Chinese Academy of Sciences, Beijing 100190, China
4 College of Energy and Power Engineering, Nanjing University of Aeronautics and Astronautics, Nanjing 210016, China; hanqn@nuaa.edu.cn
5 Aero-Engine Thermal Environment and Structure Key Laboratory of Ministry of Industry and Information Technology, Nanjing 210016, China
6 Strength Research Department, Research & Innovation Center, Mitsubishi Heavy Industries, Ltd., Takasago 676-8686, Japan; hao.li.j8@mhi.com (H.L.); hiroshi.ishikawa.j5@mhi.com (H.I.)
* Correspondence: shihj@mail.tsinghua.edu.cn; Tel.: +86-(10)-6277-2731

Abstract: An experimental method that can quantify relative slip was developed using digital image correlation (DIC) in order to evaluate the sliding portion. The bridge-type test setup was designed to establish the fretting contact condition. The relative displacements between the contact surfaces were determined by DIC methods. Based on the evolution and distribution of relative slip, the transitions from gross slip to partial slip on the contact surface were found throughout all tests. This result indicated that the fretting scar was closely correlated to relative slip. The variation of relative slip corresponding to the stick-slip state was consistent with the tangential force coefficient. Besides, the load amplitude was an important factor for fretting fatigue damage, which can affect the stick-slip state.

Keywords: fretting fatigue; digital image correlation; relative slip; tangential contact stiffness

Citation: Su, Y.; Rui, S.-S.; Han, Q.-N.; Shang, Z.-H.; Niu, L.-S.; Li, H.; Ishikawa, H.; Shi, H.-J. Estimation Method of Relative Slip in Fretting Fatigue Contact by Digital Image Correlation. *Metals* **2022**, *12*, 1124. https://doi.org/10.3390/met12071124

Academic Editor: Tilmann Beck

Received: 24 May 2022
Accepted: 28 June 2022
Published: 30 June 2022

1. Introduction

Fretting fatigue occurs between two solid surfaces in contact with a small relative displacement (usually less than 100 μm) due to the oscillating or cyclic bulk force [1–3]. This condition leads to severe surface degradation and crack initiation, which will reduce the life of components, such as railway axles [4,5], rivet holes [6,7], bolted joints [8], and ropes [9]. Especially, the fretting damage at the blade root attachments in the turbine and compressor has been widely reported, which is caused by centrifugal and aerodynamic force [10–13]. Thus, more and more researchers' attentions are focused on the fretting fatigue at the turbine blade-disk attachments.

Many studies have shown that the fretting behavior largely depends on the slip amplitude between contact surfaces [14–16]. If the remotely applied relative displacement amplitude is large enough, all points on the contact surface will undergo sliding, which is known as the gross slip condition. Conversely, some parts of the contact surfaces remain stuck when the remotely applied relative displacement amplitude is low. Therefore, the contact surface includes the slip zone and stick zone, and this is referred to as the partial slip. At the same time, the damage caused by fretting is closely related to the slip condition [17–19]. The fretting wear mainly occurs in gross slip, while the fretting fatigue is more significant in partial slip. Vingsbo [20] pointed out that the number of load cycles first decreases and then increases with increasing slip amplitude under fretting experiments. Additionally, the minimum fretting life occurs at the transition from partial slip to gross slip. The effect of slip on fretting fatigue life was also observed by Jin and Mall [21]. In addition, experimental studies have proved that the fretting crack behavior is usually closely related to the slip or stick state of contact surfaces. For example, the fracture morphology [22,23] and in situ observation [24,25] have clearly shown that the fretting cracks nucleate at the stick-slip interface. Thus, the slip amplitude measurement under fretting tests is still needed in order to distinguish the stick or slip region on the contact surfaces.

The measurement methods of slip amplitude have been widely reported in the literature [26–31]. For example, Wittkowsky et al. [26] used an extensometer assembly to measure the relative displacement between the fretting and specimen. The clip gauge was installed on the sample and fretting pad to monitor the relative displacement between the contact surfaces by Pauw et al. [30]. Besides, Ding et al. [31] performed a fretting fatigue experiment and completed the slip measurement by the linear variable differential transformers (LVDT). However, the results obtained by the above methods cannot reflect the true relative displacement on both sides of the contact surface. Since the two contact feet of measuring equipment were connected to the sample and fretting pad, they were at a certain distance from the contact surfaces. The measurement result only represents the relative displacement of two points, rather than the distribution of the relative displacement along the contact surfaces. This not only is remote data but also includes the elastic deformation of surrounding material. Furthermore, the deformation of the complicated device was also included in the slip results. In short, it is challenging to accurately measure the small relative slip between the contact surfaces under fretting conditions.

Recently, aside from the conventional methods mentioned above, the relative displacement between the fretting contact surfaces has also been studied by digital image correlation (DIC). Based on the gray level of the image, the DIC method is a noncontact full-field deformation optical measurement technique [32,33] Juoksukangas et al. [34,35] performed cantilever beam bending tests to simulate the fretting contact state. The DIC technique was used to quantify the local relative displacement field at the fretting contact interface and minimize the effect of test device compliances. Crevoisier et al. [36] obtained displacement jump in bolted assembly by the DIC method. Additionally, the in situ frictional properties of a fretting surface, such as friction coefficient and secant stiffness, were identified. Besides, Kartal [29] and Pauw [30] employed the DIC method to describe the hysteresis loop during macroscale fretting fatigue tests. These results show that the tangential contact stiffness depends highly on the normal pressure and contact area. However, their works were limited to an overall slip value of the entire contact surface. The distribution and evolution of the slip along the contact surface were rarely reported. It is impossible to give the stick-slip state of the contact surface under partial slip condition. Therefore, in order to obtain the ideal displacement data, it is necessary to develop a novel measurement method based on the DIC technique with a higher pixel and accuracy.

This paper aims to grasp the distribution and evolution of the stick-slip zone between the contact surfaces under fretting conditions. The DIC technique was employed to measure the relative displacement field between the contact interfaces. Then a method that can be used to evaluate the sliding part of the contact surface under fretting fatigue was established. For this purpose, Section 2 introduces fretting fatigue experiments to simulate the fretting fatigue behavior of the blade-disk attachments in a turbine engine. Additionally, a novel optical system was built in Section 2. Section 3 presents the hysteresis loops of relative displacement and tangential force. The slip values and tangential contact stiffness during the fretting fatigue tests were determined in Section 4. Then the distribution of the stick-slip state along the contact surfaces was obtained under the fretting conditions. Besides, the effect of cyclic load amplitude on fretting properties was investigated. Finally, the conclusions are summarized in Section 5.

2. Materials and Experiment

For the experimental work, the bridge-type device was suitable for simulating the fretting fatigue conditions in the laboratory. A camera was used to implement DIC for the measurement of the relative displacement between the contact surfaces. In the following sections, detailed information on the experimental configuration will be given.

2.1. Material and Sample

The material chosen for this paper was SUS410 stainless steel. This alloy is widely used in a turbine or compressors due to its excellent mechanical properties [37]. The typical mechanical properties are listed in Table 1, which are obtained from tensile tests based on the ASTM E08 standard [38]. All specimens and fretting pads were made from the same plate of stainless steel.

Table 1. Mechanical properties of SUS410 steel in this work.

Young's Modulus (GPa)	Proportionality Limit (MPa)	0.2% Proof Stress (MPa)	Tensile Strength (MPa)	Elongation (%)	Poisson's Ratio
212.31	438.29	539.52	715.44	28.78	0.26

The geometries of the specimen and fretting pad are shown in Figure 1. These dimensions are prepared according to the JSME standard [39]. As can be seen from the figure, the fretting pad has two contact feet, and each of which has a 3.1 mm flat surface with rounded edges of R = 0.2 mm. This foot is in contact with the flat side of the dog-bone specimen (see Figure 2). The thicknesses of the dog-bone specimen and fretting pad are 6 mm. There are no edges in the thickness direction to ensure complete contact. Thus, the area of the contact surface is 3.1×6 mm^2. All specimens and fretting pads were machined by wire electrical discharge machining, and the contact surfaces were both ground and polished to reduce the finish surface roughness (Ra = 0.4 μm).

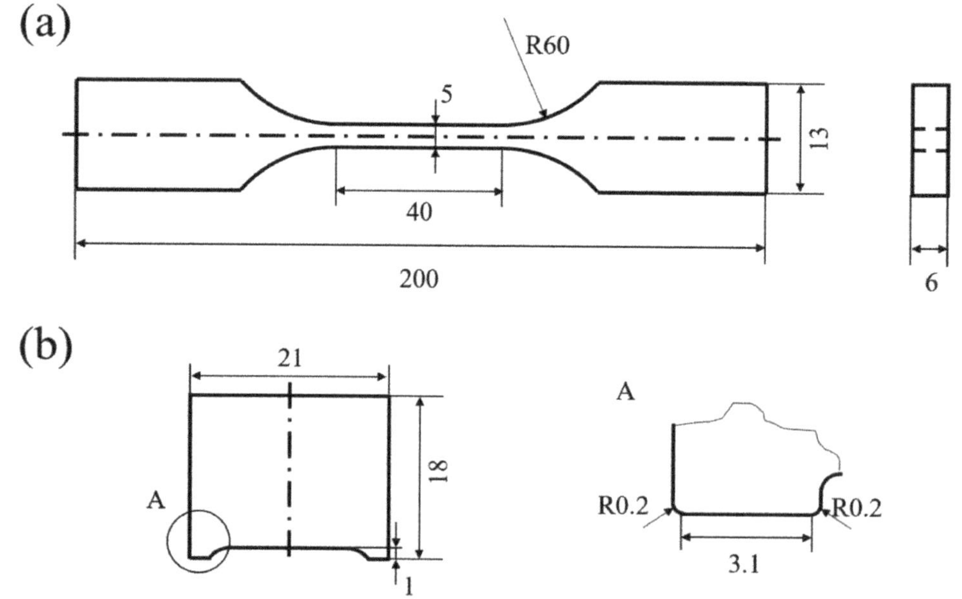

Figure 1. Dimensions of the (**a**) specimen and (**b**) fretting pad (all dimensions are in mm).

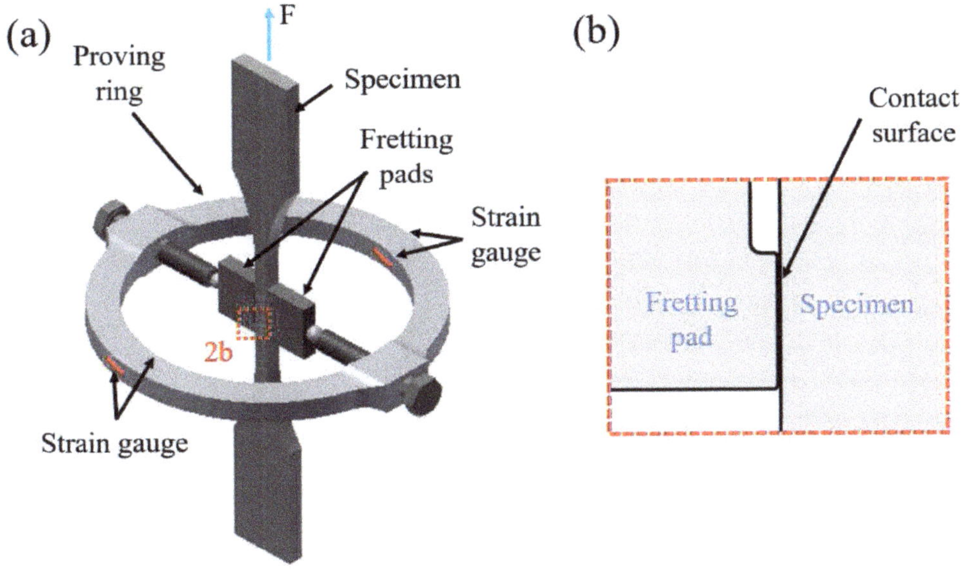

Figure 2. Schematic diagram of (**a**) the fretting fatigue experimental setup and (**b**) the contact surface between the fretting pad and specimen.

2.2. Fretting Fatigue Test

An experimental setup was designed to simulate the fretting contact condition, which is based on the JSME standard [39]. A schematic of the bridge-type fretting fatigue test apparatus is shown in Figure 2. Two fretting pads were pressed on the side of the dog-bone specimen. The proving ring and screw were used to apply the normal contact force between the specimen and fretting pads. The dog-bone specimen was mounted on the servo-hydraulic fatigue testing machine (Instron 8801). Then, the upper end of the specimen was fixed by the chuck, and the cyclic load was applied to the specimen along the length direction by the movement of the lower gripper. Hence, a total of four flat-on-flat fretting contact surfaces (Figure 2b) were formed between the specimen and fretting pads. This is a typical contact form for the fir-tree-type blade-disk attachment in the turbine engine. Since the contact was in plane strain state, the bending of the fretting pads was ignored. As shown in Figure 2a, four strain gauges were used to monitor the normal contact force, which are attached at the proving ring. These results of strain gauges revealed that the force value was changed only slightly during the testing. A strain gauge was installed between the fretting pad contact feet, and the tangential force (Ft) was calculated from the corresponding strain values. More detailed information about measurement methods can be found in the literature [39,40]. In this fretting fatigue test, these strain values were collected at a frequency of 1000 Hz.

In this paper, the fretting fatigue test was carried out at room temperature. These test parameters are listed in Table 2. The normal contact pressure in the horizontal direction was chosen to be 90 MPa, according to the deformation of the proving ring, during which the screw is tightened. This roughly presents the normal contact pressure in the actual designs of the blade-disk attachment. A sinusoidal cyclic load with a frequency of 20 Hz was axially applied to the specimen. The stress ratio of the applied cyclic load was set as −1. In order to investigate the effect of cyclic load amplitude (Fa) on the fretting fatigue behavior, three different load levels (8.5, 7.3, and 6.5 kN) were used in our work. The nominal stress amplitudes corresponding to three different load amplitudes were 283.33, 243.33, and 216.67 MPa. Before the test, all contact surfaces were cleaned with acetone. Additionally, marks were made in the vertical and horizontal directions of the specimen. Then, the specimen position was adjusted in the camera field of view in order to ensure alignment. Each test was conducted until the specimen broke or the number of cycles reached 200,000.

Table 2. Loading parameters of the fretting fatigue test in this work.

Stress Ratio	Frequency	Waveform	Cyclic Load Amplitude (*Fa*)	Average Normal Pressure
−1	20 Hz	Sine wave	8.5, 7.3, 6.5 kN	90 MPa

2.3. Digital Image Correlation (DIC)

The DIC method was used to measure the full-field displacement at the contact region during the above fretting fatigue testing, which is a well-known established technique in experimental mechanics [41,42]. The complete test system is shown in Figure 3, which includes the fretting fatigue setup, optical device, and strain measurement equipment. It can be seen from the figure that the fretting fatigue setup was installed on the fatigue testing machine, and was axially loaded with periodical force. Two lights were used to improve image quality. The speckle pattern was spraying on the surface of the specimen and fretting pads, and the diameter of one black spot was approximately 50 μm. This change in speckle was monitored during each experiment by a camera with a resolution of 2448 × 2050 pixels (5 megapixels). Since four contact surfaces were similar, only the lower contact region (Figure 4a) was observed during the measurement. A rectangular field of interest was recorded at a rate of 1 Hz. As shown in Figure 4b, a contact surface was found in this observation view. The frequency of the fretting fatigue test was reduced from 20 to 0.05 Hz during DIC measurement. Therefore, about 20 image data points were measured in each fatigue cycle. The tangential force was measured at two frequencies (20 and 0.05 Hz). The results show that the difference in tangential force was relatively slight. Then, the displacement field can be calculated by the correlation algorithms based on the deformed image. Table 3 summarizes the postprocessing parameters. A subset of 101 × 101 pixels was used in this work, and the step was set as 15 pixels. Before the formal test, two images were continuously recorded when the fretting fatigue setup was at rest in order to determine the displacement measurement accuracy. Then, repeated measurements were carried out for the same static state. Based on the above postprocessing parameters, the measurement error of vertical displacement is less than 1 μm.

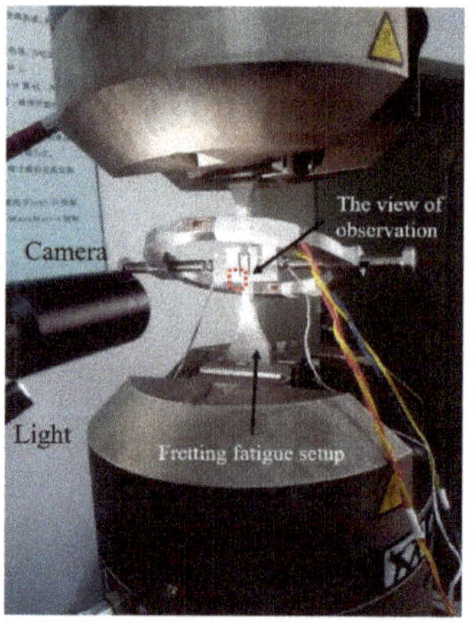

Figure 3. Photograph of the complete experimental setup in this study (The magnified area of the red wireframe is the field of view observed by the camera).

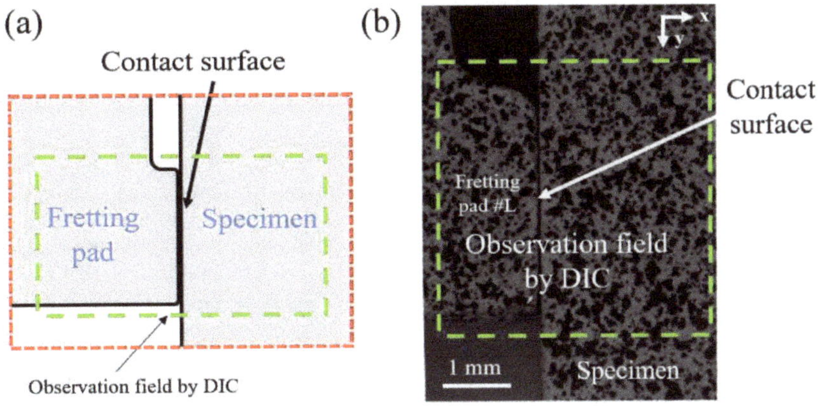

Figure 4. (**a**) Schematic diagram of the observation field for the DIC measurement and (**b**) observation field captured by the camera during the test.

Table 3. DIC measurement parameters used in this work.

Camera Pixel	Field of View	Sampling Rate	Subset	Step
5 megapixels (2448 × 2050)	8.8 × 6.6 mm^2	20 images/cycle	101 × 101 pixels	15 pixels

3. Experimental Results

3.1. Fretting Scar

As a result of the fretting damage, fretting scars were observed on all contact surfaces of the specimen and fretting pad after the experiment. Notice that the marks on the left and right contact surfaces were similar due to the symmetry of the fretting fatigue experimental setup. Therefore, only the fretting scar on the left contact surface is shown in Figure 5. For the 6.5 kN case, the contact surface includes the stick and slip zone, as illustrated in Figure 5c. The center of the contact surface remains sticky. Additionally, the severe wear marks only exist at the edge of the stick zone, which is obviously in partial slip condition. In addition, it can be seen that the severe wear marks exist on the entire contact surface under the cyclic load amplitude of 8.5 and 7.3 kN, which is known as the gross slip condition. These fretting scars indicate that the reduced load amplitude is an effective measure to improve the fretting fatigue properties. Similar results were found by Cortez and Massingham [43,44].

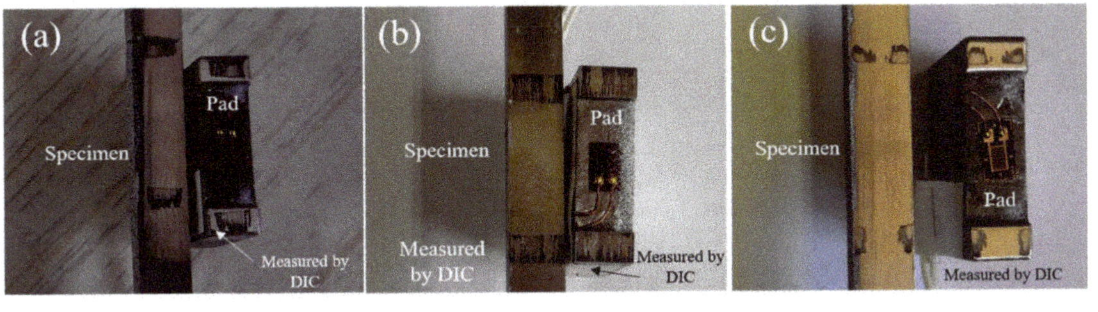

Figure 5. Fretting scar on the specimen and fretting pad for (**a**) Fa = 8.5 kN, (**b**) *Fa* = 7.3 kN, and (**c**) *Fa* = 6.5 kN.

3.2. Tangential Force Coefficient

The tangential force coefficient (TFC) is a significant parameter as it affects the fretting response [45]. It was calculated using Equation (1):

$$\text{TFC} = \frac{Ft}{Fn} \tag{1}$$

where Ft is the tangential force amplitude and Fn is the normal contact force. They were obtained according to the strain during the test.

Figure 6 shows the evolution of the TFC during the test. In all load amplitude cases, the TFC at the initial stage was low, and their values were 0.1–0.2. This indicates that the slip occurs between the contact surfaces, which is caused by the high surface quality at the original state. Then, the TFC increases rapidly as the number of cycles increases. This large range of variation reveals that the wear appeared on the surface and the roughness deteriorated due to the slip on the contact surfaces at the beginning of the test. When the number of cycles reached about 1000, the TFC was gradually stabilized. This result reflects that the contact condition has become a stable state. On the other hand, the effect of cyclic load amplitude on the TFC response was also discussed in this paper. When the Fa reduced from 8.5 to 7.3 kN, the steady-state value of the TFC did not change, and their values were both between 0.71 and 0.72. However, as the load amplitude reduced to 6.5 kN, the steady-state value of the TFC was 0.5. Hence, for the 6.5 kN case, the contact surfaces were in a partial slip state. In turn, the gross slip condition appeared at the 8.5 kN case. It is worth noting that, based on the TFC alone, the state of the slip plane cannot be determined. However, according to the fretting scar, the gross slip occurred at the 6.5 kN test.

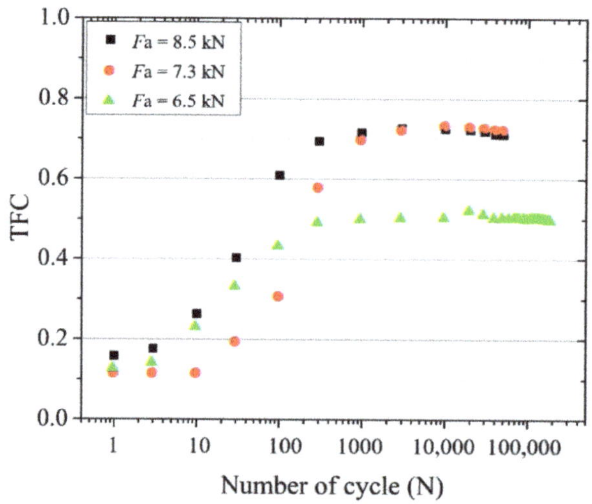

Figure 6. Evolution of TFC with the cycle number during the test.

3.3. Displacement Field

The vertical displacement (v) field of the observation area obtained by DIC postprocessing is depicted in Figure 7a. The cyclic load value was 7.3 kN. As shown in the figure, the dog-bone specimen was located on the right side of the area, and the fretting pad appeared on the left side. Thus, the contact surface was observed in this area. As expected, a bandlike form displacement field appeared in the observation area of the specimen. Since the cyclic load was applied to the specimen through the movement of the lower chuck, the vertical displacement near the lower side was higher. There is a nondata area near contact surfaces due to the divided calculation area and the limitation of a postprocessing algorithm. Besides, the vertical displacement of the fretting pad is lower than that of the specimens.

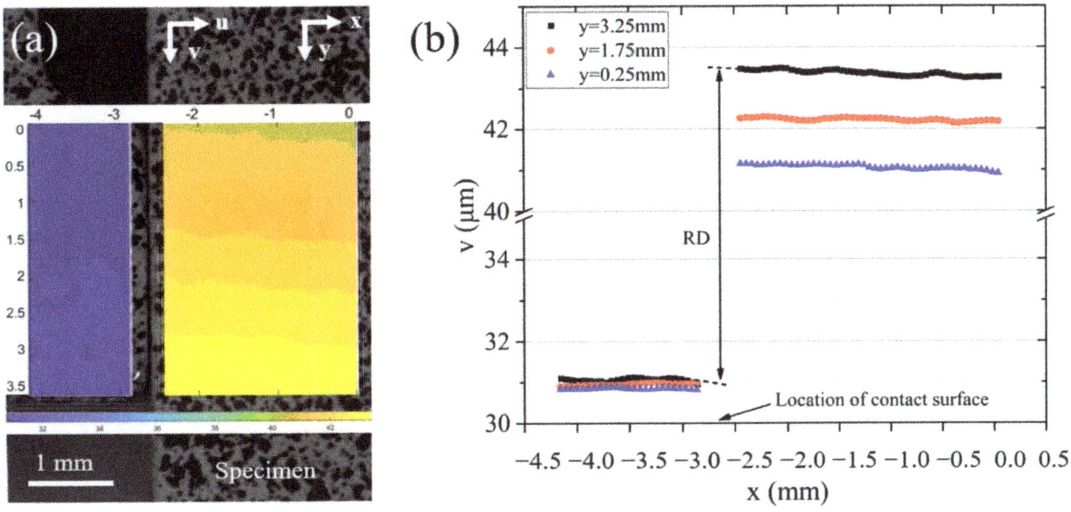

Figure 7. (**a**) Vertical displacement cloud of the observation area and (**b**) distribution of vertical displacement along the x-direction for the 7.3 kN case (cycle 1).

In order to describe the relative displacement between the contact surfaces, the distribution of vertical displacement along the x-direction is shown in Figure 7b. It can be seen that the displacement variation of the fretting pad along the x-direction was lower than the specimen. Additionally, the vertical displacement increases with the y-coordinate increases, which is consistent with the cyclic load applied by the movement of the lower chuck. Besides, there is a value jump due to the contact surface. Despite the lack of data near contact surfaces, the displacement still exhibits a linear law, as shown in Figure 7b. Therefore, the vertical displacement of the specimen and fretting pad on the contact surfaces can be determined by linear fitting. Furthermore, the relative displacement (RD) between the contact surfaces can be obtained in this study as follows:

$$\text{RD} = v_{\text{specimen}} - v_{\text{pad}} \tag{2}$$

where v_{specimen} and v_{pad} are the vertical displacement of the specimen and fretting pad on the contact surfaces, respectively.

At the same time, the displacement field was obtained along the thickness direction, but the displacement change was smaller than that of the contact surface. Consider that this work focuses on the fretting state of the contact surface. Therefore, the distribution in the thickness direction was not discussed in detail.

3.4. Hysteresis Loops

Figure 8 shows the hysteresis loops of relative displacement and tangential force during the experiment for the 8.5 kN case. The tangential force was measured by the strain, and the relative displacement was determined based on the DIC measurement results mentioned above. As shown in Figure 8c, the tangential stiffness was defined as the slope of the hysteresis loop at the stick stage, which is as follows:

$$\text{k} = \Delta Ft/\Delta \text{RD} \tag{3}$$

where ΔFt and ΔRD are the increments of tangential force and relative displacement.

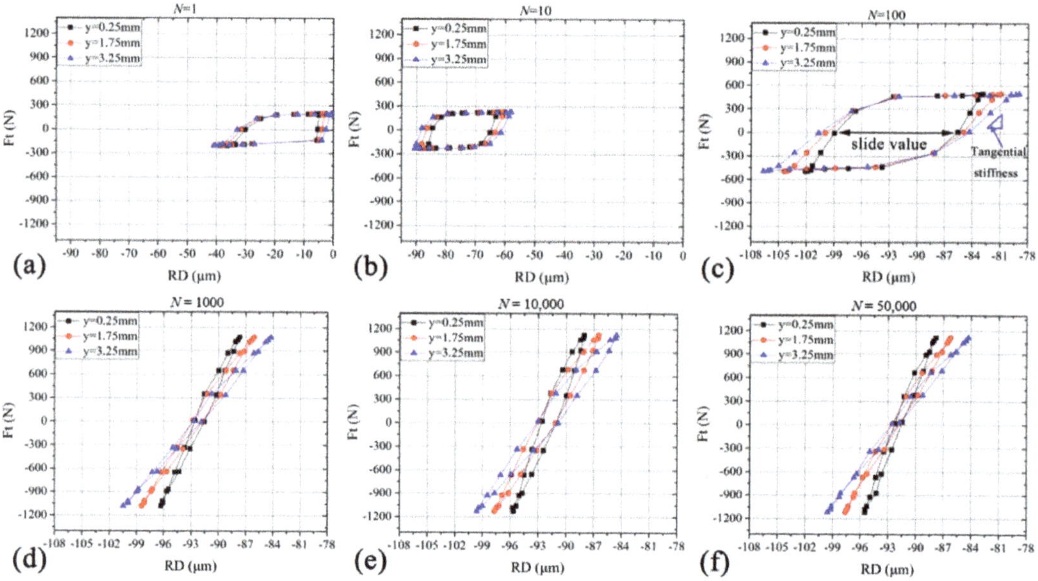

Figure 8. Hysteresis loops of tangential force (*Ft*) in a function of relative displacement (RD) for the 8.5 kN case: **(a)** *N* = 1, **(b)** *N* = 10, **(c)** *N* = 100, **(d)** *N* = 1000, **(e)** *N* = 10,000, and **(f)** *N* = 50,000.

For cycle 1, the contact surface was in gross slip condition due to the parallelogram-like hysteresis loops at different positions. When the tangential frictional force was reduced to zero, the relative displacement cannot be fully recovered. With the increasing number of cycles, the tangential force gradually increased and stabilized at 1175 N after 10,000 cycles. Additionally, note that the hysteresis loops change from parallelogram to elliptical. In addition, the relative displacement between the contact surfaces increases when the y-coordinate increases. This reveals that the movement closer to the lower chuck was greater. Especially, the curve-enclosed area of the hysteresis loop represents the frictional energy loss during the experiment. It is worth noting that severe wear occurred during the early stage of the experiment. Subsequently, the energy loss decreases and stabilizes as the number of cycles increases.

The effect of cyclic load amplitude on the hysteresis loop was also explored in this paper. Figures 9 and 10 show the hysteresis loops for the 7.3 and 6.5 kN cases. For all three cases, the hysteresis loops exhibit the transformation from a parallelogram to an ellipse with the increasing load cycle. Additionally, these loops have a similar evolution process at different heights, as shown in Figures 9 and 10. However, for the 6.5 kN case, the shape of hysteresis loops became a straight line after 10,000 cycles. Further, the tangential force amplitude was stabilized at 950 N, which was lower than the 1175 N for other load conditions. These results reflect that the local contact surface was in stick condition. In addition, since the fretting pad cannot be completely symmetrical, self-equilibrium will occur on the contact surface, which leads to complicated changes in the relative displacement in the early stage. As such, the mean value of the RD was positive during the first 10 cycles and then became negative for the 7.3 kN case, as shown in Figure 9a,b. However, in all cases, the mean values were all negative at the stable stage. This is consistent with previous works [29,34].

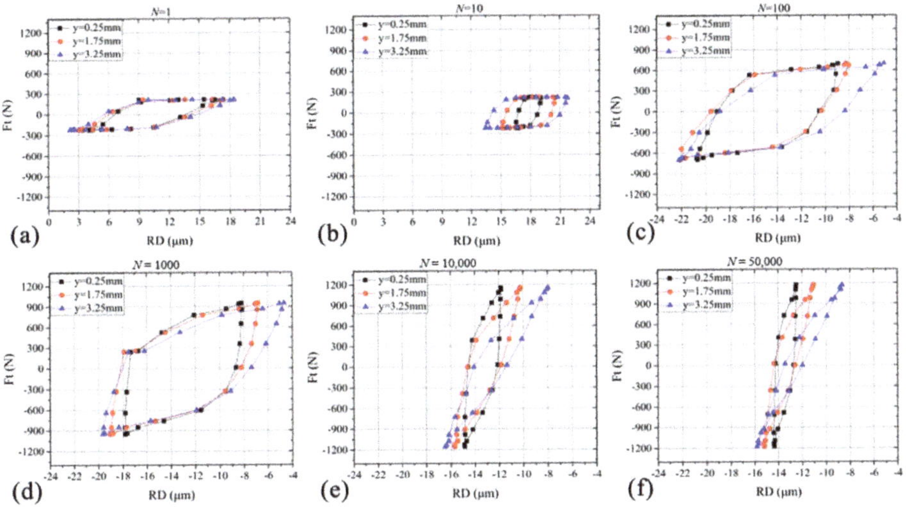

Figure 9. Hysteresis loops of tangential force (*Ft*) in a function of relative displacement (RD) for the 7.3 kN case: (**a**) *N* = 1, (**b**) *N* = 10, (**c**) *N* = 100, (**d**) *N* = 1000, (**e**) *N* = 10,000, and (**f**) *N* = 50,000.

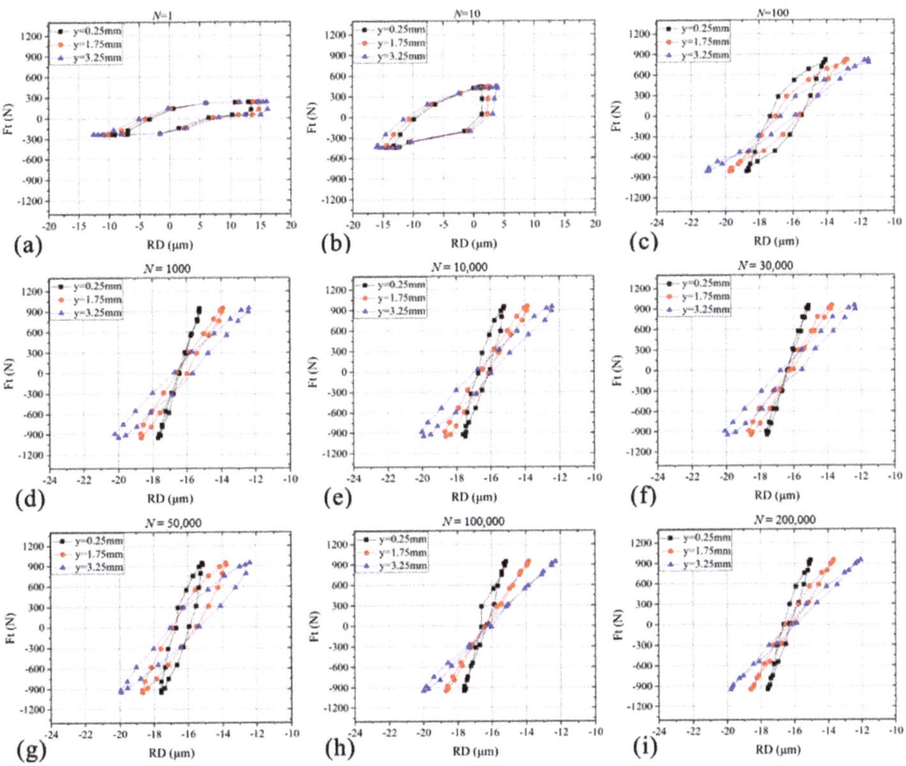

Figure 10. Hysteresis loops of tangential force (*Ft*) in a function of relative displacement (RD) for the 6.5 kN case: (**a**) *N* = 1, (**b**) *N* = 10, (**c**) *N* = 100, (**d**) *N* = 1000, (**e**) *N* = 10,000, (**f**) *N* = 30,000, (**g**) *N* = 50,000, (**h**) *N* = 100,000, and (**i**) *N* = 200,000.

4. Discussion and Analysis

4.1. Definition of Slip Value and Tangential Stiffness

Many studies have confirmed that slip amplitude and tangential contact stiffness are important parameters of contact properties. For example, Madge et al. investigated the effect of slip amplitude on fretting fatigue life [46,47]. Allara et al. used contact stiffness to evaluate the fretting fatigue response of turbine blades [48]. Meanwhile, lip value and tangential stiffness can be determined through the hysteresis loop, and their definitions can be found in Figure 8c. In this paper, these parameters were calculated by the hysteresis loops under different test conditions mentioned above.

4.2. Evolution of Relative Slip with the Cycle Number

Relative slip is an important factor that characterizes the properties of fretting contact surfaces, and it is helpful to determine whether the contact point is stick or slip state. To illustrate the evolution law of relative slip, its change with the cycle number during the experiment at different cyclic load amplitudes is shown in Figure 11. It can be seen that there is a large relative slip at the beginning of the experiments. Then, for the 8.5 kN case, these values decrease sharply and stabilize as the number of cycles increases. This phenomenon indicates that the larger slide occurred in the initial stage. Additionally, with the further cycles, the slide between the contact surfaces decreases rapidly. This is because the debris adheres to the interface surfaces, which deteriorates the surface quality. When the state of the contact surface is stable, the variation of relative slip with the number of cycles becomes smaller. This is consistent with the evolution law of the TFC described above (see Figure 6). In addition, three curves of relative slip at different height positions (y-direction) are also shown in Figure 11a. Although the relative slip at y = 0.25 mm was lower than others, the evolution law of relative slip was still similar for the three curves. A more detailed discussion about the distribution of relative slip along the contact surfaces can be found later in Section 4.3.

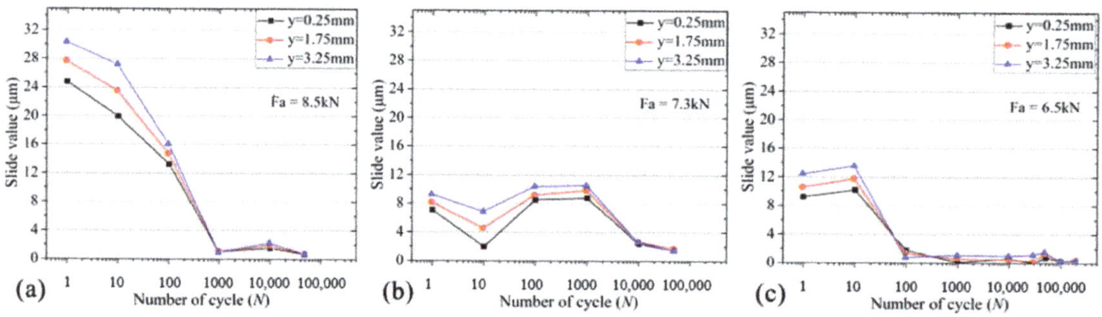

Figure 11. The evolution of relative slip with the cycle number during the test for the (**a**) 8.5 kN, (**b**) 7.3 kN, and (**c**) 6.5 kN cases.

In order to investigate the effect of the load amplitude on the fretting fatigue behavior, the relative slips at the 7.3 and 6.5 kN cases were also compared in Figure 11b,c. Obviously, there is a similar evolution law of relative slip as the cyclic load amplitude decreases. In all tests, the relative slip started from a relatively high value, but decreased rapidly and stabilized. However, this figure also reflects the difference in relative slip between the three tests. For instance, the relative slip at the beginning of the experiments for the 8.5 kN case was between 24 and 30 μm, values that were highest in both cases. When the load amplitude decreased, the relative slip at 1 cycle was reduced from about 28 to 10 μm. The highest relative slip at the initial stage indicates that there is more severe wear between the contact surfaces at 8.5 kN. This result is consistent with the fretting scar (see Figure 5) on the contact surfaces. It should be noted that the relative slip was no longer decreased when the load amplitude was reduced to 6.5 kN, and their values remained at about 10 μm in cycle 1. However, the relative slip was stable after 100 cycles for the 6.5 kN case, instead of 10,000 cycles in the 7.3 kN case. Hence, a longer period of slide occurred on the contact surfaces, since the cyclic load amplitude increased from 6.5 to 7.3 kN. At the same time, it is worth noting that the relative slip increased slightly at the beginning of the experiment, as shown in Figure 8b. It is possible that self-equilibration occurred early in the experiment because of the incomplete symmetry of the fretting pad. Nonetheless, the overall trend is clear.

4.3. Distribution of Relative Slip along the Contact Surface

Figure 12a shows the distribution of relative slip along the y-direction throughout all tests at 8.5 kN in order to further recognize the slip or stick state of the contact position. It can be seen from the figure that all relative slips increase as the y-coordinate increases. Before 100 cycles, the relative slip of the contact surfaces all exceeded 5 μm. This indicates that all points on the contact surfaces were in a slip state, which is known as gross slip condition. It provides an explanation for the dramatic increase in TFC at the beginning of the experiment. Then, the relative slip decreased rapidly as the experiment ran. Since the higher relative slip was near the lower chuck (y = 3.25 mm), the upper side of the contact surface (y = 0.25 mm) became stuck earlier. In contrast, other positions on the contact surfaces were still in a slip state. Therefore, the contact condition changes from a gross slip to a typical partial slip at this moment. Furthermore, when the number of cycles was higher than 10,000, the relative slip of all points on the interface surfaces was less than 2 μm, as shown in Figure 12a. This result shows that the entire surface was in a sticky condition after 10,000 cycles.

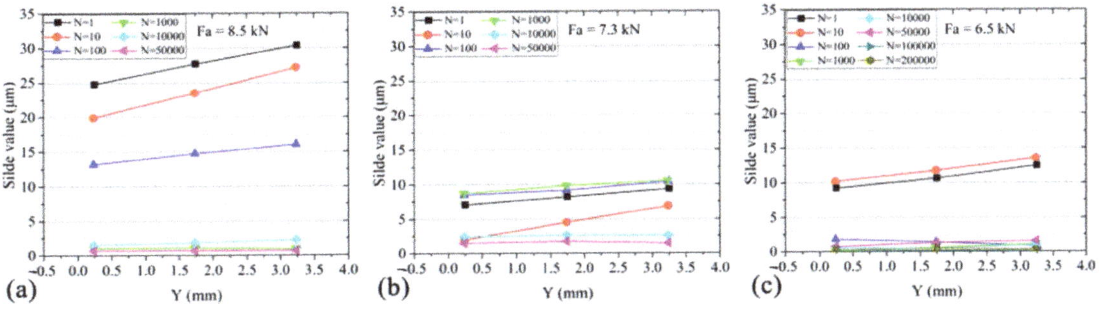

Figure 12. Distribution of relative slip along the contact surface for the (**a**) 8.5 kN, (**b**) 7.3 kN, and (**c**) 6.5 kN cases.

The effect of load amplitude on the slide or stick state is also clearly shown in Figure 12. Despite that the relative slip at the same number of cycles was reduced, the upward trend along the contact surfaces was also observed at 7.3 and 6.5 kN. Besides, the fretting contact condition changes from the initial gross slip to a partial slip when the number of cycles increases. For example, when the load amplitude was reduced to 6.5 kN, the relative slip at different locations was below 5 μm after 100 cycles. This indicates that the contact surface was in gross slip for a shorter time at 6.5 kN, compared with the 7.3 kN case. Besides, it should be noted that all points on the surfaces became sticky when the relative slip was stabilized below 2 μm.

4.4. Tangential Contact Stiffness

In order to further explore the contact properties under the fretting fatigue condition, the tangential contact stiffness obtained from the hysteresis loops is shown in Figure 13. Clearly, the tangential stiffness increased with the number of cycles at *Fa* = 6.5 kN during the experiment, as shown in Figure 13a. One of the possible reasons for this effect was the increase in contact area throughout the test. Figure 13b shows the variation in tangential stiffness versus the y-coordinate for the 6.5 kN case. It can be seen from the figure that there was a higher tangential stiffness at $y = 0.25$ mm (near the lower chuck). This nonuniform distribution of stiffness can reflect the variation of contact pressure along the interface surface. In detail, the contact pressure decreases as the y-coordinate increases. This is consistent with the increase in relative slip along the contact surfaces for all tests. These results about tangential contact stiffness were also investigated by Kartal et al. [29].

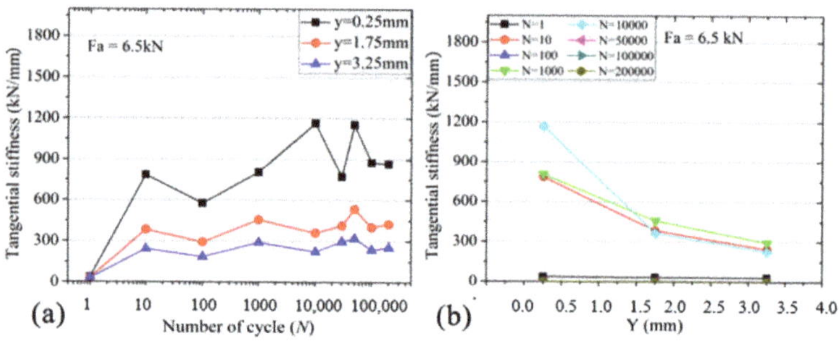

Figure 13. (**a**) Evolution of tangential stiffness with the number of cycles and (**b**) distribution of tangential stiffness along the contact surface for the 6.5 kN case.

5. Conclusions

In this study, an experimental method for accurately measuring the relative displacement between contact surfaces under the fretting fatigue condition was developed. The displacement field between the contact surfaces was measured by the DIC method. Through the distribution of displacement along the x-direction, it can be found that there was a value jump in the vertical displacement on the contact surface. Thus, the relative displacement between the specimen and fretting pad was determined by the linear fitting. The hysteresis loops of relative displacement and tangential force started from a parallelogram, then turned into an ellipse and stabilized into a straight line throughout all tests. Consequently, the DIC method was a feasible approach to acquire the distribution and evolution of relative slip in the fretting fatigue condition.

At the beginning of the test, the higher relative slip indicates that severe wear occurs on the contact surfaces, which is obtained from the hysteresis loops. Additionally, it was found that the value decreases markedly and stabilizes as the number of cycles increases. At the same time, the fretting condition of the contact surface exhibits a transition from gross slip to partial slip. Besides, the decreases in surface quality caused by wear lead to an increase in TFC. Hence, the evolution rule of the relative slip is in good agreement with the TFC. In order to investigate the effect of cyclic load amplitude on fretting fatigue properties, tests were carried out for three load amplitudes. Results show that the relative slip at the initial stage dropped sharply when the load amplitude was reduced from 8.5 to 7.3 kN. Compared with the 7.3 kN case, the partial slip condition appears earlier on the contact surface at $Fa = 6.5$ kN, even if they have a similar relative slip at the initial stage. Therefore, the reduction of cyclic load amplitude plays an important role in the mitigation of fretting fatigue damage, which can affect the stick or slip state of the contact surfaces. The tangential contact stiffness obtained from hysteresis loops indicates the variation of contact pressure along the interface surface, which is consistent with the distribution of relative slip along the contact surface for all tests.

Author Contributions: Study concept and design: Y.S., Q.-N.H. and H.L. Acquisition of data: Y.S. Analysis and interpretation of data: Y.S. and S.-S.R. Drafting of the manuscript: Y.S. and S.-S.R. Critical revision of the manuscript for important intellectual content: L.-S.N. and H.-J.S. Statistical analysis: Z.-H.S. Obtained funding: H.-J.S. and H.I. Study supervision: H.-J.S., H.L., and H.I. All authors have read and agreed to the published version of the manuscript.

Funding: This work is financially supported by Mitsubishi Heavy Industries, Ltd., Japan, and the National Natural Science Foundation of China (Grant Nos. 12172193, 11632010, 11672151, 91860101), and National Science and Technology Major Project (Grant No. J2019-VI-0002-0115, 2017-VI-0003-0073).

Institutional Review Board Statement: Not applicable.

Informed Consent Statement: Not applicable.

Data Availability Statement: Not applicable.

Conflicts of Interest: The authors declare that they have no known competing financial interests or personal relationships that could have appeared to influence the work reported in this paper. The authors declare no conflict of interest.

References

1. Waterhouse, R.B. *Fretting Fatigue*; Elsevier Science & Technology: Amsterdam, The Netherlands, 1981.
2. Nowell, D. *An Analysis of Fretting Fatigue*; University of Oxford: Oxford, UK, 1988.
3. Hills, D.A. Mechanics of fretting fatigue. *Wear* **1994**, *175*, 107–113. [CrossRef]
4. Foletti, S.; Beretta, S.; Gurer, G. Defect acceptability under full-scale fretting fatigue tests for railway axles. *Int. J. Fatigue* **2016**, *86*, 34–43. [CrossRef]
5. Ekberg, A. Fretting fatigue of railway axles—A Review of predictive methods and an outline of a finite element model. *Proc. Inst. Mech. Eng. Part F J. Rail Rapid Transit* **2004**, *218*, 299–316. [CrossRef]
6. Moraes, J.; Rao, H.M.; Jordon, J.; Barkey, M. High cycle fatigue mechanisms of aluminum self-piercing riveted joints. *Fatigue Fract. Eng. Mater. Struct.* **2017**, *41*, 57–70. [CrossRef]
7. Huang, L.; Bonnen, J.; Lasecki, J.; Guo, H.; Su, X. Fatigue and fretting of mixed metal self-piercing riveted joint. *Int. J. Fatigue* **2016**, *83*, 230–239. [CrossRef]
8. Oskouei, R.; Ibrahim, R. Improving fretting fatigue behaviour of Al 7075-T6 bolted plates using electroless Ni–P coatings. *Int. J. Fatigue* **2012**, *44*, 157–167. [CrossRef]
9. Peterka, P.; Krešák, J.; Kropuch, S.; Fedorko, G.; Molnar, V.; Vojtko, M. Failure analysis of hoisting steel wire rope. *Eng. Fail. Anal.* **2014**, *45*, 96–105. [CrossRef]
10. Rajasekaran, R.; Nowell, D. Fretting fatigue in dovetail blade roots: Experiment and analysis. *Tribol. Int.* **2006**, *39*, 1277–1285. [CrossRef]
11. Barella, S.; Boniardi, M.; Cincera, S.; Pellin, P.; Degive, X.; Gijbels, S. Failure analysis of a third stage gas turbine blade. *Eng. Fail. Anal.* **2011**, *18*, 386–393. [CrossRef]
12. Su, Y.; Han, Q.N.; Qiu, W.; He, Z.; Shang, Y.B.; Shi, H.J.; Niu, L.S. High temperature in-situ SEM observation and crystal plasticity simulation on fretting fatigue of Ni-based single crystal superalloys. *Int. J. Plast.* **2019**, *127*, 102645. [CrossRef]
13. Han, Q.; Rui, S.; Qiu, W.; Su, Y.; Ma, X.; He, Z.; Cui, H.; Zhang, H.; Shi, H. Subsurface crack formation and propagation of fretting fatigue in Ni-based single-crystal superalloys. *Fatigue Fract. Eng. Mater. Struct.* **2019**, *42*, 2520–2532. [CrossRef]
14. Zhou, Z.; Vincent, L. Mixed fretting regime. *Wear* **1995**, *181–183*, 531–536. [CrossRef]
15. Pasanen, A.; Lehtovaara, A.; Rabb, R.; Riihimäki, P. Friction behavior of quenched and tempered steel in partial and gross slip conditions in fretting point contact. *Wear* **2009**, *267*, 2200–2207. [CrossRef]
16. Sabelkin, V.; Mall, S. Investigation into relative slip during fretting fatigue under partial slip contact condition. *Fatigue Fract. Eng. Mater. Struct.* **2005**, *28*, 809–824. [CrossRef]
17. Fouvry, S.; Kapsa, P.; Vincent, L. Quantification of fretting damage. *Wear* **1996**, *200*, 186–205. [CrossRef]
18. Zhou, Z.; Nakazawa, K.; Zhu, M.; Maruyama, N.; Kapsa, P.; Vincent, L. Progress in fretting maps. *Tribol. Int.* **2006**, *39*, 1068–1073. [CrossRef]
19. Han, Q.N.; Rui, S.S.; Qiu, W.; Ma, X.; Su, Y.; Cui, H.; Zhang, H.; Shi, H. Crystal orientation effect on fretting fatigue induced geometrically necessary dislocation distribution in Ni-based single-crystal superalloys. *Acta Mater.* **2019**, *179*, 129–141. [CrossRef]
20. Vingsbo, O.; Söderberg, S. On fretting maps. *Wear* **1988**, *126*, 131–147. [CrossRef]
21. Jin, O.; Mall, S. Effects of slip on fretting behavior: Experiments and analyses. *Wear* **2004**, *256*, 671–684. [CrossRef]
22. Rengaraj, B.; Baba, S.; Okazaki, M. Influence of Crystal Orientation on Cyclic Sliding Friction and Fretting Fatigue Behavior of Single Crystal Ni-Base Superalloys. In *Superalloys 2016: Proceedings of the 13th Intenational Symposium of Superalloys*; John Wiley & Sons, Inc.: Hoboken, NJ, USA, 2016.
23. Su, Y.; Han, Q.N.; Zhang, C.C.; Shi, H.; Niu, L.S.; Deng, G.J.; Rui, S.S. Effects of secondary orientation and temperature on the fretting fatigue behaviors of Ni-based single crystal superalloys. *Tribol. Int.* **2018**, *130*, 9–18. [CrossRef]
24. Cai, Z.B.; Zhu, M.H.; Yang, S.; Xiao, X.B.; Lin, X.Z.; Yu, H.Y. In situ observations of the real-time wear of PMMA flat against steel ball under torsional fretting. *Wear* **2011**, *271*, 2242–2251. [CrossRef]
25. Han, Q.N.; Qiu, W.; He, Z.; Su, Y.; Ma, X.; Shi, H.J. The effect of crystal orientation on fretting fatigue crack formation in Ni-based single-crystal superalloys: In-situ SEM observation and crystal plasticity finite element simulation. *Tribol. Int.* **2018**, *125*, 209–219. [CrossRef]

26. Wittkowsky, B.U.; Birch, P.R.; Dominguez, J.; Suresh, S. An apparatus for quantitative fretting fatigue testing. *Fatigue Fract. Eng. Mater. Struct.* **1999**, *22*, 307–320. [CrossRef]
27. Popov, V.L. *Contact Mechanics and Friction*; Springer: Berlin/Heidelberg, Germany, 2010.
28. Kartal, M.E.; Mulvihill, D.M.; Nowell, D.; Hills, D.A. Determination of the Frictional Properties of Titanium and Nickel Alloys Using the Digital Image Correlation Method. *Exp. Mech.* **2010**, *51*, 359–371. [CrossRef]
29. Kartal, M.; Mulvihill, D.; Nowell, D.; Hills, D. Measurements of pressure and area dependent tangential contact stiffness between rough surfaces using digital image correlation. *Tribol. Int.* **2011**, *44*, 1188–1198. [CrossRef]
30. De Pauw, J.; De Waele, W.; Hojjati-Talemi, R.; De Baets, P. On the use of digital image correlation for slip measurement during coupon scale fretting fatigue experiments. *Int. J. Solids Struct.* **2014**, *51*, 3058–3066. [CrossRef]
31. Ding, J.; Bandak, G.; Leen, S.; Williams, E.; Shipway, P. Experimental characterisation and numerical simulation of contact evolution effect on fretting crack nucleation for Ti–6Al–4V. *Tribol. Int.* **2009**, *42*, 1651–1662. [CrossRef]
32. Pan, B.; Qian, K.; Xie, H.; Asundi, A. Two-dimensional digital image correlation for in-plane displacement and strain measurement: A review. *Meas. Sci. Technol.* **2009**, *20*, 062001. [CrossRef]
33. Blaber, J.; Adair, B.S.; Antoniou, A. Ncorr: Open-Source 2D Digital Image Correlation Matlab Software. *Exp. Mech.* **2015**, *55*, 1105–1122. [CrossRef]
34. Juoksukangas, J.; Lehtovaara, A.; Mäntylä, A. Applying the digital image correlation method to fretting contact for slip measurement. *Proc. Inst. Mech. Eng. Part J J. Eng. Tribol.* **2016**, *231*, 509–519. [CrossRef]
35. Juoksukangas, J.; Lehtovaara, A.; Mäntylä, A. A comparison of relative displacement fields between numerical predictions and experimental results in fretting contact. *Proc. Inst. Mech. Eng. Part J J. Eng. Tribol.* **2016**, *230*, 1273–1287. [CrossRef]
36. De Crevoisier, J.; Swiergiel, N.; Champaney, L.; Hild, F. Identification of in situ frictional properties of bolted assemblies with digital image correlation. *Exp. Mech.* **2012**, *52*, 561–572. [CrossRef]
37. Cuevas Arteaga, C.; Rodríguez, J.A.; Clemente, C.M.; Segura, J.A.; Urquiza, G.; Hamzaoui, Y.E. Estimation of useful life in turbines blades with cracks in corrosive environment. *Eng. Fail. Anal.* **2013**, *35*, 576–589. [CrossRef]
38. ASTM. *Standard Test Methods for Tension Testing of Metallic Materials*; ASTM International: West Conshohocken, PA, USA, 2016; Volume E8/E8M-162016.
39. JSME. *JSME Standard Method of Fretting Fatigue Testing*; Japan Society Mechanical Engineers: Tokyo, Janpan, 2002; Volume JSME S 015-20022002.
40. Nix, K.J.; Lindley, T.C. The Application of Fracture Mechanics to Fretting Fatigue. *Fatigue Fract. Eng. Mater. Struct.* **1985**, *8*, 143–160. [CrossRef]
41. Pan, B. Recent Progress in Digital Image Correlation. *Exp. Mech.* **2010**, *51*, 1223–1235. [CrossRef]
42. Chu, T.C.; Ranson, W.F.; Sutton, M.A. Applications of digital-image-correlation techniques to experimental mechanics. *Exp. Mech.* **1985**, *25*, 232–244. [CrossRef]
43. Massingham, M.; Irving, P. The effect of variable amplitude loading on stress distribution within a cylindrical contact subjected to fretting fatigue. *Tribol. Int.* **2006**, *39*, 1084–1091. [CrossRef]
44. Cortez, R. Investigation of variable amplitude loading on fretting fatigue behavior of Ti-6Al-4V. *Int. J. Fatigue* **1999**, *21*, 709–717. [CrossRef]
45. Mubarak Ali, M.; Ganesh Sundara Raman, S.; Pathak, S.D.; Gnanamoorthy, R. Influence of plasma nitriding on fretting wear behaviour of Ti–6Al–4V. *Tribol. Int.* **2010**, *43*, 152–160. [CrossRef]
46. Madge, J.; Leen, S.; McColl, I.; Shipway, P. Contact-evolution based prediction of fretting fatigue life: Effect of slip amplitude. *Wear* **2007**, *262*, 1159–1170. [CrossRef]
47. Madge, J.; Leen, S.; Shipway, P. The critical role of fretting wear in the analysis of fretting fatigue. *Wear* **2007**, *263*, 542–551. [CrossRef]
48. Allara, M. A model for the characterization of friction contacts in turbine blades. *J. Sound Vib.* **2008**, *320*, 527–544. [CrossRef]

metals

Article

A Combined Numerical–Analytical Study for Notched Fatigue Crack Initiation Assessment in TRIP Steel: A Local Strain and a Fracture Mechanics Approach

Peter I. Christodoulou and Alexis T. Kermanidis *

Laboratory of Mechanics and Strength of Materials, Department of Mechanical Engineering, University of Thessaly, 38334 Volos, Greece; pchristod@mie.uth.gr
* Correspondence: akermanidis@mie.uth.gr

Abstract: In the fatigue design of metallic components using the safe-life approach, fatigue crack initiation as a development of slip systems at the nanoscale, followed by microstructurally short crack growth, is critical for the onset of structural failure. The development of reliable analytical tools for the prediction of crack initiation, although very complex due to the inherent multiscale fatigue damage processes involved, is important for promoting a more sophisticated design but, more importantly, enhancing the safety in regard to fatigue. The assessment of fatigue crack initiation life at the root of a V-shaped notch is performed by implementing a local strain and a fracture mechanics concept. In the low cycle fatigue analysis, the finite element method is used to determine the local stress–strain response at the notch root, which takes into account elastoplastic material behavior. Fatigue crack initiation is treated as the onset of a short corner crack by incremental damage accumulation and failure of a material element volume at the notch root. The finite element results are compared against established methodologies such as the Neuber and strain energy density methods. In the fracture mechanics approach, fatigue crack initiation is treated as the onset and propagation of a corner crack to a finite short crack. Fatigue experiments in two different transformation-induced plasticity (TRIP) steels were conducted to evaluate the analytical predictions and to determine the physical parameters for the definition of crack initiation. The analytical results show that the finite element method may be successfully implemented with existing fatigue models for a more accurate determination of the local stress–strain behavior at the notch tip in order to improve the assessment of fatigue crack initiation life compared to the established analytical methodologies.

Keywords: fatigue crack initiation; short crack; Neuber's rule; strain energy density (SED); finite element analysis (FEA); TRIP steel

Citation: Christodoulou, P.I.; Kermanidis, A.T. A Combined Numerical–Analytical Study for Notched Fatigue Crack Initiation Assessment in TRIP Steel: A Local Strain and a Fracture Mechanics Approach. *Metals* **2023**, *13*, 1652. https://doi.org/10.3390/met13101652

Academic Editor: Haitao Cui

Received: 31 August 2023
Revised: 18 September 2023
Accepted: 23 September 2023
Published: 27 September 2023

1. Introduction

Conducting a reliable assessment of fatigue crack initiation life in structural components containing stress raisers is a critical engineering problem for material selection and design optimization against fatigue failure. Major simplifications and empirical fatigue concepts are often used to simulate the complex nature of damage processes [1], while there is no adequate and consistent validation against experimental data to ensure the reliability of analytical predictions. Hence, most methodologies rely on predictions of the total fatigue life of notched specimens [2].

A common approach in the prediction of fatigue crack initiation is the assumption of low-cycle fatigue conditions. With that approach, the notch root fatigue damage is simulated with the damage of a macroscopic specimen subjected to similar cyclic plastic straining conditions [3]. For the determination of the notch stress–strain state involving local plasticity, Neuber's rule [4] and the strain energy density (SED) criterion [5] are the most widely used empirical models. A review of these models was performed in [6], where the elastoplastic stress–strain behavior at notch roots under monotonic and cyclic loading was examined. According to [7], the SED model is more reliable compared to Neuber' s rule when predicting elastic–plastic notch stresses and strains. On the other hand, Sharpe et al. [8] claimed that Neuber's rule provides better results in cases where plane stress conditions are present. They suggested that estimations using the SED criterion and Neuber' s rule will give lower and upper limits on the local strain, which can be used to bound an uncertainty field for the life prediction. When yielding at the notch takes place, more accurate results of the local notch response may be obtained with an finite element analysis (FEA) [9].

As the fraction of fatigue life in a component is higher when the crack is short, investigating the growth of short cracks initiating from a notch from is of practical significance, especially when the crack's length is typically small compared to the plastic zone ahead of the notch. In many design situations, Linear Elastic Fracture Mechanics (LEFM) analyses allow for a direct comparison of fatigue crack growth behavior between engineering components and laboratory specimens using the stress intensity factor range, ΔK. Although this is an accepted methodology in the case of long cracks, for short cracks, limitations of continuum mechanics lead to the underestimation of the fatigue crack growth rate [10]. The limitations in LEFM or anisotropic material behavior of microstructurally short cracks requires more complex fatigue crack growth analyses due to the length of the crack being comparable to the microstructural scale of the material (e.g., grain size) or to the plastic zone size at the crack tip. An analytical model involving crystal plasticity based on the local stored energy at the crack tip for simulating short crack growth in a polycrystalline material was presented in [11]. For physically small cracks (typically smaller than 0.5 mm) with a length exceeding the microstructural scaling, limitations may occur when the short crack is embedded within the notch plasticity [12]. To overcome this, modified LEFM concepts can be used for short cracks embedded in the inelastic field induced by the presence of the notch. Characteristic paradigms are the correlative model by Leis [12] where a correlation of LEFM concept has been proposed when the crack is completely submerged in the notch tip plastic zone and, thus, bulk plasticity conditions dominate the growth behavior; the concept of El Haddad et al. [13] which introduces a threshold crack length as minimum for applicability of LEFM; or models based on the determination of short crack growth parameters (e.g., [14]). Modified J-integral solutions for short cracks were also extended to plastically strained notches of different severities [15]. As the crack closure is another important factor resulting in the anomaly of short crack growth behavior, thus influencing the crack initiation, the effective stress intensity factor range (ΔK_{eff}) was used by Tanaka and Nakai to correlate the early stage of fatigue crack growth rate from a notch when the crack was short, with the long crack growth rate data [16], while in [17], a plasticity corrected stress intensity factor range was implemented to determine the extent of fatigue initiated physically short cracks growth emanating from notches.

The aforementioned methods, by being macroscopic, do not account for certain microstructural influences, which control the material's plastic deformation behavior. For instance, the phase transformation effect contributes to the plastic straining under external loading and is an important parameter of the TRIP steel behavior concerning crack initiation. From the available research findings, it was shown that the transformation of RA in front of the crack tip, inside the monotonic plastic zone [18], acts in favor of suppressing the fatigue crack initiation and/or propagation through the plastic relaxation induced by the transformation effect [19]. Although such effects are important in crack initiation analyses, the complexity of satisfactorily combining damage mechanisms at a multiscale level (atomic/microscopic and macroscopic) is high.

Despite the above complexities involved in the simulation, the main problem is the strong sensitivity of the fatigue crack initiation phase on the local stress–strain values at the notch root, specifically in the presence of local plasticity. In [20], it was shown that a strong correlation exists between the applied notch stress range and the experimental fatigue life, leading to short or long crack initiation. Finite element simulations in combination with analytical/empirical theories have provided more reliable results of the local stress–strain behavior and, hence, the crack initiation estimation. In [21], an SWT-based model was developed to predict the fatigue crack initiation of an additively manufactured notched maraging steel under bending–torsion, while in [22], a finite element model in combination with a physically based model was implemented to predict crack initiation in a stent.

The objective behind the present study was to develop a suitable numerical–analytical methodology to estimate the fatigue life for crack initiation at a V-notch under cyclic loading. The methodology effectively combines classical fatigue theory with a numerical simulation in order to accurately determine the local cyclic stress/strain state at the notch root. In the methodology, both a local strain approach and a fracture mechanics concept were investigated for estimation of crack initiation. The analysis was performed on two different TRIP steels, materials for which a similar analysis is not reported in the recent literature, and the numerical results are compared against the experimental results of fatigue crack initiation on the steels under examination.

2. Materials and Methods

2.1. Materials

The materials used in the investigation were a hot-rolled and a cold-rolled Al-containing low alloy TRIP steel, in sheet form, with the thickness and chemical composition given in Table 1.

Table 1. State and chemical composition (wt.%) of TRIP steels.

Steel	Condition	Thickness (mm)	C	Mn	Al	Si	P
(A)	Hot rolled	1.75	0.18	1.61	1.45	0.7	-
(B)	Cold-rolled	1.45	0.202	1.99	1.07	0.35	0.009

A thorough experimental investigation, including microstructural characterization, tensile, LCF, and HCF tests, for the two steels was performed in [23], where the microstructural characteristics and static and cyclic mechanical properties were evaluated.

2.2. Mechanical Testing

Constant-amplitude high cycle fatigue tests were carried out using an INSTRON 8801 servo-hydraulic machine (Norwood, MA, US) with a 100 kN load capacity at σ_{max} = 200 MPa, a stress ratio of R = 0.1, and a frequency of 25 Hz. Notched fatigue specimens with a single 60° V-shape notch were prepared according to ASTM E466, with the geometry shown in Figure 1. The notch configuration results in an elastic stress concentration factor value of K_t = 3.5, based on the Noda et al. approach [24].

The 60° V-shaped side notch was introduced in the samples by using the sinker Electrical Discharge Machining (EDM) method. The EDM method, as a non-deforming cutting process, has the advantage of avoiding any plastic deformation of the steel during the removal of material. Thus, no significant transformation effect (TRIP effect) at the notch root caused by the machining process was expected.

Fatigue crack initiation and propagation at the notch root was monitored using the replica technique. Careful preparation of the notch area was conducted in order to remove imperfections (burrs) from the machining process by grinding with fine #1000–4000# grit papers to ensure that no surface effects were induced upon crack initiation. The notched fatigue tests were interrupted at regular cycle intervals, and a small tensile load was applied to open the crack surfaces. Extreme care was taken when applying this tensile load in order not to exceed the test's maximum force (P_{max}), which could possibly introduce an overload in the loading history. Special replica foils of 0.1 mm in thickness and wetted with acetone were used to take replicas from both sides of the notch root and were examined in an optical microscope. Cyclic stressing was continued for a small interval of cycles, and the above procedure was repeated until a first crack was observed in the replica through the microscopic observation. When the first crack was observed, the procedure was continued by making the interruption intervals more frequent. The length of the fatigue crack was measured using the replica images taken from microscope and processed with image analysis software (ImageJ: Open Source, U.S. National Institutes of Health, Bethesda, MD, USA). It should be noted that a different series of fatigue crack initiation tests on the same materials were performed in [23], but, in this case, they were used to evaluate the difference in material behavior between the two steels.

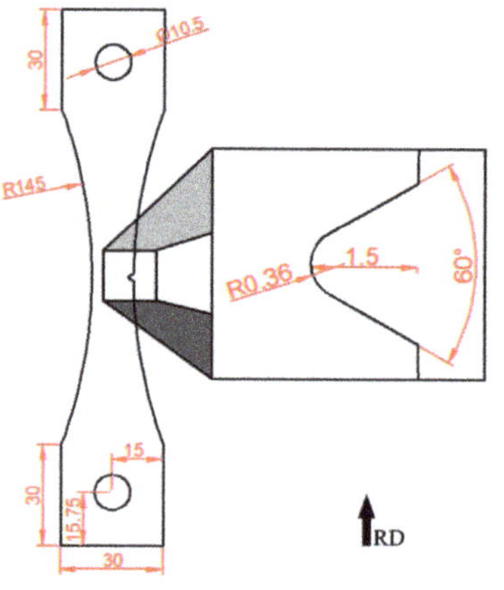

Figure 1. Geometry of fatigue test specimen and detail of V-notch (dimensions in mm).

2.3. Fatigue Crack Initiation Analysis

2.3.1. Local Strain Approach

A local strain (LS) approach using the typical Neuber and strain energy density (SED) methods was implemented to relate the crack initiation life at the notch root to the crack initiation life of the smooth laboratory specimen. The principle behind the LS approach, which is depicted in Figure 2, is that a smooth specimen tested under cyclic strain control (LCF conditions) can simulate the fatigue damage at the notch root of an engineering component. An equivalent fatigue damage is assumed to occur at the notch root and in the smooth specimen when both are subjected to identical stress–strain histories (constant plastic strain range, $\Delta_{\varepsilon \text{pm}}$).

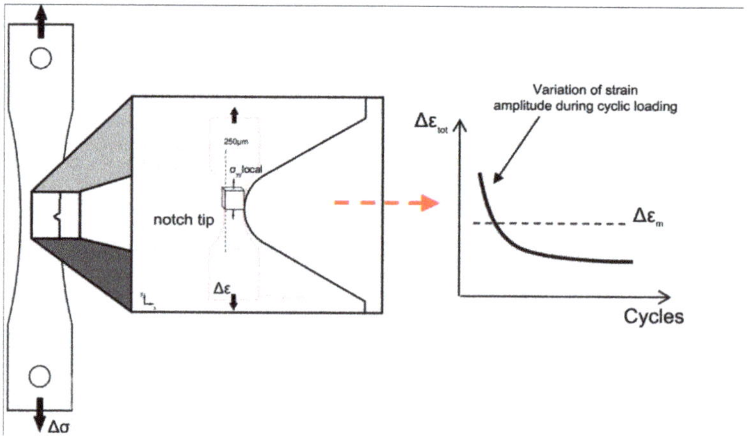

Figure 2. LS approach showing material element ahead of the notch tip subjected to LCF conditions.

The LS approach uses the strain–life (ε-N) curve for the prediction of life cycles, and the estimation of crack initiation life is based on the criterion of the smooth specimen failure by assuming an equivalent strain state of the material volume in the notched specimen. This hypothesis contains specific shortcomings due to (i) the size effect involved, (ii) the fact that the material element volume is subjected to deformation constraints from the adjacent material, (iii) the deformation state in the material element is not strain controlled as in the case of the low cycle fatigue test. Strain triaxiality at the notch tip is not expected to be an influential factor in the fatigue behavior examined due to the selection of the corner element promoting plane stress conditions. This assumption, although introducing specific limitations in the analysis, is considered acceptable to approach this type of problem.

Assumptions

A fatigue analysis for the prediction of crack initiation life adopts certain hypotheses for simulating the involved damage mechanisms. Crack initiation at the notch tip is associated with failure of an elementary material volume (Figure 2) adjacent to the notch root. The material volume is subjected to low cycle fatigue conditions and is located at the corner of the notch tip, at the free surface of the specimen.

The criterion for selecting the specific element location and size is based on the fractographic observations that are discussed in Section 3.2. According to the experimental findings, cracks originate as corner cracks at the tip of the notch and then propagate radially before coalescence occurs into a single through-thickness crack. The assumption of the critical length of 250 μm width for the corner element satisfies the conditions of an isotropic, homogeneous medium for the implementation of fatigue life equations. That is because it significantly exceeds the material's grain size, which, as determined from [23], is 9.47 μm. Using this concept, fatigue failure of the material element designates the initiation of a 250 μm corner crack at the tip of the notch on the outer specimen surface. Similar lengths have been observed by the experiments to lead to the transition of the initial corer crack to the through-thickness crack, as is later shown by the fractographic observations.

2.3.2. Analytical Methods

Neuber showed that, for a shear-strained prismatic body with an arbitrary nonlinear stress–strain law, the product of the notch stress and strain, σ and ε, can be estimated from theoretical stress concentration factor, K_t; the far-field applied stress range, S; and the elastic modulus of the material, E:

$$\varepsilon\sigma = \frac{(K_t S)^2}{E} \tag{1}$$

Generally, this is rewritten in terms of stress and strain ranges for the case when the stress range remote to the notch is linear elastic:

$$\Delta\varepsilon\Delta\sigma = \frac{(K_t \Delta S)^2}{E} \tag{2}$$

Topper et al. [1] showed that smooth specimen fatigue data could be used to adequately predict the fatigue lives of notched members made from 2024 and 7075 aluminum alloys under fully reversed loading. They suggested the use of the fatigue notch factor, K_f, instead of the theoretical stress concentration factor, K_t, for cyclic loading, modifying Equations (1) and (2) as follows:

$$\varepsilon\sigma = \frac{(K_f S)^2}{E} \tag{3}$$

$$\Delta\varepsilon\Delta\sigma = \frac{(K_f \Delta S)^2}{E} \tag{4}$$

Although Neuber's rule was derived for monotonic loading, it was applied to fatigue loading by Manson and Hirschberg in [25], who suggested using the cyclic stress–strain curve (Equation (6)) instead of the monotonic curve given by the Ramberg–Osgood equation (Equation (5)):

$$\varepsilon = \frac{\sigma}{E} + \frac{\sigma}{K}^{1/n} \quad (5)$$

$$\varepsilon_\alpha = \frac{\sigma_\alpha}{E} + \frac{\sigma_\alpha}{K'}^{1/n'} \quad (6)$$

where K and K′ are the static and cyclic strength coefficient, n and n′ are the monotonic and cyclic strain hardening exponent, and ε_α and σ_α are the strain and stress amplitude, respectively. Thus, by simultaneously solving Equation (3) for maximum stress and strain ($\varepsilon = \varepsilon_{max}$; $\sigma = \sigma_{max}$) and Equation (5) for $\varepsilon = \varepsilon_{max}$ and $\sigma = \sigma_{max}$, the notch root stress–strain at maximum loading can be estimated.

Modifying the Ramberg–Osgood relationship for cyclic loading by replacing the strain and stress amplitude with the strain and stress ranges, respectively, and assuming a Masing factor of 2 (symmetric deformation behavior in tension and compression), Equation (6) yields the following:

$$\Delta\varepsilon = \frac{\Delta\sigma}{E} + 2\left(\frac{\Delta\sigma}{2K'}\right)^{1/n'} \quad (7)$$

By simultaneously solving Equations (4) and (7), the cyclic notch stress and strain amplitudes can be estimated.

An alternate approximate method is the strain energy density (SED) criterion proposed by Molski and Glinka [5]. In this approach, it is assumed that the strain energy density at the notch root does not change significantly if the localized plasticity is surrounded by predominantly elastic material (Figure 3).

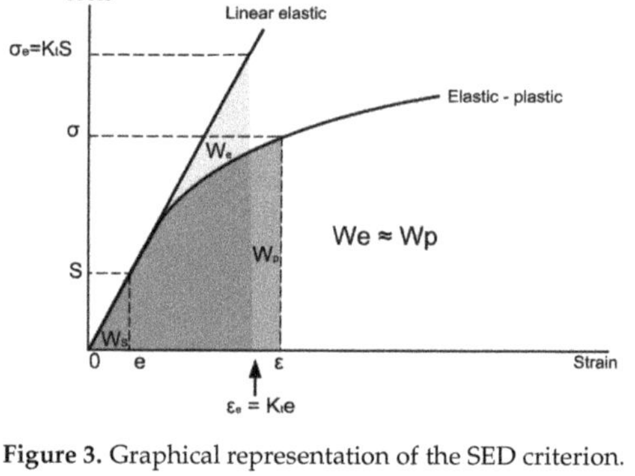

Figure 3. Graphical representation of the SED criterion.

Based on this assumption, computation of the strain energy density at the notch root will yield identical results for either the elastic or the elastic–plastic material law. The SED criterion has the following form:

$$W_e = W_p \; \frac{1}{2}\frac{(K_tS)^2}{E} = \int_0^\varepsilon \sigma_{(\varepsilon)} d\varepsilon \rightarrow \frac{(K_tS)^2}{E} = \frac{\sigma^2}{E} + \frac{2\sigma}{n+1}\left(\frac{\sigma}{K}\right)^{1/n} \quad (8)$$

where S is the gross-section nominal stress, and $\sigma_{(\varepsilon)}$ is the notch root stress as a function of ε. For a given nominal stress, S, the notch stress and strain during monotonic loading can be calculated simultaneously solving Equations (5) and (8).

With the use of Equation (5) for the maximum stress and strain (ε_{max} and σ_{max}) and Equation (7) for $\sigma = \sigma_{max}$ and $S = S_{max}$, the notch root stress–strain at maximum load can be calculated.

For cyclic loading, Equation (8) can be rewritten in terms of stress and strain ranges, while the material monotonic deformation properties (K and n) can be replaced by the cyclic deformation properties (K′ and n′), resulting in the following equation:

$$\frac{(\Delta\sigma)^2}{E} + \frac{4\Delta\sigma}{n'+1}\left(\frac{\Delta\sigma}{2K'}\right)^{1/n'} = \frac{(K_t S)^2}{E} \quad (9)$$

By simultaneously solving Equations (5) and (9), the cyclic notch stress and strain amplitudes can be calculated.

The analytical expressions provided above consider the notch stress–strain response for completely reversed straining, $R = \varepsilon_{min}/\varepsilon_{max} = -1$. In strain histories with a mean strain, usually mean stress relaxation occurs due to the presence of plastic deformation. To account for the mean stress effect in fatigue life prediction for the elastic strain component, the Smith–Watson and Topper parameter (SWT parameter) or the Morrow's mean stress method is often used. The SWT parameter [26] was used in the present analysis in the strain–life equation as follows:

$$\sigma_{max}\varepsilon_{\alpha} = \frac{(\sigma_f')^2}{E}(2N_f)^{2b} + \sigma_f'\varepsilon_f'(2N_f)^{b+c} \quad (10)$$

where $\sigma_{max} = \sigma_m + \sigma_{\alpha}$ and ε_{α} is the maximum stress and strain amplitude in the alternating strain history, σ_m is the mean stress, σ_f', ε_f', b, c are cyclic properties taken from [23] and N_f is the fatigue life. Equation (10) assumes that for different combinations of strain amplitude, ε_{α} and mean stress, σ_m, the product $\sigma_{max}\varepsilon_{\alpha}$ remains constant for a given fatigue life. If σ_{max} is zero, then Equation (10) predicts infinite life, which implies that tensile loading must be present for fatigue fracture to occur.

Calculations of the local stress–strain history may be combined with an appropriate notched fatigue analysis in order to estimate fatigue crack initiation as the number of cycles required to initiate the crack at the location of interest (notch tip). In this approach, the calculated fatigue life is strongly dependent on the material element length and the associated strain values, which usually lack certain accuracy when elastoplastic material behavior is involved. For this purpose, a numerical analysis was performed in order to obtain a greater accuracy for strains at the position of the material element. A suitable finite element model was developed for the determination of the local maximum strains and stresses of a material element volume at the notch tip during a far-field loading. The criteria for the selection of the material volume length are analyzed in the following section. The numerical results are combined with the analytical methods of Neuber and SED and incorporated into the SWT model to obtain the critical number of cycles for fatigue crack initiation. The flowchart of the implemented methodology is presented in the diagram in Figure 4.

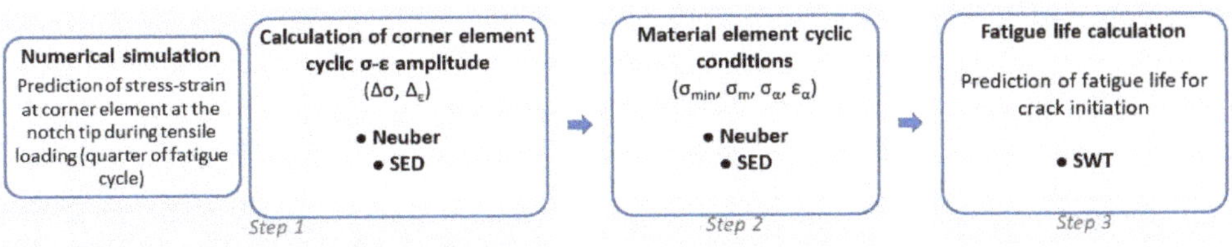

Figure 4. Flowchart of the steps used in the fatigue analysis for the prediction of fatigue life corresponding to crack initiation.

The monotonic and cyclic material properties, namely σ_y, n, K, n′, K′, σ_f', b, ε_f', c, and K_f, are taken from [23] and are used as input parameters in the analysis. The predicted fatigue lives in cycles for each material examined are presented in the next section.

2.4. Numerical Simulation of the Notch Stress–Strain Behavior

Finite Element Analysis

The finite element model used to obtain the representative local stress–strain behavior at the notch root was first presented in [23]. In that work, the elastoplastic stress–strain behavior of the material element was numerically simulated under an external monotonic loading using the Abaqus finite element (FE) software version 6.12-3. A maximum far-field tensile stress of 200 MPa was applied, corresponding to the quarter of a cycle in the fatigue-loading history. For the simulation, a 3D finite element model was constructed, and a step-loading analysis was performed (Figure 5).

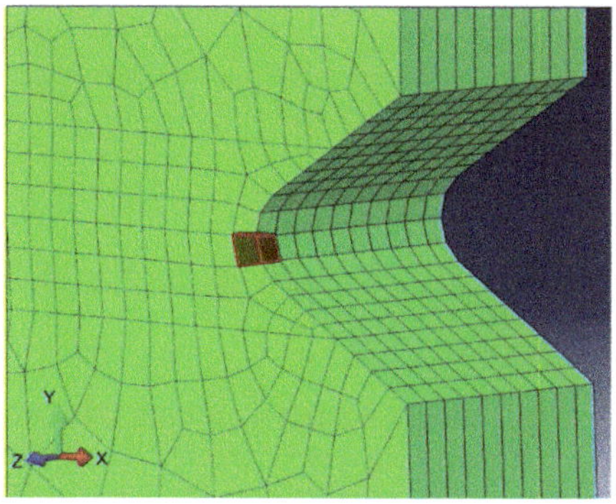

Figure 5. Finite element model to simulate the notch stress–strain behavior, showing corner element and meshing at the vicinity of the notch.

The values of the Poisson's ratio and Young's Modulus used in the analysis are $\nu = 0.3$ and E = 205.9 GPa, respectively. After yielding, nonlinear behavior is taken into account using an isotropic hardening behavior based on the calculated true stress–strain plastic curve data presented in [23]. For greater accuracy at the vicinity of the notch, a geometric nonlinearity was also taken into account.

The element width at the notch tip was set to 250 μm, which corresponds to the characteristic crack length considered for crack initiation in Section 3.1. A mesh optimization was performed in the present study by increasing the element size along with the increasing distance from the notch tip in order to reduce the computational time.

A mesh sensitivity validation was also performed with the use of finer mesh. The element width in the vicinity of the notch was reduced to 25 μm, and the monotonic simulation was repeated. The deviation of predicted notch stress–strain values with the finer mesh was lower than 4%, compared to the results of the 250 μm width. The latter findings suggest that a finer mesh will have a negligible effect on evaluation of results.

2.5. Fracture Mechanics Approach

Assumptions

In the LS approach described previously, crack initiation is considered as the onset of a 250 μm long crack due to the failure of a material volume at the notch tip, which is subjected to LCF conditions. A more detailed approach combining the LS approach with a fracture mechanics concept is attempted in this section and is presented as a case study for material (B). The fracture mechanics approach is based on the prediction of the number of cycles required for a physically short crack (not microstructurally short crack) to initiate at the root of the notch and propagate incrementally to the final length of 250 μm. It is considered that crack initiation is associated with the onset of a corner crack at the free edge of the notch (Figure 5) due to the fatigue failure of a material element at the free edge, as was experimentally observed (see Section 3.2). A more realistic approximation of fatigue crack initiation must consider that the occurring fatigue damage takes place at a smaller scale; therefore, the dimensions of the material element examined in the previous section are reduced to 50 μm × 50 μm. A simplified model is assumed, where, following the onset of crack initiation, the corner crack propagates under mode I until it reaches a final length of 250 μm. The calculation steps presented in Figure 4 are used to predict the number of cycles for failure of the corner element (crack initiation) of 50 μm. Afterwards, a fracture mechanics analysis is implemented to calculate the number of cycles for the growth of the crack until a final length of 250 μm from the notch tip (crack propagation).

3. Results

3.1. Fatigue Crack Initiation

The static and cyclic mechanical properties of the investigated materials (A) and (B), were evaluated in [23] and were used in the present study as input parameters in the proposed methodology. Fatigue crack initiation is considered to be the fatigue stage where the length of the growing crack at the notch root reaches 250 μm. With the length of 250 μm, the crack is significantly larger than the fictitious mean grain size of 6.5 μm, considering both materials A and B; therefore, the assumption of continuum mechanics is not violated (Figure 6). In [23], the average ferrite grain size was measured to be 3.62 μm and 9.47 μm for steels (A) and (B), respectively.

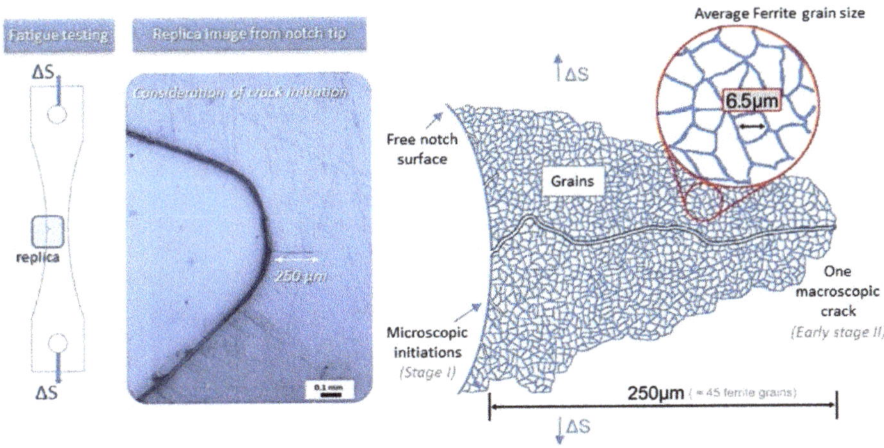

Figure 6. Schematic representation of stages I (shear mode) and II (tensile mode) considered in the present study for the assumption of crack initiation.

In the findings of Jacques et al. [27], in situ measurements of stress and strain partitioning in TRIP steels during monotonic loading revealed that a representative microstructural "element" typically involves ~100 grains, and, at that level of magnification, the average mechanical properties of the constituent phases are relevant.

The number of cycles corresponding to crack initiation life (N_{ini}) for notched specimens is presented in Table 2 as the average number of cycles at which the first 250 μm crack at the free surface is detected. In Figure 7, the fatigue crack growth curves for materials (A) and (B) are displayed.

Table 2. Fatigue crack initiation with EDM method (σ_{max} = 200 MPa).

Steel	N_f (Cycles)		N_{ini} (Cycles)	
(A)	87,300 140,420	113,860	53,080 79,000	66,040
(B)	168,860 181,450	175,150	106,060 111,870	108,970

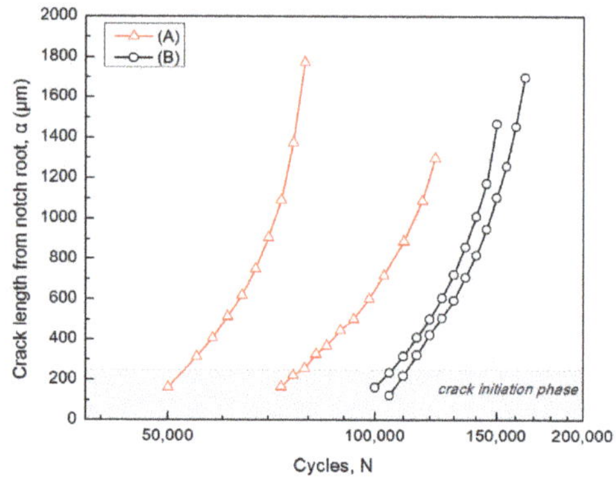

Figure 7. Fatigue crack growth curves of the examined TRIP steels (σ_{max} = 200 MPa).

From the results of Table 2, it can be seen that steel (B) is the more damage tolerant material, with a fatigue life of 175,150 cycles, followed by steel (A), with 113,860 cycles before failure. Regarding the fatigue crack initiation stage, steel (B) exhibits a higher crack initiation resistance, which corresponds to a life of 108,970 cycles, while steel (A) has a crack initiation life of 66,040 cycles.

The difference in the crack initiation behavior of steels (A) and (B) is a material and microstructure-related problem, which was examined in [23] and was not the subject of investigation in the present study. The experimental results obtained here are in agreement with the fatigue crack initiation and fatigue life trends observed in [23].

In Figure 8, replica images showing the existence of cracks initiating at the notch tip are displayed for both materials for the case of a maximum stress of σ_{max} = 200 MPa. A crack with a length of 254 μm was detected in steel (A) at 79,000 cycles, while the more damage-tolerant steel (B) showed a crack length of 233 μm at a life of 105,000 cycles (Figure 8b).

Crack initiation resistance is primarily associated with microstructural aspects controlling the start of cyclic slip or twinning mechanisms under cyclic strains, which is macroscopically associated with the materials fatigue limit [28]. In notched fatigue problems, it may also be relevant to the materials' LCF behavior [27]. Additionally, the transformation mechanism due to the localized plastic strains at the root of the notch is also expected to have an impact on crack initiation [18,28]. For the above reasons, providing an explanation for the differences in crack initiation behavior of the TRIP steels requires a more rigorous investigation.

3.2. Fractographic Observations

Stereoscopic observations of the fracture surface of notched specimens revealed that fatigue cracks initiate at the free edge adjacent to the notch due to less constraint for the development of slip systems. The fractographic details of Figure 9 are characteristic of the majority of specimens examined. In Figure 9a, a specimen with typical fatigue segment characteristics, including the crack initiation section, fatigue crack growth regime, and fast fracture region, is shown. In the fracture surface of Figure 9b, the crack initiates at the free edge surface of the notch area and is advancing radially from the corner edge, towards the mid-thickness direction. The first crack was observed at one specimen surface, and after a certain number of cycles, a second crack on the other edge was observed, propagating through thickness and leading eventually to crack coalescence and the formation of a single through-thickness crack that propagates until the failure of the specimen. A characteristic example is shown in Figure 9b, where replica observations in steel (B), at 120,000 cycles revealed the existence of a 211 μm crack only at one side (side A) of the specimen. The above observations agree with the hypothesis used in the finite element analysis for crack initiation simulation, taking into account the corner crack element with a reference size of 250 μm at the free edge of the V-notch.

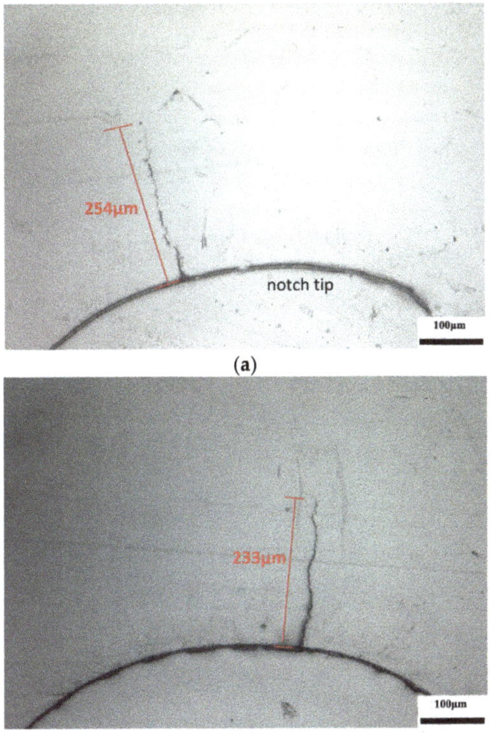

(**a**)

(**b**)

Figure 8. Replica images from notch tip at (**a**) 79,000 cycles for steel (A) and (**b**) 105,000 cycles for steel (B). Loading axis is in the horizontal direction.

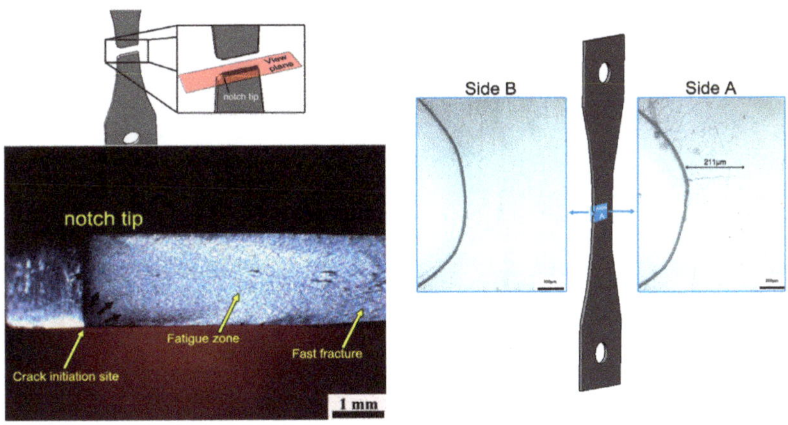

(**a**) (**b**)

Figure 9. (**a**) Stereoscopic image from fracture surface of notched steel (A) (N_f = 122,420 cycles at σ_{max} = 200 MPa; the 250 μm crack detected at 87,700 cycles). (**b**) Lateral replica images in steel (B) at 110,000 cycles showing the onset of an edge surface crack (N_f = 188,800 cycles at σ_{max} = 200 MPa).

3.3. *Analytical Results with LS Approach*

In [23], a stepped load increment was applied up to a far-field tensile stress of 200 MPa, and the calculated stress–strain behavior of the corner element at the notch tip, as well as the distribution of normal stresses and strains in the loading direction, is reproduced and presented in Figure 10 for convenience. As shown in the results of Figure 10, local yielding at the notch tip takes place.

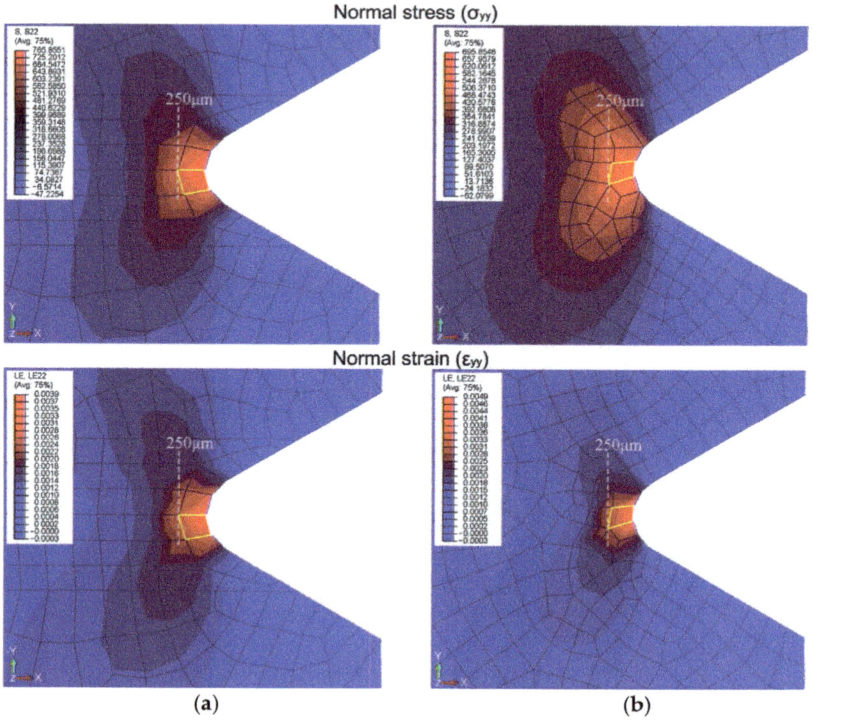

Figure 10. Normal stress and strain distribution at the vicinity of the notch for (**a**) steel (A) and (**b**) steel (B) (taken from [23]).

The maximum stresses and strains at the notch root can be calculated using the Neuber method or the SED criterion by simultaneously solving Equations (1) and (5) and Equations (5) and (8), respectively, for incremental monotonic loading.

In Figure 11, the analytical local stress–strain results obtained using the Neuber and SED method are compared with the numerical results obtained from the finite element model. By examining the results presented in Figure 11, it can be noticed that the analytical results underestimate the numerical findings. In material (A), for a maximum notch strain of 0.0034, the predicted notch stress with the Neuber or SED criterion is underestimated by 33.4% (FEA prediction = 676 MPa; Neuber/SED = 450 MPa), while for material (B), the difference is 33.5%.

As discussed earlier, Neuber's rule may overestimate the notch root strains under multiaxial stress state conditions, while it seems to provide more reliable results for plane stress conditions. A comparison of the analytical approximations with the FE analysis showed that the Neuber method and the SED criterion provide almost identical results, which are conservative compared to the results of the FE model and, hence, may be better used as first estimations [9]. However, the underestimation of notch root strains by quite large margins was also reported elsewhere [9].

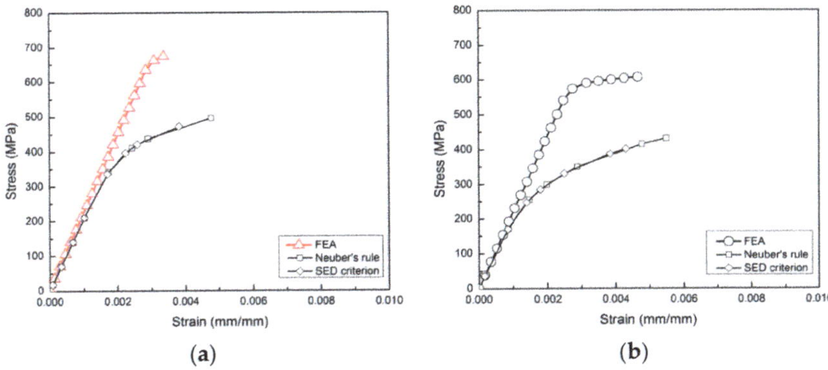

Figure 11. Notch root strains from FEA, Neuber's rule, and SED criterion under monotonic tensile loading for (**a**) material (A) and (**b**) material (B).

Fatigue Crack Initiation Assessment

The fatigue predictions from Table 3 are very sensitive to the calculated stress–strain history at the notch root. Small differences in notch stress–strain of the order of 5–6% may result in significant differences in the predicted fatigue life of the order of 34% in Table 3. The SED criterion predicts, in general, higher values of notch stress and strain compared to Neuber's method, leading to more conservative fatigue life predictions. Both methods overestimate the experimental results. The plane stress conditions corresponding to the element examined in Figures 5 and 10 do not seem to favor the prediction of the Neuber method here possibly due to other factors associated with the microstructure and TRIP effect. The above trend is in line with the observations presented in the literature, since Neuber's rule considers the stress state to be uniaxial at the notch root, which leads to less conservative results.

Table 3. Estimated fatigue life for the formation of a 250 μm crack using FEA results.

	(A)		(B)	
	Neuber	**SED**	**Neuber**	**SED**
Notch stress amplitude, σ_α (MPa) [a]	290	312.5	275.5	285
Notch strain amplitude, ε_α [a]	0.00141	0.00153	0.00149	0.00156
Notch mean stress, σ_m (MPa)	385.7	363.2	327.4	317.9
Numerical solution of maximum notch stress (FE), σ_{max} (MPa)	676	676	603	603
Product of $\varepsilon_\alpha \sigma_{max}$	0.9525	1.0311	0.8964	0.9417
Estimated fatigue life, N_f (cycles) [b]	53,800	37,890	104,500	77,500

[a] Calculated with Equations (4) and (7) for Neuber and Equations (7) and (9) for SED criterion. [b] Solved with Equation (10).

A comparison of predictions using, in the fatigue analysis, the numerical notch stress–strain results, and the results from the Neuber or SED method are shown in Table 4, using the experimental findings as verification. The numerical results improve the fatigue analysis and bring the analytical solutions closer to the experimental results; however, a slight underestimation of the number of cycles for fatigue crack initiation is present.

Table 4. Comparison of estimated fatigue life for crack initiation with experimental data.

Method	(A)		(B)	
	Neuber	SED	Neuber	SED
Estimated fatigue life without FE calculations, N_f (cycles) [a]	180,000	130,000	650,000	560,000
Estimated fatigue life with FE calculations, N_f (cycles)	53,800	37,890	104,500	77,500
Experimental investigation, N_{ini} (cycles)	66,040		108,970	

[a] Calculated with Equations (3), (5) and (8) implemented in Equation (10).

3.4. *Analytical Results with Fracture Mechanics Approach*

3.4.1. Crack Initiation (Onset of 50 μm Crack at the Notch Tip)

A description of the finite element analysis carried out and the notch plastic zone analysis for the specimen geometry used is presented. In Figure 12a, the finite element model is presented with a finer mesh (25 μm element size) to expand the notch stress-strain analysis at a characteristic distance of 50 μm from the notch tip for a far field tensile stress of S = 200 MPa. In Figure 12b,c the distribution of local stress and strain with the distance from the notch tip for material (B) is shown. The maximum stress and strain calculated at a distance of 50 μm from the notch root are σ_{yy} = 630 MPa and ε_{yy} = 0.007128, showing a 4.5% percent increase in the maximum stress compared to the element of 250 μm. Using these values in the analytical model described in Section 2.3.2 and shown in Figure 4, the cycles for failure of the material element are 80,300 and 60,500 cycles, using the Neuber and SED criterion, respectively (Table 5).

Table 5. Estimated fatigue life considering onset for crack initiation at 50 μm for steel (B).

Method	Neuber	SED
Estimated fatigue life with FE calculations, N_f (cycles)	80,300	60,500

3.4.2. Crack Propagation

In the case examined, the assumed crack length of 50 μm is significantly larger than the fictitious average grain size (6.5 μm in Figure 6), and, therefore, the problem does not fall within the microstructurally short crack regime. We considered the cyclic plastic zone at the tip of the crack based on the following analytical expression [29]:

$$r_{cy} = \frac{1}{8\pi}\left(\frac{\Delta K}{\sigma_{y0.2}}\right)^2 \tag{11}$$

where ΔK is the stress intensity factor range; $\sigma_{y0.2}$ the cyclic yield strength; and the resulting plastic zone size is r_{cy} = 17.65 μm, which is smaller compared to the considered crack length. On the other hand, the plastic zone at the notch root for a remote stress of 200 MPa is approximately 300 μm, taking into account the yield strength of 515 MPa [23] of steel (B) as and the result shown in Figure 12c, meaning that the stage of short crack propagation takes place inside the plasticity of the notch. In the work of Leis [12], a correlation of the LEFM concept was proposed for bulk plasticity conditions that dominate the growth behavior, e.g., the case of a crack completely submerged in the notch tip plastic zone. Since the notch inelastic field does not transmit the far field stress and stress cycle to the crack in the same manner as would an elastic field, it is reasonably assumed that, for a notched specimen subjected to a far field constant stress amplitude (load controlled), there is a material volume at the tip of the notch, with its deformation being displacement controlled (Figure 13).

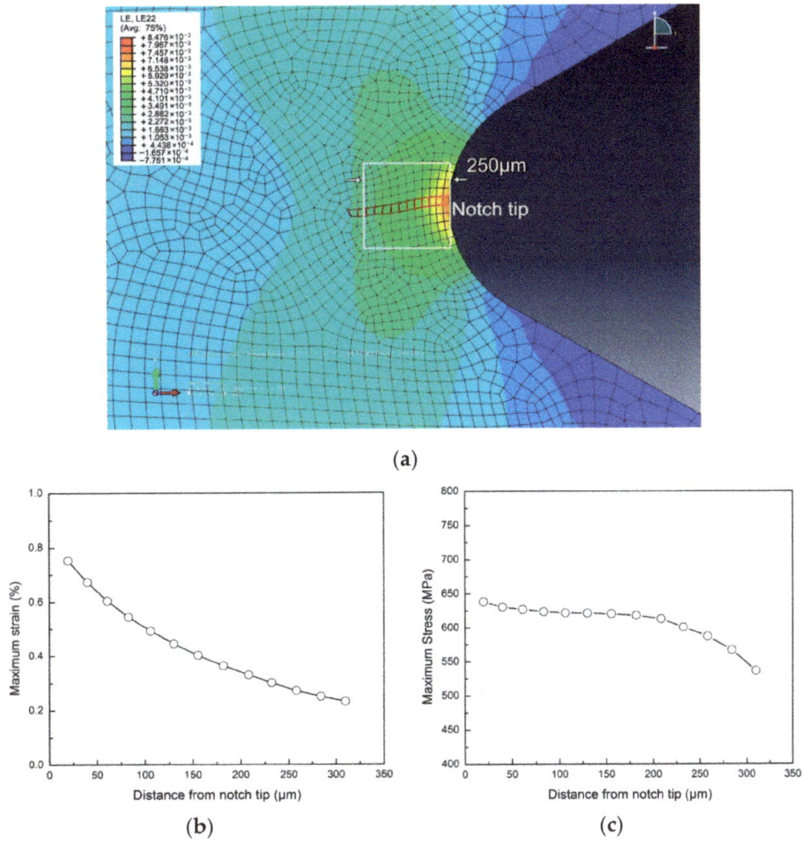

Figure 12. (**a**) Normal strain distribution in material (B) with finer mesh (25 μm element size), (**b**) maximum strain, and (**c**) maximum stress distribution vs. distance from notch tip for a remote stress of 200 MPa.

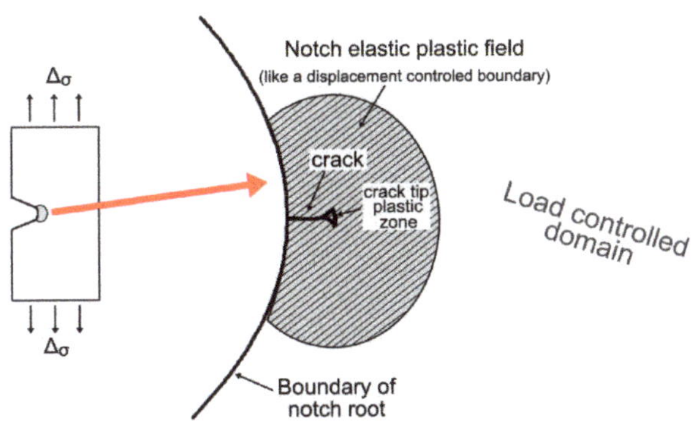

Figure 13. Displacement control of crack contained within notch plastic zone.

Under this viewpoint, it is postulated by Leis [12] that the driving force for the above crack to grow in the notch inelastic field can be approximated by the following equation:

$$K_{mx} = 1.12\varepsilon_{mx}E_{m}\sqrt{\pi\alpha} \tag{12}$$

where "K_{mx}" is an equivalent stress intensity factor, ε_{mx} is the cyclically stable maximum strain in the material element at the depth of interest (in the absence of the crack), E_m is the monotonic modulus, 1.12 is the free surface correction factor, and α is the length of the crack measured from the notch root. In Equation (12), the product $\varepsilon_{mx} * E_m$ is primarily used to estimate a "pseudostress" in keeping with the linear elastic nature of LEFM. In [30], the application of Equation (12) in a pearlitic rail steel resulted in the accurate prediction of the cracking behavior for cracks as small as 50 μm.

Since the total fatigue crack would be fully embedded in the notch tip plastic zone propagating under LCF conditions, the parameter ε_{mx} of Equation (12) can be estimated as an average maximum strain within the length scale of 50 μm to 250 μm in Figure 12b. From Figure 12b, we can see that the average maximum strain at the ligament of 50 μm to 250 μm is $\varepsilon_{mx} = 0.00417$. Using Equation (12) with a maximum strain value of $\varepsilon_{mx} = 0.00417$, E_m = 205.9 GPa, and the initial crack length based on the size of the corner element $\alpha = 0.00005$ m, the calculated stress intensity factor is $K_{mx} = 12.05\ \text{MPa}\sqrt{m}$.

The number of cycles for the crack to propagate from an initial length, α_{in} = 50 μm, to a final length, α_f = 250 μm (Figure 14), for a uniaxial; mode I problem under a stress ratio, R = 0.1; and a maximum far-field stress of σ_{max} = 200 MPa can be calculated by integrating a Paris–Erdogan crack growth law [31]:

$$\frac{d\alpha}{dN} = C(\Delta K)^m = C\left(Y\Delta\sigma\sqrt{\pi\alpha}\right)^m \rightarrow N_f = \frac{1}{CY^m\Delta\sigma^m\pi^{m/2}}\left[\frac{a_f^{1-m/2} - a_{in}^{1-m/2}}{1-\frac{m}{2}}\right] \quad (13)$$

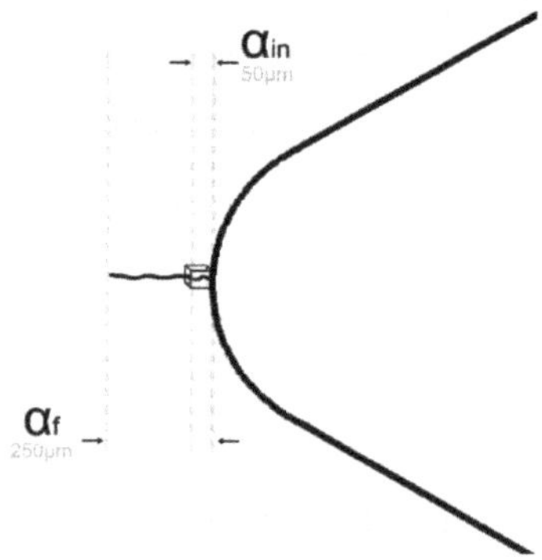

Figure 14. Initial and final crack length for the growth of the short crack.

In Equation (13), C and m are the Paris–Erdogan fatigue crack growth parameters; Y is the stress intensity factor correction; and a_f and a_{in} are the final and initial crack length, respectively.

The parameters $C = 1.81005 \times 10^{-13}$ and m = 4.10051 for the steel (B) were taken from [32] and are used here under the assumption of being applicable to the short crack problem. The "pseudostress" amplitude is $\Delta\sigma$ = 772 MPa, calculated from $\varepsilon_{mx} * E_m$ for R = 0.1, and the correction factor was taken as Y = 1.12. Using the above values in Equation (13), the number of fatigue cycles for the crack to propagate to the critical length of 250 μm is 12,254 cycles. By adding the cycles calculated in Table 5 for the onset of crack initiation, we see that the total fatigue life for development of a 250 μm crack at the root of the notch is 92,554 and 72,754, with the Neuber and SED method, respectively (Table 6).

Table 6. Predicted fatigue life for the formation of a 250 μm crack in material (B).

Method	Neuber		SED
Predicted Fatigue Life with FE Calculations, for Failure of a 50 μm Element, N_{ini} (Cycles)	80,300		60,500
Crack advancement from 50 μm to 250 μm, N_{prop} (cycles)		12,254	
Total fatigue life, N_f (cycles)	92,554		72,754
Experimental investigation, N_{ini} (cycles)		108,970	

4. Discussion

Conducting a fatigue crack initiation assessment for a corner element at the notch root, taking into account that the sheet steel materials have a thickness of under 2 mm, allows for the consideration of a uniaxial loading scenario for the material element under examination. It should be noted that, in the TRIP steels examined, possible phase transformation effects [33] under cyclic loading at the notch tip, which may promote local strain hardening of the material or produce certain crack closure mechanisms, were not taken into account in the analysis. RA transformation is microstructurally accompanied by volume expansion [34] and may cause a certain stress relaxation at the tip of a stress raiser due to the elastic surrounding material, thus extending the fatigue life prior to crack initiation. Additionally, phase transformation at the crack tip has been shown to be beneficial for fatigue crack growth in numerous studies [34]. The main mechanism contributing to the retardation of crack growth is the stress relaxation imposed by the phase transformation during the volume change [34].

Furthermore, while the material element at the free edge is subjected to uniaxial loading (σ_x, $\sigma_z = 0$), as the damage progressively grows towards the mid-thickness and away from the notch influence, plane strain conditions are expected, and therefore the RA transformation rate can be influenced by the stress triaxiality. Stress triaxiality effects can influence RA transformation and, hence, the associated mechanical behavior of TRIP steels.

By comparing the results of Table 6, focusing on steel (B), it can be seen that, by using the numerical results within the fracture mechanics approach, a good correlation with experimental data is obtained, but a slight underestimation of the fatigue crack initiation life is observed compared to the analysis involving the numerical simulation with the LS approach.

5. Conclusions

A coupled numerical–analytical methodology was implemented for the assessment of fatigue crack initiation life in TRIP steel, using a V-notched specimen. Fatigue crack initiation has been defined as the number of fatigue cycles required for the development of a short fatigue corner crack with finite length based on experimental fractographic observations. Depending on the definition of the finite short crack length, the prediction of crack initiation life is very sensitive to the local stress–strain field in the close vicinity of the notch. The results indicate that incorporating the numerical solution using the finite element method may lead to a more accurate determination of the stress–strain response at the free edge, the corner location of the notch tip, with regard to using solely established empirical–analytical solutions. Experimental observations of fatigue crack initiation in two different TRIP steels confirmed that fatigue crack initiation occurs as a corner crack and develops progressively as a through-thickness crack. The analytical findings of the present study highlight the importance of the introduction of an appropriate finite element analysis in fatigue methods for a more accurate determination of the local notch stress–strain behavior and, accordingly, a more reliable assessment of fatigue crack initiation life of steel components. Furthermore, the finite element method combined with a fracture mechanics approach for the assessment of crack initiation at the notch root is a promising option to extend the analysis by also including fatigue crack growth mechanisms that may influence the rate of short crack growth, typical for the fatigue crack growth behavior in TRIP steels (e.g., phase transformation effects).

Author Contributions: P.I.C. developed the methodology of the research, carried out the experiments, and validated the results. He also set up the finite element model and produced the numerical results, performed the comparisons between experimental and analytical results, and participated in the writing and editing of the article. A.T.K. supervised the overall concept and methodology of the performed research. He worked on the analysis and discussion of results and contributed to the writing, review, and editing of the article. All authors have read and agreed to the published version of the manuscript.

Funding: This research received no external funding.

Data Availability Statement: Research data can be available upon request.

Conflicts of Interest: The authors declare no conflict of interest.

References

1. Topper, T.; Wetzel, R.; Morrow, J. Neuber's rule applied to fatigue of notched specimens. *J. Mater.* **1969**, *4*, 200–209.
2. Shuai, Z.; Zhu, S.-P.; He, J.-C.; Liao, D.; Correia, J.; Macek, W.; Branco, R.; Wang, Q. Fatigue life prediction of notched components under size effect using strain energy reformulated critical distance theory. *Int. J. Fatigue* **2023**, *175*, 107805. [CrossRef]
3. Murthy, A.R.; Vishnuvardhan, S.; Anjusha, K.V.; Gandhi, P.; Singh, P.K. Prediction of fatigue crack initiation life in SA312 Type 304LN, austenitic stainless steel straight pipes with notch. *Nucl. Eng. Technol.* **2022**, *54*, 1588–1596. [CrossRef]
4. Neuber, H. Theory of Stress Concentration for Shear-Strained Prismatical Bodies with Arbitrary Nonlinear Stress-Strain Law. *J. Appl. Mech.* **1961**, *28*, 544–550. [CrossRef]
5. Molski, K.; Glinka, G. A method of elastic-plastic stress and strain calculation at a notch root. *Mater. Sci. Eng.* **1981**, *50*, 93–100. [CrossRef]
6. Fatemi, A.; Zeng, Z. Elasto-plastic stress and strain behaviour at notch roots under monotonic and cyclic loadings. *J. Strain Anal. Eng. Des.* **2001**, *36*, 287–300.
7. Glinka, G.; Ott, W.; Nowack, H. Elastoplastic Plane Strain Analysis of Stresses and Strains at the Notch Root. *J. Eng. Mater. Technol.* **1988**, *110*, 195–204. [CrossRef]
8. Sharpe, J.W.N.; Yang, C.H.; Tregoning, R.L. An Evaluation of the Neuber and Glinka Relations for Monotonic Loading. *J. Appl. Mech.* **1992**, *59*, S50–S56. [CrossRef]
9. Mowbray, D.F.; McConnelee, J.E. Applications of Finite Element Stress Analysis and Stress-Strain Properties in Determining Notch Fatigue Specimen Deformation and Life. *Astm. Stp.* **1971**, *519*, 151–169.
10. Suresh, S.; Ritchie, R.O. Propagation of short fatigue cracks. *Int. Mater. Rev.* **1984**, *29*, 445–475. [CrossRef]
11. MacLachlan, D.W.; Karamitros, V.; Dunne, F.P. Mechanistic modelling of fatigue nucleation and short crack growth in polycrystalline alloys. *J. Mech. Phys. Solids* **2023**, *177*, 105314. [CrossRef]
12. Leis, B.N. Displacement controlled fatigue crack growth in inelastic notch fields: Implications for short cracks. *Eng. Fract. Mech.* **1985**, *22*, 279–293. [CrossRef]
13. El Haddad, M.H.; Smith, K.N.; Topper, T.H. A Strain Based Intensity Factor Solution for Short Fatigue Cracks Initiating from Notches. In *Fracture Mechanics: Proceedings of the Eleventh National Symposium on Fracture Mechanics: Part I, ASTM STP 677*; Smith, C.W., Ed.; Virginia Polytechnic Institute and State University: Blacksburg, VA, USA, 1979; pp. 274–289.
14. Natkowski, E.; Durmaz, A.R.; Sonnweber-Ribic, P.; Münstermann, S. Fatigue lifetime prediction with a validated micromechanical short crack model for the ferritic steel EN 1.4003. *Int. J. Fatigue* **2021**, *152*, 106418. [CrossRef]
15. Zhu, X.-K. Effects of large plastic deformation and residual stress on the path independence of J-integral for cracks in ductile materials. *Eng. Fract. Mech.* **2023**, *277*, 108945. [CrossRef]
16. Tanaka, K.; Nakai, Y. Propagation and non-propagation of short fatigue cracks at a sharp notch. *Fatigue Fract. Eng. Mater. Struct.* **1983**, *6*, 315–327. [CrossRef]
17. Meng, L.; Yang, M.; Chen, X.; Hu, Y.; Feng, M. Physically short fatigue crack growth from notch described by plasticity-corrected stress intensity factor. *Int. J. Mech. Sci.* **2020**, *176*, 105544. [CrossRef]
18. Cheng, X.; Petrov, R.; Zhao, L.; Janssen, M. Fatigue crack growth in TRIP steel under positive R-ratios. *Eng. Fract. Mech.* **2008**, *75*, 739–749. [CrossRef]
19. Nam, S.W.; Chang, Y.W.; Lee, S.B.; Kim, N.J. Fatigue Strength of Formable Ultra High-Strength TRIP-Aided Steels with Bainitic Ferrite Matrix. *Key Eng. Mater.* **2007**, *345*, 247–250.
20. Braun, M.; Fischer, C.; Baumgartner, J.; Hecht, M.; Varfolomeev, I. Fatigue Crack Initiation and Propagation Relation of Notched Specimens with Welded Joint Characteristics. *Metals* **2022**, *12*, 615. [CrossRef]
21. Branco, R.; Prates, P.A.; Costa, J.D.; Ferreira, J.A.M.; Capela, C.; Berto, F. Notch fatigue analysis and crack initiation life estimation of maraging steel fabricated by laser beam powder bed fusion under multiaxial loading. *Int. J. Fatigue* **2021**, *153*, 106468. [CrossRef]

22. Dogahe, K.J.; Kurz, I.; Binkele, P.; Schmauder, S.; Mlikota, M.; Bozic, Z. Physically-based modelling of the fatigue crack initiation life of stent components under cyclic loading employing the Finite-Element-Method (FEM). *Int. J. Fatigue* **2023**, *171*, 107594. [CrossRef]
23. Christodoulou, P.I.; Kermanidis, A.T.; Krizan, D. Fatigue behavior and retained austenite transformation of Al-containing TRIP steels. *Int. J. Fatigue* **2016**, *91*, 220–231. [CrossRef]
24. Noda, N.A.; Nisitani, H. Stress concentration of a strip with a single edge notch. *Eng. Fract. Mech.* **1987**, *28*, 223–238.
25. Manson, S.; Hirschberg, M. Crack initiation and propagation in notched fatigue specimens. In Proceedings of the First International Conference on Fracture, Sendai, Japan, 12–17 September 1966.
26. Smith, K.N.; Watson, P.; Topper, T.H. A stress-strain function for the fatigue of metals. *J. Mater.* **1970**, *4*, 767–778.
27. Jacques, P.J.; Furnémont, Q.; Lani, F.; Pardoen, T.; Delannay, F. Multiscale mechanics of TRIP-assisted multiphase steels: I. Characterization and mechanical testing. *Acta Mater.* **2007**, *55*, 3681–3693. [CrossRef]
28. Wang, K.; Song, K.; Xin, R.; Zhao, L.; Xu, L. Cyclic microstructure analysis, crack propagation and life prediction of Inconel 750H considering the slip fracture energy. *Int. J. Plast.* **2023**, *167*, 103660. [CrossRef]
29. Dugdale, D.S. Yielding of steel sheets containing slits. *J. Mech. Phys. Solids* **1960**, *8*, 100–104. [CrossRef]
30. Leis, B.N. Microcrack initiation and growth in a pearlitic steel–Experiment and analysis. In Proceedings of the 15th National Fracture Symposium, College Park, MD, USA, 7–9 July 1982.
31. Paris, P.; Erdogan, F. A critical analysis of crack propagation laws. *J. Fluids Eng.* **1963**, *85*, 528–533. [CrossRef]
32. Gonidakis, S. The Effect of Heat Treatment Procedure in Fracture Behavior of Aluminum Containing TRIP Steel. Master's Thesis, Department of Mechanical Engineering, University of Thessaly, Volos, Greece, 2017.
33. Gui, Y.; An, D.; Han, F.; Lu, X.; Kang, G.; Zhang, X. Multiple-mechanism and microstructure-based crystal plasticity modeling for cyclic shear deformation of TRIP steel. *Int. J. Mech. Sci.* **2022**, *222*, 107269. [CrossRef]
34. Olson, G.B. Transformation Plasticity and Toughening. *Le J. De Phys. IV* **1996**, *6*, C1-407–C1-418. [CrossRef]

metals

Article

Non-Unified Constitutive Models for the Simulation of the Asymmetrical Cyclic Behavior of GH4169 at Elevated Temperatures

Xuteng Hu [1,2,*], Shuying Zhuang [1,2], Haodong Zheng [1,2], Zuopeng Zhao [1,2] and Xu Jia [1,2]

1 College of Energy & Power Engineering, Nanjing University of Aeronautics and Astronautics, Nanjing 210016, China
2 Key Laboratory of Aero Engine Thermal Environment and Thermal Structure, Ministry of Industry and Information Technology, Nanjing 210016, China
* Correspondence: xthu@nuaa.edu.cn

Abstract: The tensile, creep, fatigue and creep-fatigue tests of the nickel-based superalloy GH4169 were carried out. According to the deformation characteristics of GH4169 alloy, the Ohno-Karim kinematic model (O-K model) can be used to describe the tensile behavior. The creep constitutive model presented in this paper can be used to predict the three-stage creep characteristics of the GH4169 alloy. The modified Ohno-Karim kinematic hardening model, combined with an isotropic hardening model, can well predict the cyclic softening behavior of the material under symmetric loads and the mean stress relaxation behavior under asymmetric loads. Based on the modified Ohno-Karim kinematic hardening model, isotropic hardening model and creep constitutive model, a non-unified constitutive model was established. The creep-fatigue behavior of the GH4169 alloy under symmetric and asymmetric loads is simulated by using the non-unified constitutive model. The simulation results are very close to the experimental results; however, the prediction results of the time-dependent relaxation load are relatively small.

Keywords: nickel-based superalloy; non-unified constitutive models; creep-fatigue; cyclic softening; stress relaxation

Citation: Hu, X.; Zhuang, S.; Zheng, H.; Zhao, Z.; Jia, X. Non-Unified Constitutive Models for the Simulation of the Asymmetrical Cyclic Behavior of GH4169 at Elevated Temperatures. *Metals* **2022**, *12*, 1868. https://doi.org/10.3390/met12111868

Academic Editors: Francesco Iacoviello and Denis Benasciutti

Received: 5 October 2022
Accepted: 28 October 2022
Published: 2 November 2022

1. Introduction

The turbine disk is an important part of the aero engine. The grooves of the turbine disk are subjected to asymmetric creep-fatigue loads. Accurate description of the deformation response of materials under asymmetric load is the basis for scientifically predicting the creep-fatigue life of the groove. The deformation responses of the material are more complex under asymmetric loading. It will show not only cyclic softening or cyclic hardening behavior, but also ratcheting and mean stress relaxation behavior. The material will also be subjected to creep loads at high temperatures. The grooves of the turbine disc may contain all of the above deformation behaviors.

Chaboche [1] employed two internal variable equations of the Armstrong-Frederick nonlinear kinematic hardening model (A-F model) [2] and isotropic hardening model to describe the deformation behavior of materials well under symmetric loads. However, this method would lead to excessive ratchet strain under asymmetric loads [3]. Subsequently, Chaboche and Nouailhas [4–7] analyzed several possible forms of the constitutive equation to reduce the predicted ratcheting without disturbing the tensile plastic flow. It was of great significance that they proposed a kinematic hardening model with thresholds in the dynamic recovery term. Ohno and Wang [8,9] proposed a kinematic hardening model in terms of nonintersecting multisurfaces, called the O-W I model. Ohno and Abdel-Karim [10,11] proposed a new kinematic hardening model called the O-K model, which was developed by furnishing the A-F model with the O-W I model. The O-K model was successful in simulating multiaxial ratcheting and multiaxial cyclic stress relaxation responses.

The creep deformation behavior can be described by the kinematic hardening model with the steady-state recovery term; however, this method can only describe the second stage of creep. With the development of the creep constitutive model, many constitutive models describing the characteristics of three stages of creep have been proposed, such as the θ projection method [12–14], Kachanov-Rabotnov model [15–20], Wilshire model [21–24], etc. The description of creep-fatigue behavior can also be realized by the non-unified models, which considered that the inelastic strain can be decomposed into the time-independent plastic strain and creep strain [25–28]. The parameters of the creep constitutive model can be obtained from the creep tests so that the creep deformation behavior can be better described.

The purpose of this paper is to systematically establish the cyclic constitutive model of nickel-base superalloys GH4169 for turbine discs. The mechanical properties tests of GH4169 were carried out systematically, including the tensile test, creep test, fatigue and creep-fatigue test. The constitutive model established can describe the tensile, creep, fatigue and creep-fatigue behavior of the material and lay a foundation for predicting the life of the turbine disc in the future.

2. Experimental Procedures

2.1. Materials

The nickel-based superalloy GH4169 is commonly used in aero-engine turbine disks. This material has good comprehensive properties at high temperatures. The heat treatment process of this material is as follows: first heated to 720 °C and held for 8 h; subsequently, furnace-cooled at a rate of 50 °C/h to 620 °C and then, held for 8 h; and finally, air-cooled to a room temperature condition. The main chemical components of the GH4169 alloy provided by the manufacturer are shown in Table 1. The metallographic structure of the sample is observed under an optical microscope, as shown in Figure 1. It can be seen from the metallograph that the structure of the GH4169 alloy is uniform and the grain size is about 100 μm, with some twin structures and a small amount of inclusions.

Table 1. Chemical components of GH4169 alloy.

Element	Nb + Ta	Mo	Cr	Ni	Ti	Fe	Al	C	Si
Wt%	5.04	3.12	19.70	51.70	1.01	—	0.42	0.039	0.15

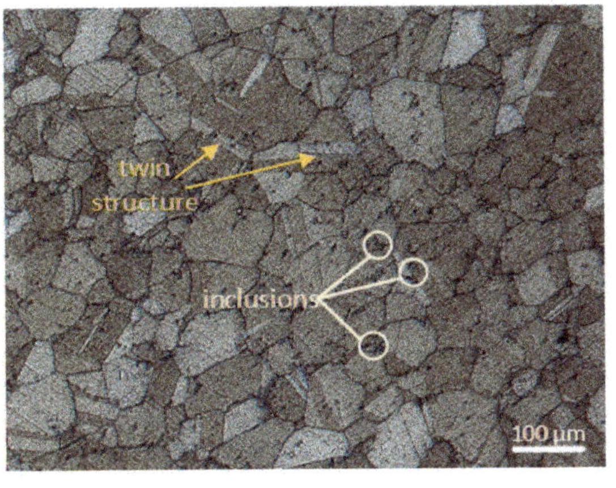

Figure 1. Metallographic structure of GH4169.

2.2. Tensile Tests and Results

The tensile tests were performed on a Zwick/Roell machine (ZwickRoell Testing Technology Co., Ltd., Shanghai, China). Prior to the test, the samples were heated to the test temperature of 600 °C with an environmental box and held for half an hour to ensure the uniform temperature inside the samples. The gauge length of the sample is set to 25 mm according to the extensometer and the deformation of the sample is monitored throughout the tensile process.

The tensile test results are an important reference for the formulation of the creep, fatigue and creep-fatigue tests. The description of tensile behavior is the basis of other tests. Three strain rates of 5×10^{-4}/s, 2×10^{-3}/s and 1×10^{-2}/s were used for the tensile tests. The engineering stress-engineering strain curves obtained from quasi-static tensile tests at different strain rates are shown in Figure 2. The curves of the same strain rate almost coincide. The curves of all the strain rates show unstable plastic flow as serrations in varying degrees. It is the typical Portevin Le Chatelier (PLC) effect. The most commonly accepted explanation for the phenomenon of the PLC effect is based on a model called dynamic strain ageing (DSA), which is defined as the interaction between the moving dislocations and diffusing solute atoms [29,30]. It also leads to the phenomenon that the tensile curve with the smallest strain rate (5×10^{-4}/s) will have the greatest stress under the same strain. The stress–strain curve exhibits a negative strain rate sensitivity. The behavior of the negative strain rate sensitivity cannot be described by the viscoplastic model. The curves of the strain rate 1×10^{-2}/s and 2×10^{-3}/s are very close, as shown in Figure 2d. The tensile strain rates of the fatigue and creep-fatigue tests are between 2×10^{-3}/s and 1×10^{-2}/s in this paper. Therefore, the total strain rate can be divided into three parts without considering the effect of viscoplasticity:

$$\dot{\varepsilon} = \dot{\varepsilon}_e + \dot{\varepsilon}_p + \dot{\varepsilon}_c \quad (1)$$

where $\dot{\varepsilon}_e$ is the elastic strain rate, $\dot{\varepsilon}_p$ is the plastic strain rate and $\dot{\varepsilon}_c$ is the creep strain rate.

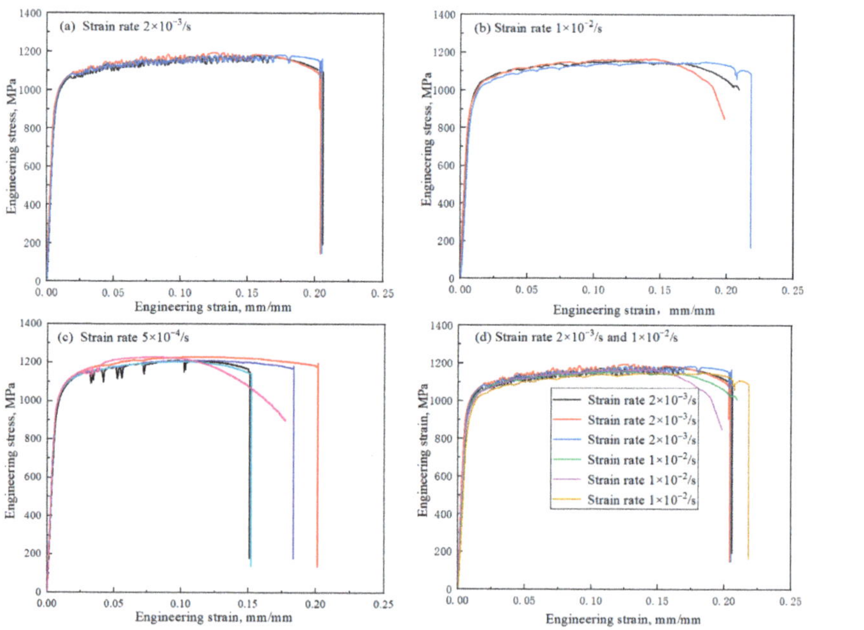

Figure 2. Tensile test results at different strain rates: (**a**) 2×10^3/s; (**b**) 1×10^{-2}/s; (**c**) 5×10^{-4}/s; (**d**) 2×10^{-3}/s and 1×10^{-2}/s.

2.3. Creep Tests and Results

The creep tests were performed on the creep testing machine Qianbang QBR-100 (Changchun Qianbang Test Equipment Co., Ltd., Changchun, China). The samples were heated to the test temperature and held for 1 h to ensure the uniform temperature inside the samples prior to the test. The temperature control system of the testing machine guarantees temperature fluctuation less than 2 °C and the load control system enables the applied load error not to exceed 0.5%.

The creep testing loads can be roughly determined according to the tensile strength of the material. The test time of creep test is generally long and the ultimate tensile strength is the maximum value that the creep test load can achieve. In this paper, 80% of the ultimate tensile strength is selected as the maximum load of the creep test at 600 °C and then, the load is gradually reduced. At the same time, to verify the generality of the creep constitutive model proposed later, this paper also supplemented the creep tests at 650 °C. Considering that the tests at 650 °C are only confirmatory, the maximum life is controlled within 350 h.

The creep curve generally shows the characteristics of three stages of creep. The primary stage is characterized by the gradual decline of the creep strain rate, with the increase in creep time at a short time. When the creep strain rate decreases to the minimum value, it indicates the beginning of the second stage of creep. The creep strain rate is nearly constant at the second stage. When the creep strain rate reaches 2 times of the minimum strain rate, it indicates the beginning of the accelerated creep stage [31]. It can be seen from Figure 3 that the creep curve of the GH4169 alloy is mainly composed of the second and accelerated creep stage. The primary stage of creep is relatively insignificant and the creep time of the first stage accounts for less than 10% of the total life. The second stage of creep occupies the main time of creep, accounting for more than 60%. The deformation of the accelerated creep stage is the largest, accounting for more than 60% of the total creep deformation. The creep fracture strain is also called creep ductility. The creep ductility of 865 MPa and 825 MPa are obviously greater than that of other loads. Perhaps due to the small number of tests, there is no obvious law for the change of the ductility with stress in this paper.

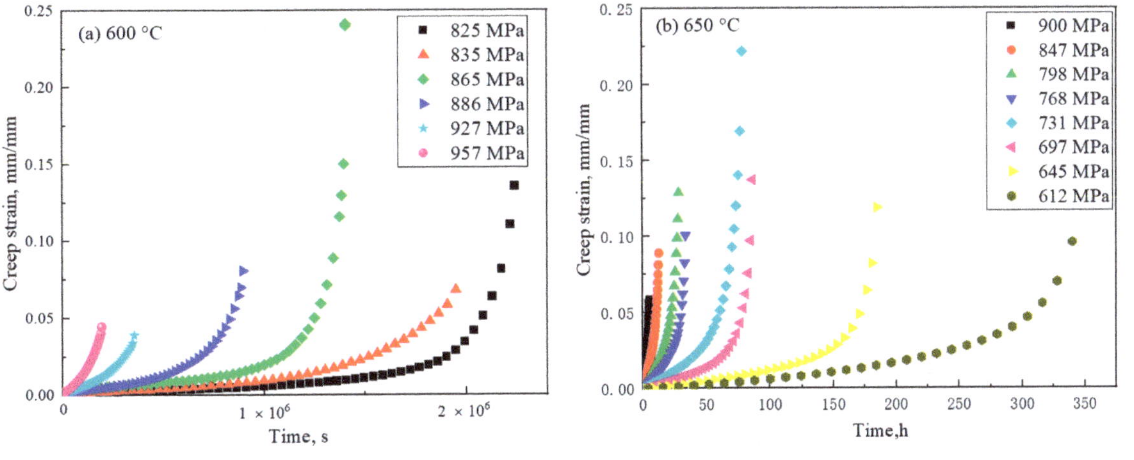

Figure 3. Creep strain–time curves of the GH4169 alloy at different temperatures: (**a**) 600 °C; (**b**) 650 °C.

2.4. Fatigue and Creep-Fatigue Tests and Results

The fatigue and creep-fatigue tests were carried out on an Instron fatigue testing machine (Instron Corp., Norwood, MA, USA). The samples were held for over half an hour at the test temperature and the temperature-controlling errors of the test section are less than 2 °C.

Materials will exhibit cyclic softening or hardening under symmetric cyclic loading. The GH4169 alloy is a kind of cyclic softening material. The hysteresis loops of the first 10 cycles under uniaxial strain-controlled cyclic with a constant strain amplitude $\varepsilon_a = 1.0\%$ are shown in Figure 4. The peak stress decreases with the number of cycles. The yield point, which is defined as the boundary point between elasticity and plasticity, decreases with the increasing number of cycles. The yield radius decreases with the increasing number of cycles in essence. This phenomenon of the softening or hardening of materials can be described by an isotropic hardening model.

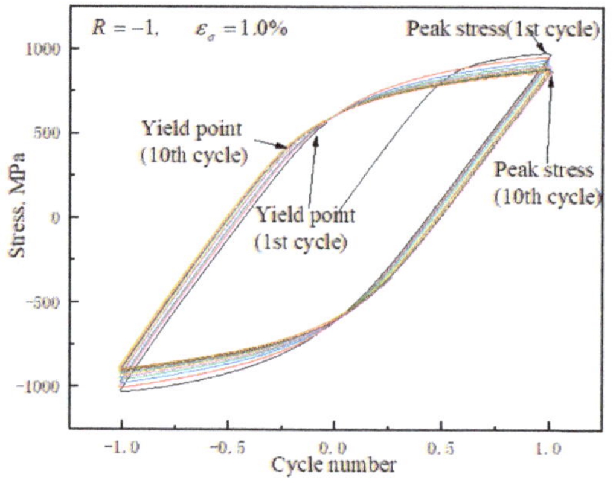

Figure 4. The hysteresis loops of the first 10 cycles.

Due to the cyclic softening effect of materials, the stress amplitude decreases with the number of cycles, as shown in Figure 5. The stress amplitude can be divided into three stages: the amplitude in the first stage decreases rapidly; the amplitude in the second stage decreases steadily; and the material in the third stage tends to show severe softening effects, and the stress amplitude drops sharply until fracture. This is caused by the crack on the sample surface and the reduction of the net-bearing area.

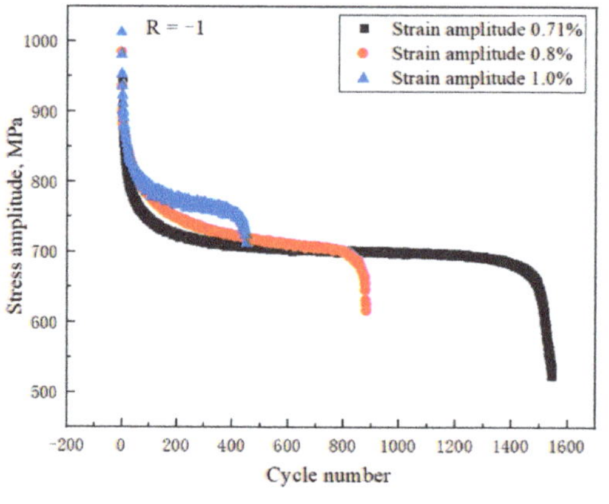

Figure 5. Results of cyclic softening characteristic test of GH4169 alloy.

The material shows a mean stress relaxation under the strain-controlled asymmetric load. The average stress relaxation curve is similar to the cyclic softening curve, which also has three stages, as shown in Figure 6.

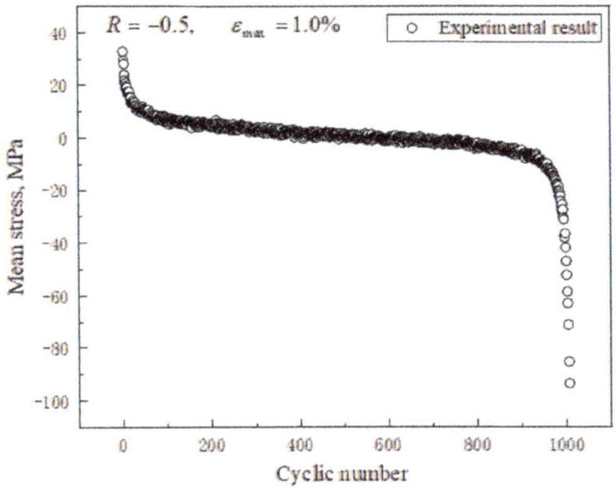

Figure 6. Results of mean stress relaxation characteristic test of GH4169 alloy.

The material exhibits not only cyclic softening and mean stress relaxation, but also time-dependent stress relaxation behavior under the strain-controlled creep-fatigue load, as shown in Figure 7a. The stress relaxation behavior has little influence on the cyclic softening behavior, as shown in Figure 7b.

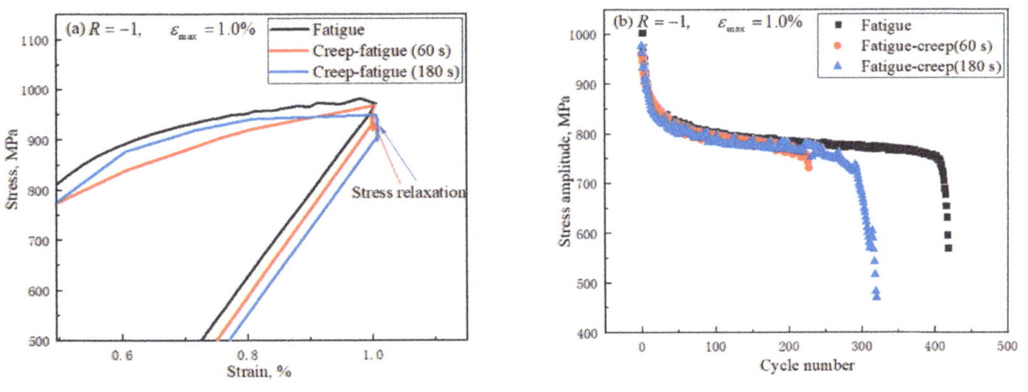

Figure 7. Stress relaxation behavior: (**a**) cyclic curve of stress relaxation at different holding times; (**b**) stress amplitude at different holding times.

3. Constitutive Model and Simulation Results of Experiments

3.1. Simulation Results of Tensile Experiments

The tensile behavior can be described by a nonlinear kinematic hardening model. The O-K model is employed to describe the mean stress relaxation behavior. In order to better describe the movement of the yield surface, Chaboche decomposed the back stress into several terms and made each term follow the O-K model:

$$\dot{\alpha} = \sum_{i=1}^{M} \dot{\alpha}_i \tag{2}$$

The O-K model can be expressed as:

$$\dot{\alpha}_i = \zeta_i r_i \left[\frac{2}{3} \dot{\varepsilon}^p - \mu_i \frac{\alpha_i}{r_i} \dot{p} - H(f_i) \left\langle \dot{\lambda}_i \right\rangle \frac{\alpha_i}{r_i} \right] \quad (3)$$

$$\dot{\lambda}_i = \frac{\alpha_i}{r_i} : \dot{\varepsilon}^p - \mu_i \dot{p} \quad (4)$$

where H denotes the Heaviside step function; $\langle\rangle$ is the Macauley bracket; ζ_i, r_i and μ_i ($0 \leq \mu_i \leq 1$) are material parameters. The parameter μ_i can control ratcheting strain and mean stress with the increasing number of cycles. In this paper, it is assumed that each component of μ_i is equal to a fixed value μ. The O-K model is reduced to the O-W I and A-F models when $\mu = 0$ and $\mu = 1$, respectively. The fitting methods of different models are different.

When $\mu = 0$, under the one-dimensional stress state, stress can be written as

$$\sigma = \sigma_{y0} + \sum_{i=1}^{M} r_i [1 - \langle 1 - \zeta_i p \rangle] \quad (5)$$

When $0 < \mu < 1$, under the one-dimensional stress state, stress can be written as

$$\sigma = \sigma_{y0} + \sum_{i=1}^{M} r_i \left[1 - \left\langle 1 - \frac{1 - \exp(-\zeta_i \mu p)}{\mu} \right\rangle \right] \quad (6)$$

When $\mu = 1$, under one-dimensional stress state, stress can be written as

$$\sigma = \sigma_{y0} + \sum_{i=1}^{M} r_i [1 - \exp(-\zeta_i p)] \quad (7)$$

where p is equivalent plastic strain.

Formulas (5) and (6) are piecewise expressions. M is the number of segments. Theoretically, the more segments there are, the more accurate the simulation result will be. In a comprehensive consideration, $M = 20$ is selected in this paper. The selected 20 data points are shown in Figure 8.

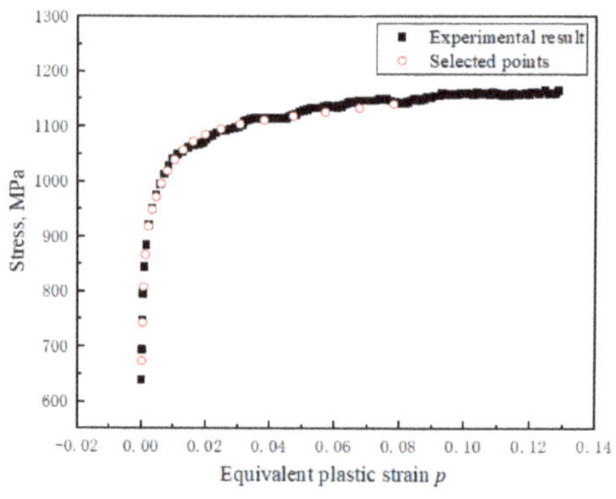

Figure 8. Selected points for parameter fitting.

Parameter fitting of the A-F model is relatively simple. Formula (7) can be directly fitted according to the data. The A-F kinematic equations are generally given by the superposition of three rules; that is, $M = 3$. The first rule describes the nonlinear behavior for the moderate plastic strains. The second one gives the nonlinearity for the very small strains, in the transition regime between elasticity and plasticity. The third kinematic rule is used to describe the approximately constant tangent stiffness for large strains. The fitting results of the A-F model are shown in Figure 9.

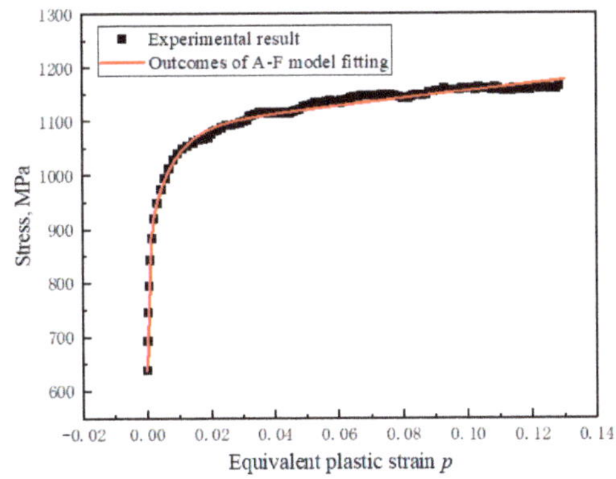

Figure 9. Outcomes of A-F model fitting.

Simulation results of tensile tests with different μ values are shown in Figure 10. The value of μ has little effect on the tensile simulation results. When the values are within a small range ($0 \leq \mu_i \leq 0.1$), the values of ζ_i, and r_i do not change much.

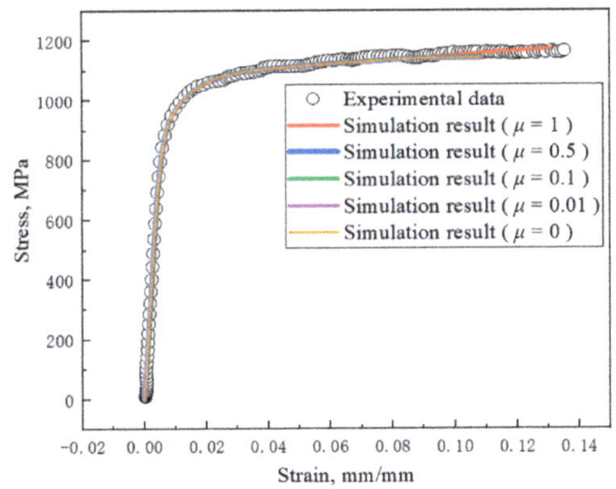

Figure 10. Simulation results of different μ values.

3.2. Simulation Results of Creep Tests

The relationship between creep strain and creep time is shown in Figure 11a. The creep strain rate remains constant during the second stage. The creep strain rate of the second stage is the minimum creep strain rate. The relationship between the creep strain rate and normalized creep time is shown in Figure 11b. The creep strain rate first decreases rapidly to the minimum value with time, indicating the beginning of the second stage of creep. When the creep strain rate reaches twice of the minimum creep strain rate, it indicates the beginning of the third stage of creep. The minimum creep strain rate plays an important role in the creep curve. The creep constitutive model is written in the form of minimum creep strain rate:

$$\varepsilon_c = c'_1\left(1 - e^{-c'_2\dot{\varepsilon}_m^{c'_3}t}\right) + \dot{\varepsilon}_m t + c'_4\left(e^{c'_5\dot{\varepsilon}_m^{c'_6}t} - 1\right) \tag{8}$$

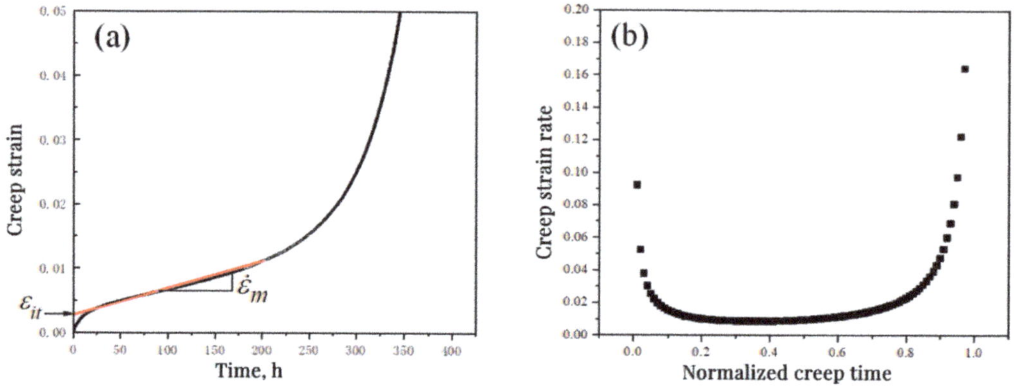

Figure 11. Typical creep curves: (**a**) creep strain–time curve; (**b**) creep strain rate–normalized creep time curve.

The first term represents the primary stage of creep, reflecting that creep strain increases to the critical value c_1' with time. The second term represents the second stage of creep and reflects the linear change of the creep strain with time. The third term represents the third stage of creep, reflecting that the creep strain increases exponentially with time.

In fact, there is a certain relationship between c_1' and creep stress; provided that the linear curve of the second stage of creep is extended to intersect with the ordinate, as shown in Figure 11a. Set the intersection as ε_{it}, which is called the intercept strain. It can be derived from Equation (8):

$$\varepsilon_{it} \approx c'_1 \tag{9}$$

If intercept strains are extracted from all the creep strain curves, it can be found that the intercept strain has a linear relationship with the nominal stress in logarithmic coordinates, as shown in Figure 12. Therefore, it can be expressed as:

$$c'_1 = b_2 \sigma^{b_1} \tag{10}$$

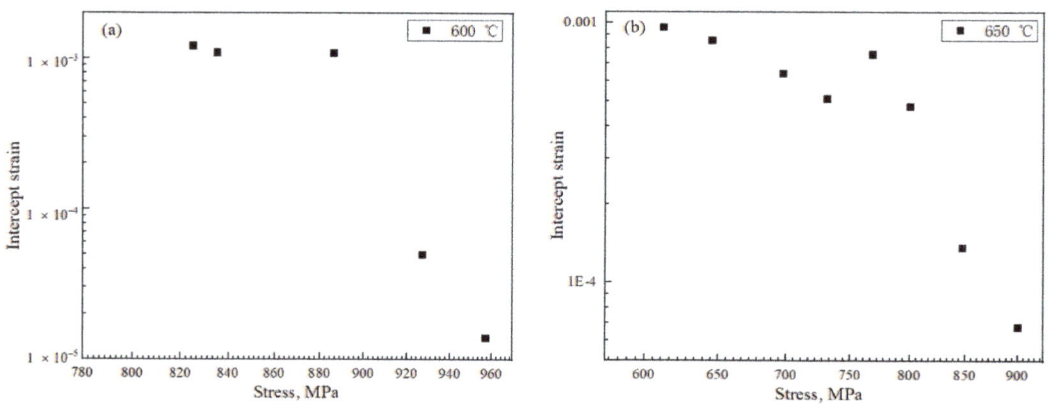

Figure 12. Relationship between intercept strain and stress: (**a**) at 600 °C; (**b**) at 650 °C.

The constitutive model of the second stage of creep is commonly described in Norton form and the minimum creep strain rate can be expressed as:

$$\dot{\varepsilon}_m = C_2\sigma^{n_2} \quad (11)$$

If Equations (10) and (11) are substituted into Equation (8), the coefficients can be sorted out and combined to obtain the following expression:

$$\varepsilon_c = c_1\sigma^{b_1}\left(1 - e^{-C_1\sigma^{n_1}t}\right) + C_2\sigma^{n_2}t + c_2\left(e^{C_3\sigma^{n_3}t} - 1\right) \quad (12)$$

where c_1, c_2, b_1, C_1, n_1, C_2, n_2, C_3 and n_3 are material constants.

In the form of Equation (12), the creep test results of a typical sample are taken for fitting; for example, a typical creep curve at 600 °C and 865 MPa, of which the fitting results are shown in Figure 13. The test values coincide completely with the calculated curve and the three-stage characteristics of the fitted curve can be clearly seen.

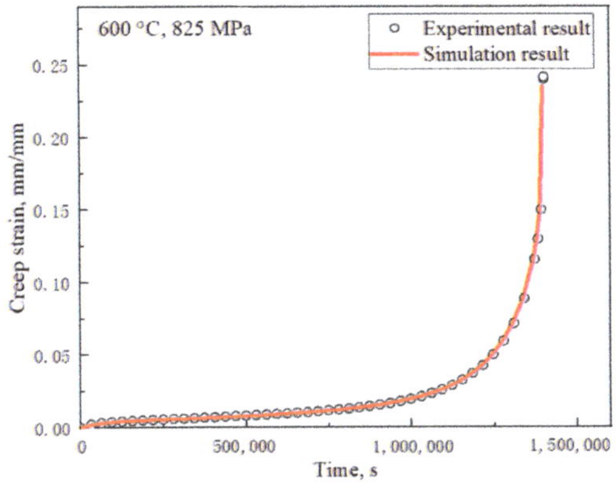

Figure 13. Simulation result of the creep test at 600 °C and 865 MPa.

The three-stage creep constitutive model of Equation (12) was adopted to fit all the creep test results of the nickel-based superalloy GH4169 at 600 °C and 650 °C, as shown in Figure 14; and the parameters are shown in Table 2. The experimental results almost coincide with the fitting results, which shows the practicability of the method.

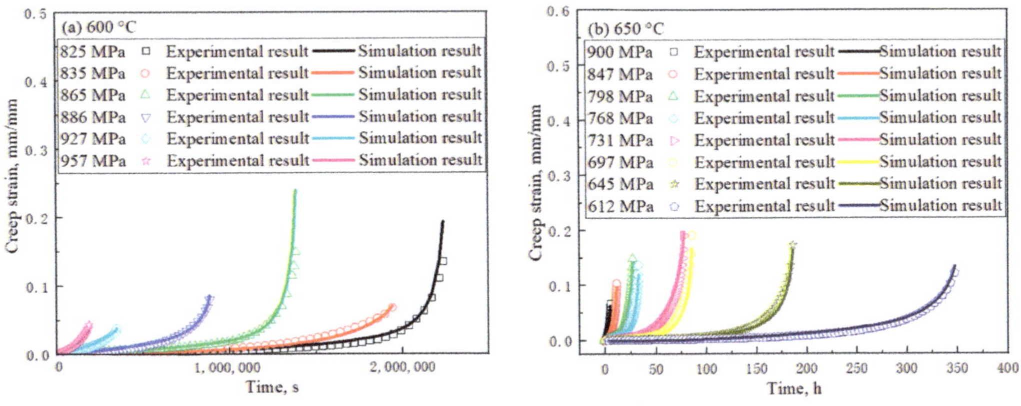

Figure 14. Simulation results of creep tests at different temperatures: (**a**) at 600 °C; (**b**) at 650 °C.

Table 2. The parameters of creep model at 600 °C.

c_1	c_2	b_1	C_1	n_1	C_2	n_2	C_3	n_3
1.0776×10^{-8}	5.8129	2.8489	5.5468×10^{-17}	3.3655	1.2997×10^{-13}	2.2323	2.7304×10^{-21}	4.7467

3.3. *Simulation of Fatigue and Creep-Fatigue Tests*

Materials tend to exhibit cyclic softening or hardening behavior under symmetric loading, which can be described by an isotropic hardening model. The GH4169 material is a typical cyclic softening material. The cyclic softening process includes three parts. The third part is the rapid crack propagation leading to the rapid curve decline, which is not a characteristic of the material itself. Therefore, the nonlinear isotropic hardening model can be expressed as:

$$\sigma_y = \sigma_{y0} + R_0 p + R_\infty (1 - e^{-bp}) \tag{13}$$

where σ_{y0} is the initial yield stress. R_0, R_∞ and b are material parameters.

To identify the isotropic hardening parameters R_0, R_∞ and b in Equation (13), parametric sensitivity studies are carried out through uniaxial strain-controlled cyclic simulations with the combined O-K model. The results of the sensitivity studies are shown in Figure 15. The parameter R_0 controls the slope of the stable stage of the curve as shown in Figure 15a. The parameter R_∞ controls the softening strength at the stabilized cycles as shown in Figure 15b and the parameter b controls the softening rate at the early stage of cyclic behavior as shown in Figure 15c.

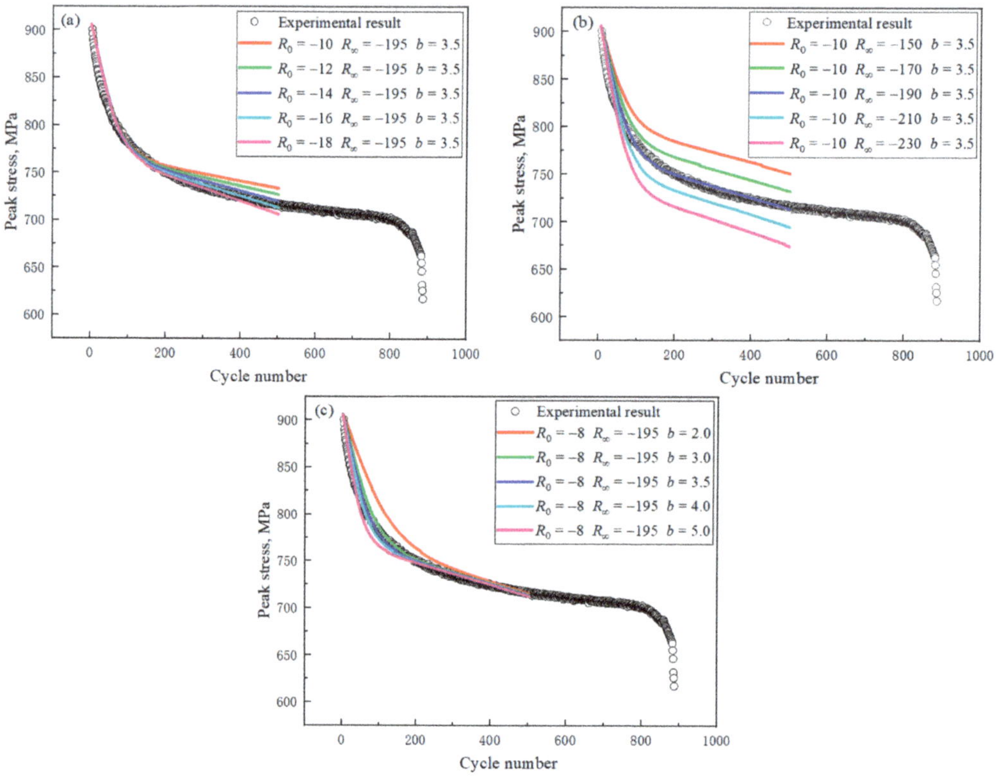

Figure 15. Results of sensitivity analysis for isotropic softening parameters: (**a**) R_0; (**b**) R_∞; (**c**) b.

The comparison between fitting results and test data is shown in Figure 16, and the specific parameters are shown in Table 3. The predicted cyclic softening behavior of the GH4169 alloy is in very good agreement with those of the tests for each cycle.

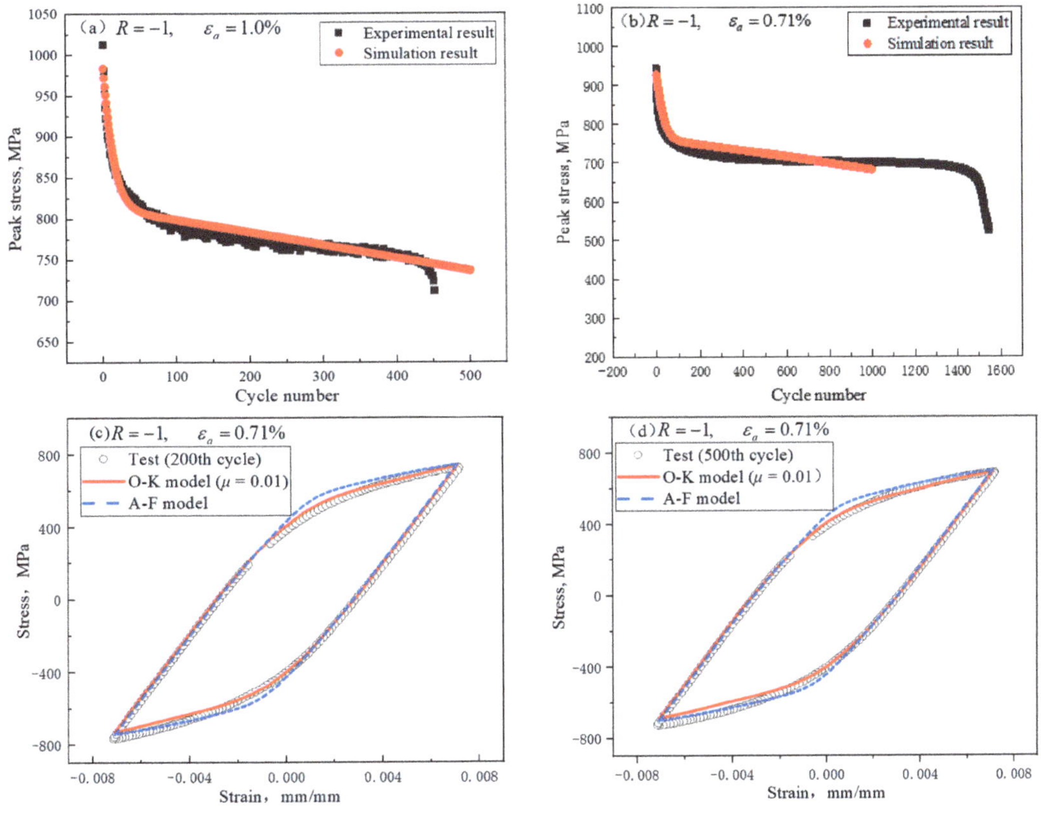

Figure 16. Comparison of symmetric cyclic behavior between tests and simulations: (**a**) $\varepsilon_a = 1.0\%$; (**b**) $\varepsilon_a = 0.71\%$; (**c**) the 200th cycle; (**d**) the 500th cycle.

Table 3. The parameters of isotropic hardening model.

σ_{y0}	R_0	R_∞	b
630	−8	−195	4

In the simulation of uniaxial ratcheting, the sensitivity studies of the O-K kinematic hardening parameter μ are carried out through uniaxial strain-controlled asymmetrical cyclic simulations with the combined isotropic hardening model, as mentioned earlier as shown in Figure 17. The parameter μ controls the mean stress relaxation rate. A larger value of μ leads to a larger mean stress relaxation. No matter what the μ value is, the simulation results do not agree with the experimental results.

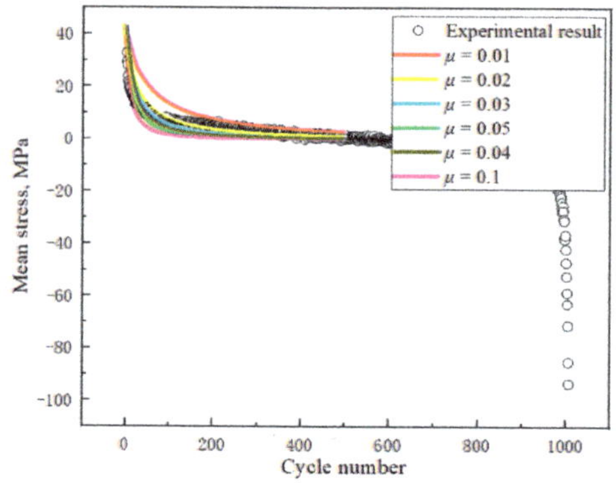

Figure 17. Results of sensitivity analysis for parameter μ.

In view of the above problems, referring to the equation of isotropic hardening, nonlinear terms are added to meet the initial stress relaxation behavior, and the new form of parameters μ is proposed:

$$\mu = \mu_0 - C_1\left(1 - e^{-C_2 p}\right) \tag{14}$$

where μ_0, C_1 and C_2 are material parameters. The simulation results are shown in Figure 18 and the material parameters are shown in Table 4.

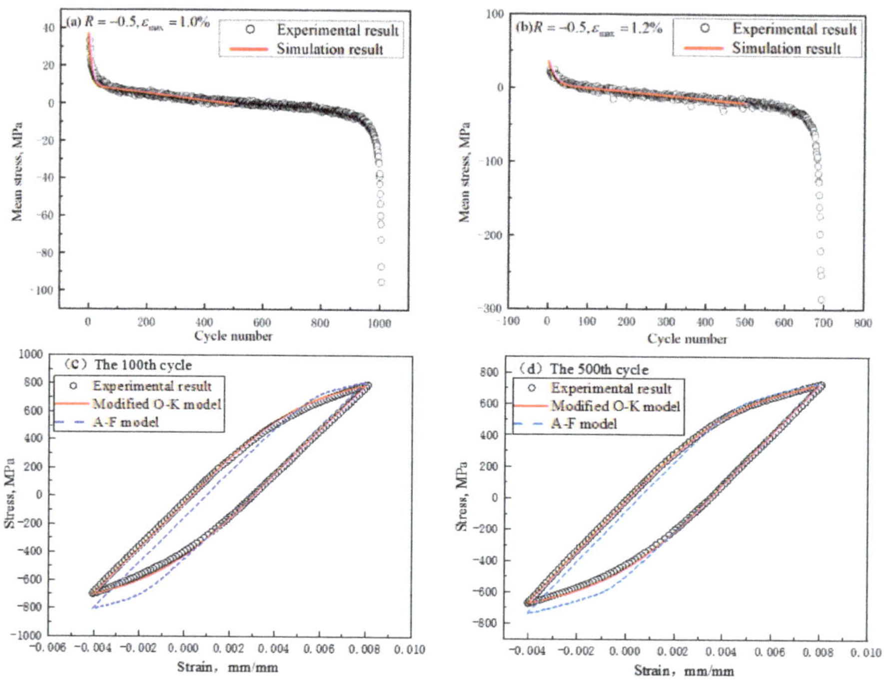

Figure 18. Comparison of stress relaxation behavior between tests and simulations: **(a)** ε_{max} = 1.0%; **(b)** ε_{max} = 1.2%; **(c)** the 100th cycle; **(d)** the 500th cycle.

Table 4. The parameters of modified O-K model.

μ_0	C_1	C_2
0.05	0.0475	5

The modified O-K model and isotropic hardening model can well describe the fatigue behavior of GH4169 under asymmetric loading, and the creep constitutive model proposed in this paper can well describe the creep behavior of GH4169. Therefore, a non-unified constitutive model can be established to describe the creep-fatigue behavior of GH4169. The simulation results are shown in Figure 19. Generally speaking, the simulation results are close to the test results; however, the amount of stress relaxation in the strain-holding phase is significantly smaller than the test results.

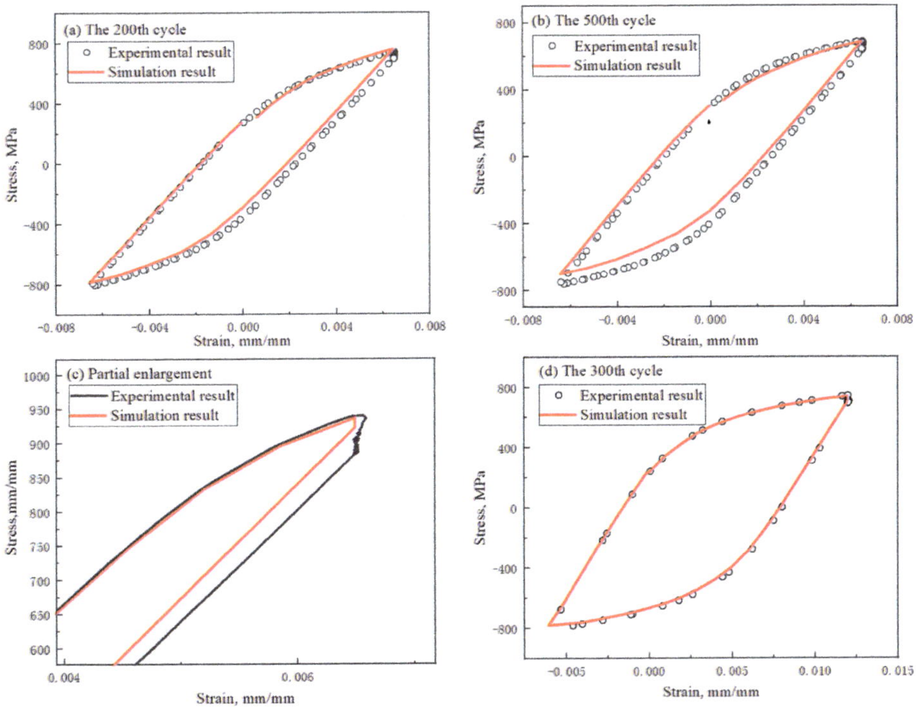

Figure 19. Comparison of creep-fatigue behavior between tests and simulations: (**a**) the 100th cycle; (**b**) the 500th cycle; (**c**) partial enlargement; (**d**) the 300th cycle.

4. Conclusions

In this paper, the tensile, creep, fatigue and creep-fatigue tests of the GH4169 alloy were carried out. The results show that the GH4169 alloy exhibits a PLC effect under a certain tensile strain rate. The tensile strain rate has little effect on the tensile curve. The creep curve of the GH4169 material is a typical three-stage creep curve. The GH4169 alloy exhibits cyclic softening under symmetric cyclic loading and a mean stress relaxation under asymmetric cyclic loading.

According to the test results, the deformation behavior of the GH4169 material was simulated. The O-K model can well describe the tensile behavior of the GH4169 material. The creep constitutive model presented in this paper can be used to predict the three-stage creep characteristics of the material. To better describe the mean stress relaxation behavior, the parameter μ is modified with reference to the isotropic hardening expression.

The modified O-K kinematic hardening model and isotropic hardening model can well predict the cyclic softening behavior of the material under symmetric loads and the mean stress relaxation behavior under asymmetric loads. Finally, the creep-fatigue behavior of the GH4169 material under symmetric and asymmetric loads is analyzed by using the non-unified constitutive model of the elastoplastic model (modified O-K model and isotropic model), coupled with the creep model. The simulation results are very close to the experimental results; however, the predicted stress relaxation in the strain-holding stage is smaller than the test results.

Author Contributions: Conceptualization, X.H. and S.Z.; methodology, X.H.; software, S.Z. and H.Z.; validation, H.Z., X.J. and Z.Z.; formal analysis, Z.Z.; investigation, X.H. and S.Z.; resources, X.H.; data curation, S.Z. and H.Z.; writing—original draft preparation, X.H.; writing—review and editing, X.H. and S.Z.; visualization, X.J.; supervision, X.H.; project administration, X.H. and Z.Z.; funding acquisition, X.H. All authors have read and agreed to the published version of the manuscript.

Funding: This research was funded by the National Science and Technology Major Project, grant number 2017-IV-0004-0041.

Data Availability Statement: The data presented in the study are available on request from the corresponding author.

Conflicts of Interest: The authors declare no conflict of interest.

References

1. Chaboche:, J.L. Viscoplastic constitutive equations for the description of cyclic and anisotropic behavior of metals. *Bull. Acad. Polon. Sci.* **1977**, *25*, 33–42.
2. Armstrong, P.J.; Frederick, C.O. *A Mathematical Representation of the Multiaxial Bauschinger Effect*; CEGB Report RD/B/N/731; Berkeley Nuclear Laboratories, R&D Department: Berkeley, CA, USA, 1966; Volume 24, pp. 1–26.
3. Chaboche, J.L.; Nouailhas, D. Constitutive modeling o ratcheting effects-Part I: Experimental facts and properties of the classical models. *ASME J. Eng. Mat. Technol.* **1989**, *111*, 384–392. [CrossRef]
4. Ohno, N. Recent topics in constitutive modeling of cyclic plasticity and viscoplasticity. *Appl. Mech. Rev.* **1990**, *43*, 283–295. [CrossRef]
5. Chaboche, J.L. On some modification of kinematic hardening to improve the description of ratcheting effects. *Int. J. Plast.* **1991**, *7*, 661–678. [CrossRef]
6. Chaboche, J.L.; Jung, O. Application of a kinematic hardening viscoplasticity model with thresholds to the residual stress relaxation. *Int. J. Plast.* **1998**, *13*, 785–807. [CrossRef]
7. Chaboche, J.L. Constitutive equations for cyclic plasticity and cyclic viscoplasticity. *Int. J. Plast.* **1989**, *5*, 247–302. [CrossRef]
8. Wang, J.D.; Ohno, N. Two Equivalent Forms of Nonlinear Kinematic Hardening: Application to Nonisothermal Plasticity. *Int. J. Plast.* **1991**, *7*, 637–650. [CrossRef]
9. Ohno, N.; Wang, J.D. Transformation of a nonlinear kinematic hardening rule to a multisurface form under isothermal and nonisothermal conditions. *Int. J. Plast.* **1991**, *7*, 879–891. [CrossRef]
10. Abdel-Karim, M.; Ohno, N. Kinematic hardening model suitable for ratcheting with steady-state. *Int. J. Plast.* **2000**, *16*, 225–240. [CrossRef]
11. Ohno, N.; Abdel-Karim, M. Uniaxial ratcheting of 316FR steel at room temperature, part II: Constitutive modeling and simulation. *ASME J. Eng. Mat. Tech.* **2000**, *122*, 35–41. [CrossRef]
12. Brown, S.; Evans, R.W.; Wilshire, B. Creep strain and creep life prediction for the cast nickel-based superalloy IN-100. *Mater. Sci. Eng.* **1986**, *84*, 147–156. [CrossRef]
13. Evans, R.W.; Wilshire, B. *Creep of Metals and Alloys;* Ztschrift Fur Met; U.S. Department of Energy Office of Scientific and Technical Information: Oak Ridge, TN, USA, 1985.
14. Brown, S.; Evans, R.W.; Wilshire, B. A comparison of extrapolation techniques for long term creep strain and creep life prediction based on equations designed to represent creep curve shape. *Int. J. Press. Vessel. Pip.* **1986**, *24*, 251–268. [CrossRef]
15. Rabotnov, Y.N. *Some Problems of the Theory of Creep*; Vestnik Moskov University: Moscow, Russia, 1948; pp. 81–91.
16. Othman, A.M.; Hayhurst, D.R.; Dyson, B.F. Skeletal Point Stresses in Circumferentially Notched Tension Bars Undergoing Tertiary Creep Modelled with Physically Based Constitutive Equations. *Proc. R. Soc. A Math. Phys. Eng. Sci.* **1993**, *441*, 343–358.
17. Dyson, B.F.; Hayhurst, D.R.; Lin, J. The Ridged Uniaxial Testpiece: Creep and Fracture Predictions Using Large-Displacement Finite-Element Analyses. *Proc. R. Soc. A Math. Phys. Eng. Sci.* **1996**, *452*, 655–676.
18. Mustata, R.; Hayhurst, D.R. Creep constitutive equations for a 0.5Cr0.5Mo0.25V ferritic steel in the temperature range 565–675 °C. *Int. J. Press. Vessel. Pip.* **2005**, *82*, 363–372. [CrossRef]
19. Xu, Q. Development of constitutive equations for creep damage behavior under multi-axial states of stress. *Adv. Mech. Behav. Plast. Damage* **2000**, *2*, 1375–1382.

20. Batsoulas, N.D. Mathematical description of the mechanical behavior of metallic materials under creep conditions. *J. Mater. Sci.* **1997**, *32*, 2511–2527. [CrossRef]
21. Wilshire, B.; Scharning, P.J.; Hurst, R. A new approach to creep data assessment. *Mater. Sci. Eng.* **2009**, *510*, 3–6. [CrossRef]
22. Whittaker, M.T.; Harrison, W.J.; Lancaster, R.J.; Williams, S. An analysis of modern creep lifing methodologies in the titanium alloy Ti6-4. *Mater. Sci. Eng. A* **2013**, *577*, 114–119. [CrossRef]
23. Abdallah, Z.; Gray, V.; Whittaker, M.; Perkins, K. A Critical Analysis of the Conventionally Employed Creep Lifing Methods. *Materials* **2014**, *7*, 3371–3398. [CrossRef]
24. Whittaker, M.T.; Harrison, W.J. Evolution of Wilshire equations for creep life prediction. *High Temp. Technol.* **2014**, *31*, 233–238. [CrossRef]
25. Kobayashi, M.; Mukai, M.; Takahashi, H.; Ohno, N.; Kawakami, T.; Ishikawa, T. Implicit integration and consistent tangent modulus of a time-dependent non-unified constitutive model. *Int. J. Numer. Methods Eng.* **2003**, *58*, 1523–1543. [CrossRef]
26. Ohguchi, K.I.; Sasaki, K.; Aso, S. Evaluation of Time-Independent and Time-Dependent Strains of Lead-Free Solder by Stepped Ramp Loading Test. *ASME J. Electron. Packag.* **2009**, *131*, 021003. [CrossRef]
27. Panteghini, A.; Genna, F. Effects of the strain-hardening law in the numerical simulation of wire drawing processes. *Comput. Mater. Sci.* **2010**, *49*, 236–242. [CrossRef]
28. Chen, W.; Wang, F.; Feng, M. Study of a modified non-unified model for time-dependent behavior of metal materials. *Mech. Mater.* **2017**, *113*, 69–76. [CrossRef]
29. Gremaud, G. Overview on dislocation-point defect interaction: The brownian picture of dislocation motion. *Mater. Sci. Eng. A* **2004**, *370*, 191–198. [CrossRef]
30. Yilmaz, A. The Portevin—Le Chatelier effect: A review of experimental findings. *Sci. Technol. Adv. Mater.* **2011**, *12*, 063001. [CrossRef]
31. Cui, L.; Yu, J.; Liu, J.; Sun, X. Microstructural evolutions and fracture behaviors of a newly developed nickel-base superalloy during creep deformation. *J. Alloys Compd.* **2018**, *746*, 335–349. [CrossRef]

metals

MDPI

Article

A 3D Polycrystalline Plasticity Model for Isotropic Linear Evolution of Intragranular Misorientation with Mesoscopic Plastic Strain in Stretched or Cyclically Deformed Metals

Shao-Shi Rui [1], Yue Su [2], Jia-Min Zhao [3], Zhi-Hao Shang [4] and Hui-Ji Shi [4,*]

1 State Key Laboratory of Nonlinear Mechanics (LNM), Institute of Mechanics, Chinese Academy of Sciences, Beijing 100190, China
2 School of Power and Energy, Northwestern Polytechnical University, Xi'an 710129, China
3 Beijing Institute of Astronautical Systems Engineering, China Academy of Launch Vehicle Technology, Beijing 100076, China
4 Applied Mechanics Laboratory (AML), School of Aerospace Engineering, Tsinghua University, Beijing 100084, China
* Correspondence: shihj@mail.tsinghua.edu.cn

Abstract: Two-dimensional electron back-scattered diffraction (2D-EBSD) mapping has been widely used for indicating the polycrystalline plasticity through intragranular misorientation parameters **KAM** and **GROD**, based on the empirically linear relationship between their average values and the mesoscopic plastic strain, in both stretched and cyclically deformed metals. However, whether the intragranular misorientation measured on the 2D-EBSD observational plane objectively reflects the 3D polycrystalline plasticity or not is a rarely reported issue. In this research, we firstly compared the **KAM** and **GROD** values measured on 2D-EBSD observational planes with different angles to loading axis of a specimen in both undeformed clamp sections and deformed gauge section, to verify whether their average values increase isotropically or not with mesoscopic plastic strain. Then, we proposed six fundamental assumptions and developed a modified 3D polycrystalline plasticity model based on the 2D polycrystalline plasticity model in our previous work. This 3D polycrystalline plasticity model can explain the isotropic linear evolution of intragranular misorientation in deformed low alloy steel with uniform equiaxial grains.

Keywords: 2D-EBSD mapping; intragranular misorientation; **KAM** and **GROD**; isotropic linear evolution; mesoscopic plastic strain; 3D polycrystalline plasticity model

Citation: Rui, S.-S.; Su, Y.; Zhao, J.-M.; Shang, Z.-H.; Shi, H.-J. A 3D Polycrystalline Plasticity Model for Isotropic Linear Evolution of Intragranular Misorientation with Mesoscopic Plastic Strain in Stretched or Cyclically Deformed Metals. *Metals* **2022**, *12*, 2159. https://doi.org/10.3390/met12122159

Academic Editors: Eric Hug and John D. Clayton

Received: 23 October 2022
Accepted: 12 December 2022
Published: 15 December 2022

1. Introduction

Two-dimensional (2D) electron back-scattered diffraction (EBSD) mapping is now a standard analysis on intragranular misorientation in both stretched and cyclically deformed metals, whose contrast can further indicate the mesoscopic plastic strain distribution [1] through intragranular misorientation parameters such as **K**ernel **A**veraged **M**isorientation (**KAM**) and **G**rain **R**eference **O**rientation **D**eviation (**GROD**) firstly proposed by Wright, S.I. et al. [2,3]. **KAM** is typically defined as the averaged misorientation between a kernel point and its surrounding points excluding those out of grain boundary, which is also called "Local **M**isorientation ($\mathbf{M}_L$)" by Kamaya, M. [4]. **GROD** is typically defined as the misorientation between an individual point and the intragranular reference point, whose average value is also called "Crystal **D**eformation ($\mathbf{C}_d$)" or "**M**odified **C**rystal **D**eformation (**MCD**)" by Kamaya, M. [5,6]. Their capability of indicating mesoscopic plastic strain is based on the following experimental phenomenon: both $\overline{\mathbf{KAM}}$ ($\mathbf{M}_{ave}$) and $\overline{\mathbf{GROD}}$ ($\mathbf{C}_d$, **MCD**) averaged over multiple grains always linearly increase with the tensile strain ε^T or cyclic strain upper ε^C_{max} applied in specimens according to abundant EBSD observation results reported in Refs. [4–9], as shown in Figure 1. To this end, average

$\overline{\mathbf{KAM}}$ and $\overline{\mathbf{GROD}}$ have been widely used for identifying the fracture modes [10–14], evaluating the fatigue [15–17] and creep [18–20] damage degree, analyzing the grain boundary induced micro-cracking [21] or fretting contact-induced micro-cracking [22], and predicting the fatigue crack tip growth rate da/dN [23] or driving force ΔK [9] as two important mesoscopic plasticity indicators.

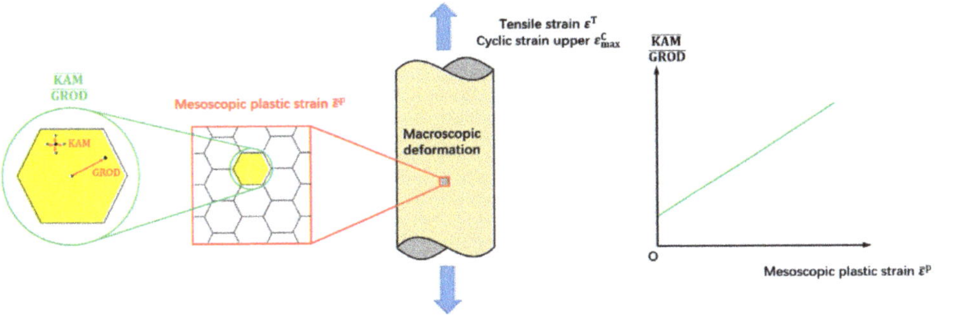

Figure 1. Linear evolution of average $\overline{\mathbf{KAM}}$ and $\overline{\mathbf{GROD}}$ with mesoscopic plastic strain in metals.

The linear evolution of average $\overline{\mathbf{KAM}}$ and $\overline{\mathbf{GROD}}$ with mesoscopic plastic strain was only reported by the above literature [4–9] as a common phenomenon in polycrystalline metals without providing any physical model explanations, which weakens its reliability in the industry application. According to the mainstream view represented by Wilkinson, A.J. [24–29], Kysar, J.W. [30–35], Pantleon, W. [36] and Raabe, D. [37,38] et al., **KAM** is regarded as an EBSD-based metric approximately measuring a special category of intragranular dislocations termed "geometrically necessary dislocation" [39,40], which can be derived from the continuum dislocation theory established by Nye, J.F. [41], Bilby, B.A. [42] and Kroner, E. [43]. However, the local geometrically necessary dislocation density cannot be linearly linked to the mesoscopic plastic strain, although it has a positive correlation with the plastic deformation in most cases. Harte, A. et al. [44] compared intragranular misorientation and plastic strain on the surface of a deformed Ni-based superalloy and revealed that the above linear relationship only works between the average misorientation and mean grain distortion in a statistical sense, while the local misorientation is the result of the number, strength and spatial distribution of intragranular slip modes and their interactions with microstructural features such as grain boundaries.

At the same time, the selection of a 2D-EBSD observational plane relative to the three-dimensional (3D) mesoscopic stress/strain principal axes is also a potential factor which may influence the $\overline{\mathbf{KAM}}$ and $\overline{\mathbf{GROD}}$ values. In other words, whether the linear evolution of average $\overline{\mathbf{KAM}}$ and $\overline{\mathbf{GROD}}$ with plastic strain reported in the above literature has the nature of isotropy or not is also an important issue to be clarified for polycrystalline metals with uniform equiaxial grains, which was rarely investigated currently. In our previous work [45], a 2D polycrystalline plasticity model considering the grain boundaries' constraint effect on intragranular lattice rotation was established. However, the above 2D polycrystalline plasticity model has not yet been generalized to the 3D case. To address the above issues, both experimental and theoretical work was carried out in this research.

On one hand, mechanical tests on low alloy steel with uniform equiaxial grains were carried out by standard round bar specimens in this research at first, and then followed by 2D-EBSD observations on those deformed or undeformed sections. Since the linear evolution law of average $\overline{\mathbf{KAM}}$ and $\overline{\mathbf{GROD}}$ has been confirmed by many studies [4–9], we mainly focused on comparing the $\overline{\mathbf{KAM}}$ and $\overline{\mathbf{GROD}}$ values measured on 2D-EBSD observational planes with different angles to loading axis in the same specimen, to verify whether they increase isotropically or not with the mesoscopic plastic strain. The isotropic evolution law is very important for averaged $\overline{\mathbf{KAM}}$ and $\overline{\mathbf{GROD}}$ because these parameters can serve as the objective plasticity indicators in actual applications only if their values are independent of the observation, including the selection of 2D-EBSD observational planes.

On the other hand, a 3D polycrystalline plasticity model was established based on six fundamental assumptions to explain the linear evolution law of averaged $\overline{\mathbf{KAM}}$ and $\overline{\mathbf{GROD}}$ reported by other studies [4–9], as well as their isotropic evolution law observed in this research. The polycrystalline plasticity model contains two scale levels: mesoscopic scale and microscopic scale. It is assumed that the Representative Volume Element (RVE, contains multiple grains) in the mesoscopic scale is homogeneous isotropic and follows the classical J_2 finite strain plasticity theory [46], while each grain in the microscopic scale follows the crystal plasticity theory. Several necessary simplifications, such as spherical grain hypothesis and minimum activated slip factors number hypothesis, were made in the above assumptions to make the mathematical derivation of $\overline{\mathbf{KAM}}$ and $\overline{\mathbf{GROD}}$ evolution easier.

2. Materials and Methods

2.1. Material and Mechanical Tests

The material used in this research is type 40Cr low alloy steel in Chinese Brand, whose chemical elements and mechanical properties are shown in Table 1. The samples for tensile mechanical tests were cut from a hot rolled plate. For higher resolution under EBSD observation, it was annealed at 850 °C for about 120 min, and followed by furnace cooling. The microstructures after heat treatment are uniform equiaxial grains made up of Body-Centered Cubic (BCC) ferrite and pearlite phases, and no initial texture exists before the plastic deformation. The metallographic figure and **I**nverse **P**ole **F**igure (**IPF**) can be referred to Ref. [9], belonging to our previous work using the same material.

Table 1. Chemical elements and mechanical properties of type 40Cr low alloy steel.

Chemical Elements (wt.%)	**C**	**Cr**	**Si**	**Mn**
	0.36%	1.56%	0.41%	1.27%
Mechanical Properties	**Yield Strength (YS)**		**Ultimate Tensile Strength (UTS)**	
	293.6 MPa		671.9 MPa	

The mechanical tests were carried out by two round bar specimens in type WDW-100® machine (Instron Inc., Norwood, MA, USA) at room temperature based on ASTM E8-08 [47] standard as shown in Figure 2a, and the loading speed for tension was set as 0.5 mm/min (statically loading). The diameters of gauge section and clamp section of round bar specimen are 7 and 14 mm, respectively, while the initial gauge length and clamp length are 25 mm and 38 mm respectively. Specimen B was continually elongated until rupture to give the full nominal stress–clamp stroke curve, while Specimen A was interrupted at nominal stress of 664.6 MPa before the obvious necking behavior appears for the following EBSD observation and analysis, as shown in Figure 2b,c. Due to the lack of extensometer, the gauge length cannot be tracked in real time, which has little effect on discussing related issues because we only care about the final gauge length in this research. The final gauge lengths of interrupted Specimen A and ruptured Specimen B are 28 and 29.3 mm, respectively, corresponding to the elongations of 12% and 17.2%.

2.2. 2D-EBSD Observations

Two-dimensional electron back-scattered diffraction observations (2D-EBSD) were carried out on five planes ①–⑤ with different angles to loading axis in both clamp sections and gauge section of interrupted Specimen A, whose exact locations are shown in Figure 3. Therein, two planes ① and ⑤ are located in clamp sections at two ends with angles of 0° and 90° to loading axis, respectively, which are undeformed during the tensile mechanical test. Meanwhile, three other planes ②, ③ and ④ are located in gauge section in the middle with angles of 0°, 45°and 90° to loading axis, respectively, which are uniformly elongated to 12% during the tensile mechanical test. The comparison between ① and ⑤ is aimed at confirming the isotropic distribution of initial intragranular misorientation before deformation, while the comparison among ②, ③ and ④ is aimed at revealing the isotropic evolution law of intragranular misorientation during deformation.

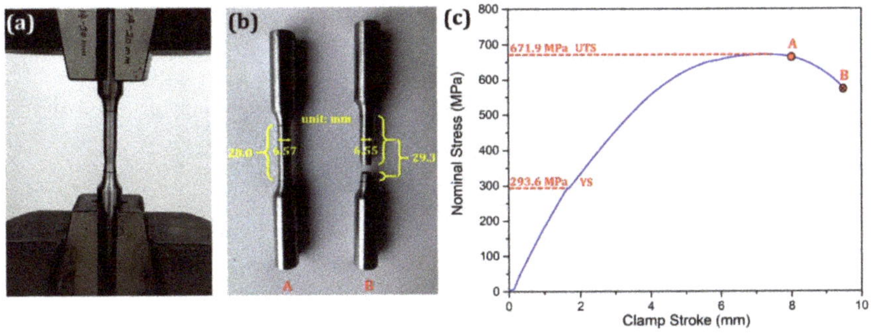

Figure 2. (**a**) Tensile mechanical tests, (**b**) interrupted Specimen A and ruptured Specimen B, as well as (**c**) nominal stress–clamp stroke curve obtained from tensile mechanical tests.

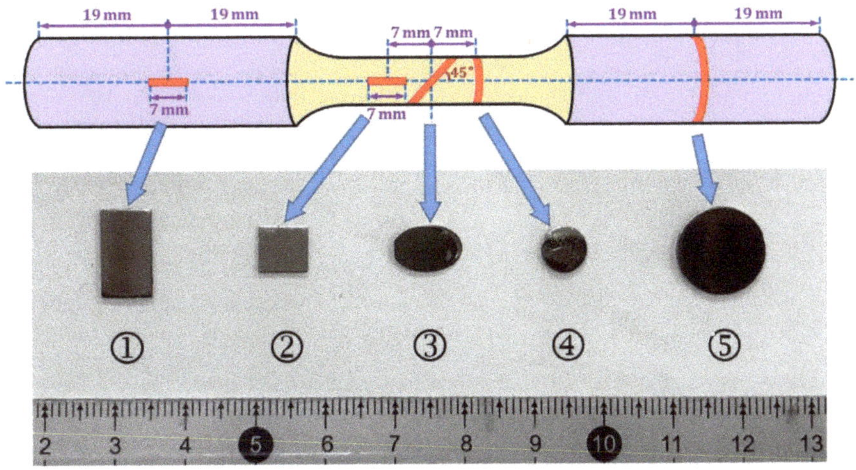

Figure 3. EBSD observational plane with various angles to loading axis in interrupted Sample A.

All the samples for the following 2D-EBSD observations were removed from the interrupted Specimen A by electro discharge cutting method. After the typical grinding procedure, these samples were firstly polished by specified cloth adhered with 3.5, 2.5, 1.5 and 0.5 μm diamond particles for 10–15 min every step and then attached in an aluminum platform and followed by 2–3 h vibratory polishing in Buehler VibroMet 2® machine (Buehler Group Inc., Lake Bluff, IL, USA) to release the sample surface residual stress. The specified polishing liquid used in Buehler VibroMet 2® machine contains tiny (~100 nm) SiO_2 particles and acid solution, which can help remove the slight deformation layer in the polished sample surface through coupled mechanical–chemical effects to ensure the Kikuchi patterns quality.

The 2D-EBSD observational region size is set as 200 μm × 200 μm, and the scanning step size is a = 0.8 μm. Then, 2D-EBSD scanning records the orientation information in the observational plane as three Euler angles (ϕ_1, ϕ, ϕ_2) point by point through Hough transformation of Kikuchi patterns as shown in Figure 4a,b. The type of field emission scanning electron microscope (SEM) used in this research is TESCAN MIRA 3 LMH®

(TESCAN ORSAY HOLDING a.s., Kohoutovice, Czech Republic). The EBSD camera equipped in SEM platform is Nordly II® (Oxford Instruments plc., Oxford, UK), and the matched EBSD data collection software is Aztec® (Oxford Instruments plc., Oxford, UK). All the data were saved in square structure as shown in Figure 4b,c, and then analyzed by OIM Analysis 6.2® (EDAX Inc., Pleasanton, CA, USA).

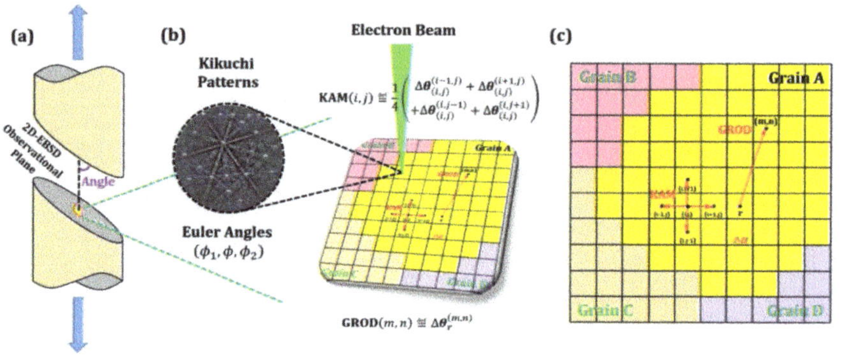

Figure 4. (**a**) Selection of 2D-EBSD observational plane, (**b**) 2D-EBSD scanning and data collection, as well as (**c**) definitions of intragranular misorientation parameters **KAM** and **GROD**.

For the convenience of calculating misorientation, the Euler angles (ϕ_1, ϕ, ϕ_2), which represent three relative rotations between local lattice coordinate ([100], [010], [001]) and global sample coordinate (RD, TD, ND) and are usually converted into the orientation matrix g as shown in Equation (1):

$$g = \begin{bmatrix} \cos\phi_2 & \sin\phi_2 & 0 \\ -\sin\phi_2 & \cos\phi_2 & 0 \\ 0 & 0 & 1 \end{bmatrix} \begin{bmatrix} 1 & 0 & 0 \\ 0 & \cos\phi & \sin\phi \\ 0 & -\sin\phi & \cos\phi \end{bmatrix} \begin{bmatrix} \cos\phi_1 & \sin\phi_1 & 0 \\ -\sin\phi_1 & \cos\phi_1 & 0 \\ 0 & 0 & 1 \end{bmatrix} \quad (1)$$

The misorientation $\Delta\theta_{\mathrm{A}}^{\mathrm{B}}$ between two different points A and B can then be calculated as shown in Equation (2), where g_{A} and g_{B} represent the orientation matrixes of points A and B, respectively. $S_k|_{k=1\sim 24}$ indicate the 24 symmetry operators of BCC lattice, which can be referred to in Refs. [22,48]. The minimum lattice rotation angle of the above 24 equivalent lattice rotation operations converting orientation of point A into orientation of point B is then set as the misorientation $\Delta\theta_{\mathrm{A}}^{\mathrm{B}}$ between them.

$$\Delta\theta_{\mathrm{A}}^{\mathrm{B}} = \min_{i,j=1\sim 24}\left[\arccos\left(\frac{\mathrm{trace}\left[(S_i \cdot g_{\mathrm{B}}) \cdot (S_j \cdot g_{\mathrm{A}})^{-1}\right] - 1}{2}\right)\right] \quad (2)$$

The parameters for quantifying intragranular misorientation in this research are **KAM** and **GROD**. **KAM** (i, j) is defined as the average misorientation between the kernel point (i, j) and its nearest neighboring points $(i-1, j)$, $(i, j-1)$, $(i, j+1)$, $(i+1, j)$ in the same grain as shown in Figure 4c and Equation (3):

$$\mathbf{KAM}(i,j) \overset{\mathrm{def}}{=} \frac{1}{4}\left(\Delta\theta_{(i,j)}^{(i-1,j)} + \Delta\theta_{(i,j)}^{(i,j-1)} + \Delta\theta_{(i,j)}^{(i,j+1)} + \Delta\theta_{(i,j)}^{(i+1,j)}\right) \quad (3)$$

where any neighboring points outside the grain boundary will be ignored. Another misorientation parameter **GROD** (m, n) is defined as the misorientation between the individual point (m, n) and the reference point r in the same grain representing its average orientation, where the detailed calculation method of average orientation can be referred to in Refs. [49,50], as shown in the Figure 4c and Equation (4):

$$\mathbf{GROD}(m,n) \stackrel{\text{def}}{=} \Delta\theta_r^{(m,n)} \quad (4)$$

After the **KAM** and **GROD** values in each EBSD scanning point were calculated, the **KAM** and **GROD** distribution maps can be drawn by the software OIM Analysis 6.2® automatically. Instead of **KAM** (i, j) and **GROD** (m, n) sensitive to the local microstructural features, $\overline{\mathbf{KAM}}$ and $\overline{\mathbf{GROD}}$ averaged over the whole 2D-EBSD observational region ($P \times Q$ scanning points) containing multiple grains, as shown in Equation (5), are widely used for measuring the mesoscopic plastic strain in the corresponding 2D-EBSD observational position because they are independent of the local microstructural features.

$$\frac{\overline{\mathbf{KAM}}}{\overline{\mathbf{GROD}}} = \frac{1}{P \times Q} \sum_{i,m=a}^{P \cdot a} \sum_{j,n=a}^{Q \cdot a} \frac{\mathbf{KAM}(i,j)}{\mathbf{GROD}(m,n)} \quad (5)$$

3. Experimental Results

The **IPF**, **KAM** and **GROD** maps given by 2D-EBSD observations on different planes with angles of 0°, 45° and 90° to loading axis, respectively, in the gauge section and two clamp sections of interrupted Specimen A are shown in Figure 5. Fe_3C layers distributed in the pearlite phase were not recognized as an independent phase separately by the EBSD data collection software; thus, both ferrite and pearlite phases were identified as the single BCC phase. From the **IPF** maps, we can see the cross profile of each grain enclosed by grain boundary (GB) on the 2D-EBSD observational plane. Most cross profiles present irregular but equiaxial shapes, and their sizes d range from several to dozens of microns. If the equiaxial grain is regarded as a sphere with diameter of D_{Grain}, the difference among those cross profile sizes d shown in the **IPF** maps should be attributed to not only the different grain sizes D_{Grain}, but also to the different distances H between 2D-EBSD observational plane and grain center (GC), as shown in both Figure 6 (illustration explaining the geometric relationship between 3D grain size D_{Grain} and cross profile size d) and Equation (6). On one hand, the larger the grain size D_{Grain} is, the larger the cross profile size d presents on the 2D-EBSD observational plane when its distance H away from GC is constant. On the other hand, the closer the distance away from GC, the larger the cross profile size d presents on the 2D-EBSD observational plane when the grain size D_{Grain} is constant.

$$d = \sqrt{D_{\text{Grain}}^2 - 4H^2} \quad (6)$$

From the **KAM** and **GROD** maps, we can see that intragranular misorientation levels are relatively low in two underformed clamp sections but high in deformed gauge section. The intragranular misorientation values averaged over the whole 2D-EBSD observational region containing multiple grains are drawn in Figure 7. The subscript "0" stands for average $\overline{\mathbf{KAM}}$ and $\overline{\mathbf{GROD}}$ values measured from undeformed clamp sections, different from those from deformed gauge section. For two undeformed clamp sections, the average $\overline{\mathbf{KAM}}$ and $\overline{\mathbf{GROD}}$ values measured on 2D-EBSD observational planes with angles of 0° and 90° to loading axis, respectively, are almost the same. This reveals the isotropic distribution law of initial intragranular misorientation. For the deformed gauge section, the average $\overline{\mathbf{KAM}}$ and $\overline{\mathbf{GROD}}$ values repeatedly measured on 2D-EBSD observational planes with angles of 0°, 45° and 90° to loading axis are also almost the same. This reveals the isotropic evolution law of intragranular misorientation during the tensile deformation. In previous works, the linear evolution law of intragranular misorientation during the tensile deformation has been widely confirmed by Refs. [4–8], which laid the foundation for indicating the mesoscopic plastic strain by average $\overline{\mathbf{KAM}}$ and $\overline{\mathbf{GROD}}$ values. Furthermore, the isotropic evolution law of intragranular misorientation confirmed by this research ensures that average $\overline{\mathbf{KAM}}$ and $\overline{\mathbf{GROD}}$ values are independent of the 2D-EBSD observational planes selection. This is quite an important property because $\overline{\mathbf{KAM}}$ and $\overline{\mathbf{GROD}}$ can only serve as objective plasticity indicators if their values are determined by mesoscopic plastic strain while not affected by observational plane selection.

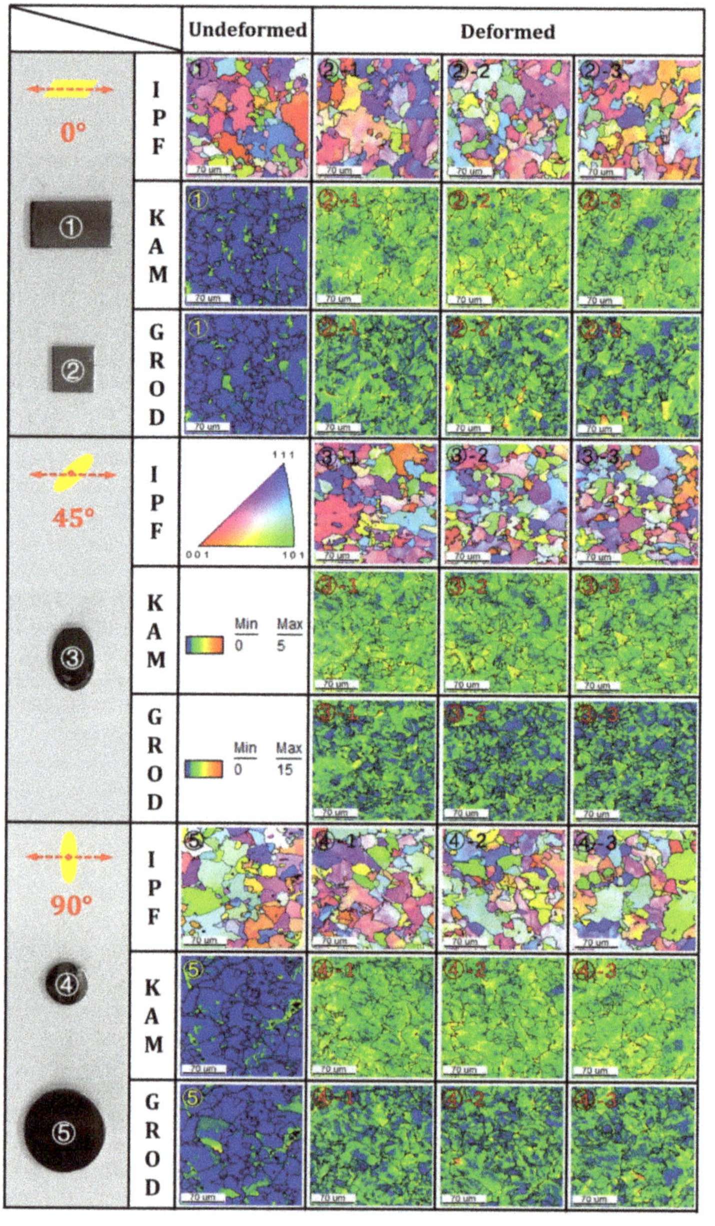

Figure 5. **IPF**, **KAM** and **GROD** maps given by 2D-EBSD observations on planes with different angles to loading axis in both undeformed clamp sections and deformed gauge section.

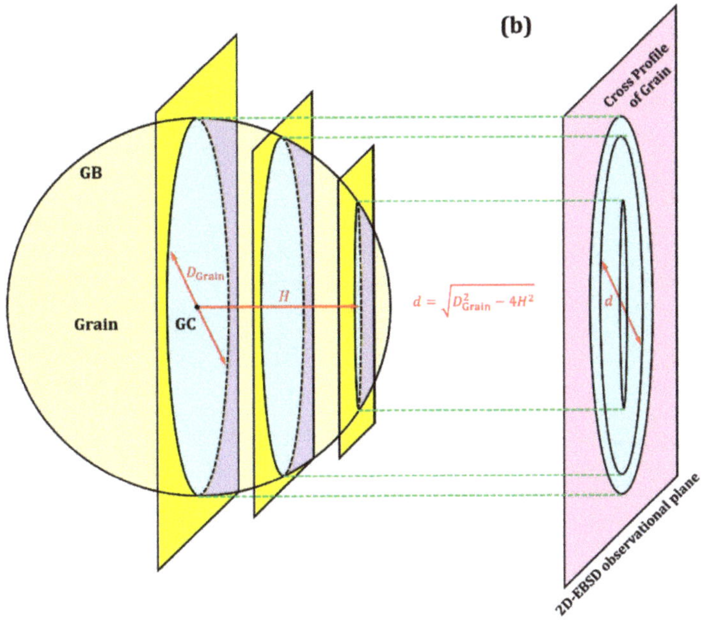

Figure 6. (**a**) The distance H between 2D-EBSD observational plane and GC, as well as (**b**) the cross profile sizes d shown on the 2D-EBSD observational plane.

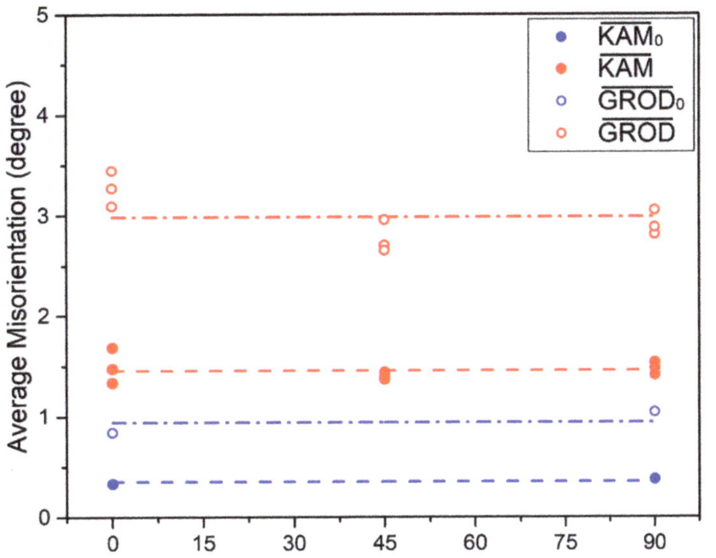

Figure 7. The average $\overline{\textbf{KAM}}$ and $\overline{\textbf{GROD}}$ values measured on 2D-EBSD observational planes with different angles to loading axis in both undeformed clamp sections and deformed gauge section.

4. Theoretical Discussions

The isotropic linear evolution law of intragranular misorientation (quantified by $\overline{\textbf{KAM}}$ and $\overline{\textbf{GROD}}$) in deformed polycrystalline metals with uniform equiaxial grains has been confirmed by this research as well as by other studies [4–8] as an important experimental phenomenon. To explain the linear evolution law of intragranular misorientation reported by other studies [4–8], our previous work [45] has established a preliminary theoretical model based on several fundamental assumptions in the 2D case. However, this 2D polycrystalline plasticity model has not yet been generalized to the 3D case and thus cannot explain the isotropic evolution law of intragranular misorientation observed in our experimental results here. To address the above issue, a modified 3D polycrystalline plasticity model explaining isotropic linear evolution law of intragranular misorientation was then developed.

4.1. Fundamental Assumptions

To establish the 3D polycrystalline plasticity model for explaining isotropic linear evolution of intragranular misorientation, six fundamental assumptions listed below should be made in advance to simplify the mathematical derivation process of $\overline{\mathbf{KAM}}$ and $\overline{\mathbf{GROD}}$ evolution law, as shown in Figure 8.

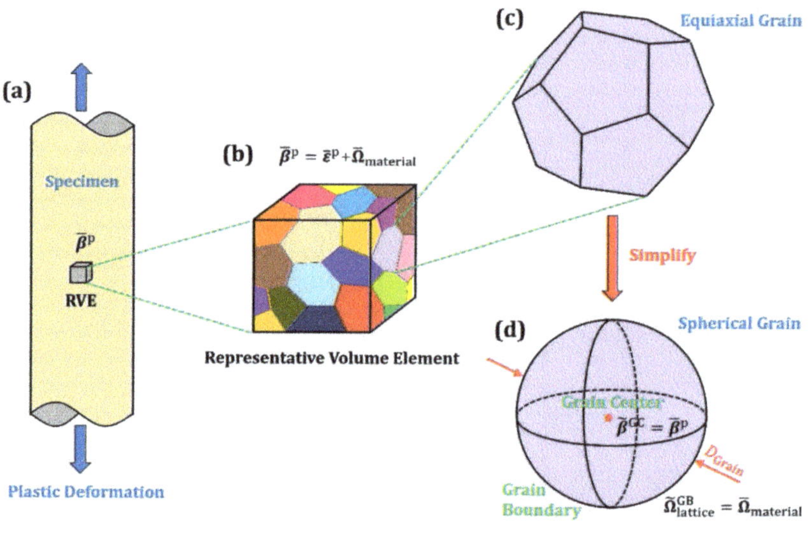

Figure 8. Several necessary simplifications for establishing the 3D polycrystalline plasticity model. (**a**) the macroscale plastic deformation applied in the specimen, (**b**) the mesoscale plastic strain and material rotation of a RVE containing multiple grains, as well as the microscale distortion in (**c**) an equiaxial grain and (**d**) a simplified spherical grain.

(1) The intragranular plastic distortion β^p follows the crystal plasticity theory. No more than five independent slip factors $(\vec{s}^{\,1}, \vec{n}^{\,1})$~$(\vec{s}^{\,5}, \vec{n}^{\,5})$ are activated to undertake the intragranular plastic distortion, which are selected from those potential slip factors of specific lattice at a given temperature with the highest five resolved shear stresses τ^1~τ^5 under the mesoscopic stress $\overline{\sigma}$ applied in RVE. The plastic strain is small enough to ensure that the additive decomposition is applicable to the distortion tensor β^p and the activated slip factors $(\vec{s}^{\,\alpha}, \vec{n}^{\,\alpha})$ can be regarded as approximately fixed during the deformation.

(2) The RVE containing multiple grains can be regarded as homogeneous and isotropic, while its mesoscopic plastic strain $\overline{\varepsilon}^p$ and mesoscopic stress $\overline{\sigma}$ follow the classical J_2 finite strain plasticity theory: $\overline{\varepsilon}^p \| \overline{\sigma}\prime$, which requires that three principal directions of deviatoric stress tensor $\overline{\sigma}\prime$ and the ratio among three principal stresses of stress tensor $\overline{\sigma}$ are fixed during the whole deformation history.

(3) The residual material distortion $\widetilde{\beta}^{GC}$ at the GC made up of microscopic plastic distortion $\beta^{p}(0)$ and residual lattice rotation $\widetilde{\boldsymbol{\Omega}}_{\text{lattice}}(0)$ is equal to the mesoscopic plastic distortion $\overline{\beta}^{p}$ of RVE, which is the same as that in Taylor's polycrystalline model.

(4) Each equaxial grain can be simplified as a sphere with the same diameter of D_{Grain}, while the distance between its GC and the 2D-EBSD observational plane is H. For each spherical grain cut by the 2D-EBSD observational plane, the ratio $2H/D_{\text{Grain}}$ is a random variable ranging from 0 to 1.

(5) The residual lattice rotation $\widetilde{\boldsymbol{\Omega}}^{GB}_{\text{lattice}}$ near the GB is close to the mesoscopic material rotation $\overline{\boldsymbol{\Omega}}_{\text{material}}$ of RVE due to the restraint from the fixed orientation relationship between the two sides of GB, as explained by Figure 9. The lattice rotation inside each grain is induced by two parts: one is induced by the overall grain rotation synchronized with the mesoscopic material rotation, and another is induced by the grain distortion accompanied with dislocations slip. Therein, the lattice rotation induced by the grain distortion must be zero near the GB; otherwise, the fixed orientation relationship between the two sides of GB will be broken (e.g., the GB misorientation angle will be changed). Taking this into account, the residual lattice rotation at the GB should be the same as the mesoscopic material rotation, since the other part must be equal to zero. A deeper physical reason is that the interior dislocations cannot be absorbed or released by those GBs at the room temperature.

(6) The intragranular residual lattice rotation $\widetilde{\boldsymbol{\Omega}}_{\text{lattice}}$ decreases from GC to GB along the grain radius r linearly and isotropically in spherical grains: $\widetilde{\boldsymbol{\Omega}}_{\text{lattice}} = \widetilde{\boldsymbol{\Omega}}_{\text{lattice}}(r)$ and $\partial^2\widetilde{\boldsymbol{\Omega}}_{\text{lattice}}/\partial r^2 = \mathbf{0}$.

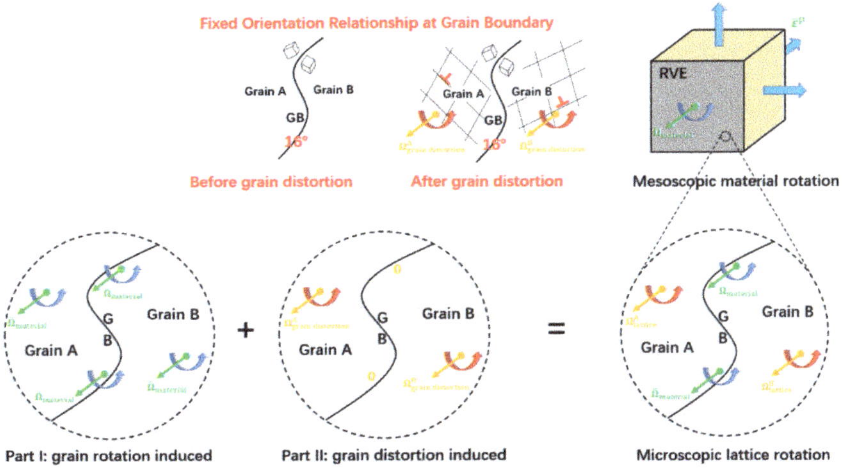

Figure 9. The explanation for the fixed orientation relationship between two sides of GB.

4.2. Establishment of 3D Polycrystalline Plasticity Model

The intragranular residual material distortion $\widetilde{\beta}$ after unloading is made up of two parts: the intragranular plastic distortion β^{p} and the intragranular residual lattice rotation $\widetilde{\boldsymbol{\Omega}}_{\text{lattice}}$. Further, the intragranular plastic distortion β^{p} is undertaken by the crystallographic slip γ_m in no more than five independent slip factors $(\vec{s}^{\,m}, \vec{n}^{\,m})$ according to Assumption (1), where the unit vector $\vec{s}^{\,m}$ represents slip direction, the unit vector $\vec{n}^{\,m}$ represents normal direction of slip plane, and m = 1–5, as shown in Equation (7). In this equation, eight quantities $\gamma_1 \sim \gamma_5$ and $\theta_1 \sim \theta_3$ are unknown, while eight independent sub-equations are given.

Therefore, a unique solution can be determined only if the five activated slip factors ($\vec{s}^{1}$, $\vec{n}^{1}$)~($\vec{s}^{5}$, $\vec{n}^{5}$) and the intragranular residual material distortion $\tilde{\beta}$ are known in advance.

$$\overbrace{\begin{bmatrix} -\varepsilon_1^{\mathrm{P}} & \gamma_3^{\mathrm{P}}+\omega_3 & \gamma_2^{\mathrm{P}}-\omega_2 \\ \gamma_3^{\mathrm{P}}-\omega_3 & -\varepsilon_2^{\mathrm{P}} & \gamma_1^{\mathrm{P}}+\omega_1 \\ \gamma_2^{\mathrm{P}}+\omega_2 & \gamma_1^{\mathrm{P}}-\omega_1 & \varepsilon_1^{\mathrm{P}}+\varepsilon_2^{\mathrm{P}} \end{bmatrix}}^{\tilde{\beta}} = \overbrace{\left(\gamma_1 \vec{s}^{1}\vec{n}^{1} + \gamma_2 \vec{s}^{2}\vec{n}^{2} + \gamma_3 \vec{s}^{3}\vec{n}^{3} + \gamma_4 \vec{s}^{4}\vec{n}^{4} + \gamma_5 \vec{s}^{5}\vec{n}^{5}\right)}^{\beta^{\mathrm{P}}} + \overbrace{\begin{bmatrix} 0 & \theta_3 & -\theta_2 \\ -\theta_3 & 0 & \theta_1 \\ \theta_2 & -\theta_1 & 0 \end{bmatrix}}^{\tilde{\Omega}_{\mathrm{lattice}}} \quad (7)$$

A coordinate system should be established in advance to help describe the evolution of various intragranular physical fields during the grain distortion, and the origin of the coordinate system is naturally located at the GC of each grain. To simplify the expression of mesoscopic plastic distortion $\overline{\beta}^{\mathrm{P}}$ of RVE, three axes x, y and z of the coordinate system are parallel to three principal directions of mesoscopic plastic strain $\overline{\varepsilon}^{\mathrm{P}}$, where the shear plastic strain components γ_1^{P}~γ_3^{P} will be equal to zero in that case. At the same time, the mesoscopic stress $\overline{\sigma}$ applied in RVE is made up of the deviatoric part $\overline{\sigma}\prime$ and the hydrostatic part, while the deviatoric part $\overline{\sigma}\prime$ should meet the requirement of $\overline{\varepsilon}^{\mathrm{P}} \| \overline{\sigma}\prime$ according to J_2 finite strain plasticity theory in Assumption (2). The detailed components of $\overline{\beta}^{\mathrm{P}}$ and $\overline{\sigma}$ are shown in Equation (8), where ε^{P} is the maximum principal plastic strain along z axis and $0 \leq k \leq 1$ is the variable controlling the ratio between another two principal plastic strains along the x and y axes. In particular, k is equal to $1/2$ in the uniaxial tension case.

$$\overline{\beta}^{\mathrm{P}} = \overbrace{\begin{bmatrix} -(1-k)\varepsilon^{p} & 0 & 0 \\ 0 & -k\varepsilon^{p} & 0 \\ 0 & 0 & \varepsilon^{p} \end{bmatrix}}^{\overline{\varepsilon}^{\mathrm{P}}} + \overbrace{\begin{bmatrix} 0 & \omega_3 & -\omega_2 \\ -\omega_3 & 0 & \omega_1 \\ \omega_2 & -\omega_1 & 0 \end{bmatrix}}^{\overline{\Omega}_{\mathrm{material}}}, \ \overline{\sigma} = \overbrace{\begin{bmatrix} -(1-k)\sigma & 0 & 0 \\ 0 & -k\sigma & 0 \\ 0 & 0 & \sigma \end{bmatrix}}^{\overline{\sigma}\prime} + \begin{bmatrix} \eta & 0 & 0 \\ 0 & \eta & 0 \\ 0 & 0 & \eta \end{bmatrix} \quad (8)$$

According to Assumption (3), the residual material distortion $\tilde{\beta}^{\mathrm{GC}}$ at the GC made up of microscopic plastic distortion $\beta^{\mathrm{P}}(0)$ and residual lattice rotation $\tilde{\Omega}_{\mathrm{lattice}}(0)$ is equal to the mesoscopic plastic distortion $\overline{\beta}^{\mathrm{P}}$ of RVE. Then, Equation (7) can be applied to the GC and generates to Equation (9).

$$\overline{\beta}^{\mathrm{P}} = \overline{\varepsilon}^{\mathrm{P}} + \overline{\Omega}_{\mathrm{material}} = \tilde{\beta}^{\mathrm{GC}} = \overbrace{\left(\gamma_1 \vec{s}^{1}\vec{n}^{1} + \gamma_2 \vec{s}^{2}\vec{n}^{2} + \gamma_3 \vec{s}^{3}\vec{n}^{3} + \gamma_4 \vec{s}^{4}\vec{n}^{4} + \gamma_5 \vec{s}^{5}\vec{n}^{5}\right)}^{\beta^{\mathrm{P}}(0)} + \overbrace{\begin{bmatrix} 0 & \theta_3(0) & -\theta_2(0) \\ -\theta_3(0) & 0 & \theta_1(0) \\ \theta_2(0) & -\theta_1(0) & 0 \end{bmatrix}}^{\tilde{\Omega}_{\mathrm{lattice}}(0)} \quad (9)$$

According to Assumptions (4) and (5), the residual lattice rotation $\tilde{\Omega}_{\mathrm{lattice}}^{\mathrm{GB}} = \tilde{\Omega}_{\mathrm{lattice}}(D_{\mathrm{Grain}}/2)$ near the GB of each spherical grain is close to the mesoscopic material rotation $\overline{\Omega}_{\mathrm{material}}$ of RVE. Then, Equation (10) connecting the mesoscopic plastic strain $\overline{\varepsilon}^{\mathrm{P}}$ of RVE to the intragranular misorientation $\tilde{\Omega}_{\mathrm{lattice}}(0) - \tilde{\Omega}_{\mathrm{lattice}}(D_{\mathrm{Grain}}/2)$ between GC and GB of each grain can be established based on Equation (9). From this equation, we can see that the intragranular misorientation will be mainly determined by mesoscopic plastic strain $\overline{\varepsilon}^{\mathrm{P}}$ rather than $\overline{\Omega}_{\mathrm{material}}$ of RVE if Assumption (5) is correct.

$$\overbrace{\begin{bmatrix} -(1-k)\varepsilon^{\mathrm{P}} & 0 & 0 \\ 0 & -k\varepsilon^{\mathrm{P}} & 0 \\ 0 & 0 & \varepsilon^{\mathrm{P}} \end{bmatrix}}^{\overline{\varepsilon}^{\mathrm{P}}} = \overbrace{\left(\gamma_1 \vec{s}^{1}\vec{n}^{1} + \gamma_2 \vec{s}^{2}\vec{n}^{2} + \gamma_3 \vec{s}^{3}\vec{n}^{3} + \gamma_4 \vec{s}^{4}\vec{n}^{4} + \gamma_5 \vec{s}^{5}\vec{n}^{5}\right)}^{\beta^{\mathrm{P}}(0)} + \left[\tilde{\Omega}_{\mathrm{lattice}}(0) - \tilde{\Omega}_{\mathrm{lattice}}(D_{\mathrm{Grain}}/2)\right] \quad (10)$$

To calculate the intragranular misorientation $\tilde{\Omega}_{\mathrm{lattice}}(0) - \tilde{\Omega}_{\mathrm{lattice}}(D_{\mathrm{Grain}}/2)$ between GC and GB determined by the mesoscopic plastic strain $\overline{\varepsilon}^{\mathrm{P}}$, the five activated slip factors

$(\vec{s}^{1}, \vec{n}^{1})$~$(\vec{s}^{5}, \vec{n}^{5})$ should be determined first. For the type 40Cr low alloy steel made up of ferrite and pearlite phases used in this research, the potential slip system of BCC lattice at room temperature is mainly the {110}<111> including twelve equivalent slip factors $(h^{\alpha}\ k^{\alpha}\ l^{\alpha})[u^{\alpha}\ v^{\alpha}\ w^{\alpha}]$ as shown in Table 2.

Table 2. Twelve equivalent slip factors of BCC lattice at room temperature.

α	1	2	3	4	5	6	7	8	9	10	11	12
$(h^{\alpha}\ k^{\alpha}\ l^{\alpha})$	(110)	(110)	$(1\bar{1}0)$	$(1\bar{1}0)$	(101)	(101)	$(10\bar{1})$	$(10\bar{1})$	(011)	(011)	$(01\bar{1})$	$(01\bar{1})$
$[u^{\alpha}\ v^{\alpha}\ w^{\alpha}]$	$[1\bar{1}1]$	$[\bar{1}11]$	[111]	$[11\bar{1}]$	$[11\bar{1}]$	$[\bar{1}11]$	$[1\bar{1}1]$	[111]	$[1\bar{1}1]$	$[11\bar{1}]$	$[\bar{1}11]$	[111]

The resolved shear stresses τ^{α} of twelve potential equivalent slip factors $(h^{\alpha}\ k^{\alpha}\ l^{\alpha})[u^{\alpha}\ v^{\alpha}\ w^{\alpha}]$ in one grain with the orientation of $\boldsymbol{g}(\phi_1, \phi, \phi_2)$ under mesoscopic stress $\overline{\sigma}$ applied in RVE can be calculated as shown in Equation (11). The five activated slip factors $\left(\vec{s}^{m}, \vec{n}^{m}\right)\Big|_{m=1\sim5}$ for undertaking the intragranular plastic distortion β^{P} are selected from the twelve potential slip factors $\left(\vec{s}^{\alpha}, \vec{n}^{\alpha}\right)\Big|_{\alpha=1\sim12}$ with the highest five resolved shear stresses among τ^{1}~τ^{12} according to Assumption (1).

$$\vec{s}^{\alpha} = g^{T}\cdot\begin{bmatrix} u^{\alpha} \\ v^{\alpha} \\ w^{\alpha} \end{bmatrix},\ \vec{n}^{\alpha} = g^{T}\cdot\begin{bmatrix} h^{\alpha} \\ k^{\alpha} \\ l^{\alpha} \end{bmatrix},\ \tau^{\alpha} = \overline{\sigma}I : \left(\vec{s}^{\alpha}\vec{n}^{\alpha}\right),\ \left(\vec{s}^{m}, \vec{n}^{m}\right)|_{m=1\sim5} = \left(\vec{s}^{\alpha}, \vec{n}^{\alpha}\right)|_{\tau^{\alpha}\ \in\ \text{highest five}\ \{\tau^{1}\sim\tau^{12}\}} \quad (11)$$

Once the intragranular misorientation $\tilde{\Omega}_{\text{lattice}}(0) - \tilde{\Omega}_{\text{lattice}}(D_{\text{Grain}}/2)$ between GC and GB of each grain under the mesoscopic plastic strain $\overline{\varepsilon}^{\mathrm{P}}$ is determined, the relative lattice rotation vector $\vec{\boldsymbol{R}}(r)$ can be then introduced to describe the distribution of intragranular misorientation $\tilde{\Omega}_{\text{lattice}}(r) - \tilde{\Omega}_{\text{lattice}}(D_{\text{Grain}}/2)$ along the grain radius r according to Assumption (6), as shown in Equation (12)

$$\vec{\boldsymbol{R}}(r) = \overbrace{[\theta_1(r) - \theta_1(D_{\text{Grain}}/2)]\vec{e}_1}^{R_1(r)} + \overbrace{[\theta_2(r) - \theta_2(D_{\text{Grain}}/2)]\vec{e}_2}^{R_2(r)} + \overbrace{[\theta_3(r) - \theta_3(D_{\text{Grain}}/2)]\vec{e}_3}^{R_3(r)} = \left(1 - \frac{2r}{D_{\text{Grain}}}\right)\vec{\boldsymbol{R}}(0) \quad (12)$$

4.3. Linear Evolution Law of Intragranular Misorientation

After establishing the 3D polycrystalline plasticity model, the next issue is to clarify the linear evolution law of intragranular misorientation with the mesoscopic plastic strain $\overline{\varepsilon}^{\mathrm{P}}$ of RVE based on Equations (10) and (11), while the intragranular misorientation distribution $\vec{\boldsymbol{R}}(r)$ is controlled by $\vec{\boldsymbol{R}}(0)$ at the GC completely according to Equation (12). Therein, Equation (10) is a system of linear equations; thus, both the crystallographic slip amounts γ_1~γ_5 and relative lattice rotation components $R_1(0)$~$R_3(0)$ at the GC will be proportional to the maximum principal plastic strain ε^{P} of RVE. However, the change rates of lattice rotation components with the maximum principal plastic strain $C_{1\sim3} = \mathrm{d}R_{1\sim3}(0)\ /\mathrm{d}\varepsilon^{\mathrm{P}}$ will be influenced by the variable k controlling the ratio between another two principal plastic strains, as well as the lattice orientation ϕ_1, ϕ, ϕ_2 controlling the grain activated slip factors $\left(\vec{s}^{m}, \vec{n}^{m}\right)\Big|_{m=1\sim5}$.

The analytical expression of $R_{1\sim3}(0) = C_{1\sim3}\cdot\varepsilon^{\mathrm{P}}$ is too complex to be derived from Equation (10) directly. Therefore, the numerical solutions to Equation (10) were achieved by MATLAB, instead of analytical approach. Figure 10 displayed the numerical solutions to Equation (10) under four special cases: (a) $\phi_1 = \phi = \phi_2 = \pi/4$; (b) $\phi_1 = \phi = \pi/4$, $\phi_2 = \pi/5$; (c) $\phi_1 = \pi/5$, $\phi = \phi_2 = \pi/4$; (d) $\phi = \pi/5, \phi_1 = \phi_2 = \pi/4$. For each case, k value was set as 0.0, 0.5 and 1.0, respectively. The calculation results shown in Figure 10 help us confirm that the lattice rotation components $R_{1\sim3}(0)$ at the GC are proportional to the maximum principal plastic strain ε^{P} of RVE, while the lattice rotation change rates $\mathrm{d}R_{1\sim3}(0)/\mathrm{d}\varepsilon^{\mathrm{P}}$ are functions of both variable k and Euler angles: $C_{1\sim3} = C_{1\sim3}(k,\ \phi_1, \phi, \phi_2)$.

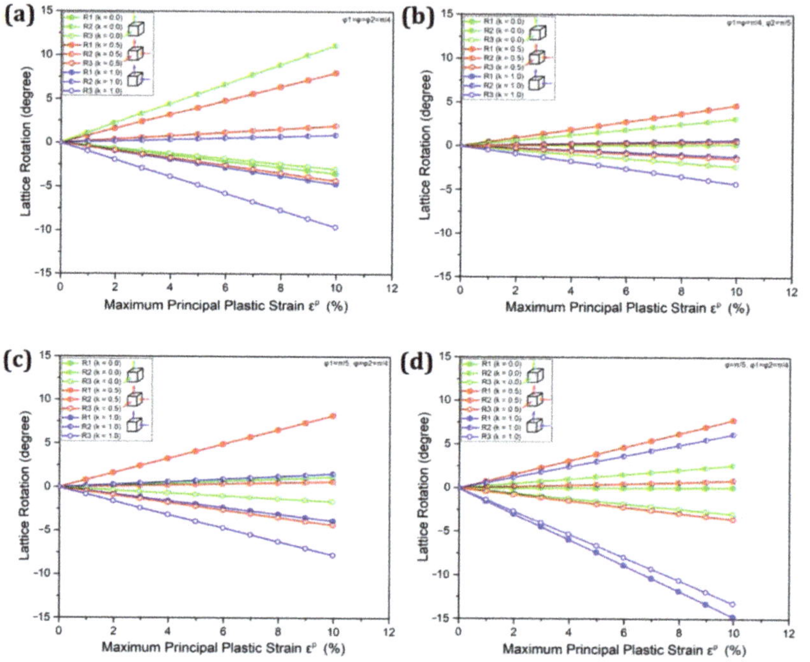

Figure 10. Linear relationship between relative lattice rotation components $R_1(0)$~$R_3(0)$ and maximum principal plastic strain ε^{P} in cases of: (**a**) $\phi_1 = \phi = \phi_2 = \pi/4$; (**b**) $\phi_1 = \phi = \pi/4,\ \phi_2 = \pi/5$; (**c**) $\phi_1 = \pi/5,\ \phi = \phi_2 = \pi/4$; (**d**) $\phi = \pi/5, \phi_1 = \phi_2 = \pi/4$ (k = 0.0, 0.5, 1.0).

After confirming the linear relationship between relative lattice rotation components $R_{1\sim3}(0)$ at the GC and the maximum principal plastic strain ε^{P} of RVE, the linear evolution law of relative lattice rotation vector $\vec{\boldsymbol{R}}(0)$ at the GC can be then determined as shown in Equation (13). The ratio among C_1, C_2 and C_3 is fixed once the variable k and Euler angles ϕ_1, ϕ, ϕ_2 are fixed (the fixed k corresponds to the proportional loading case of RVE), which means that the direction of relative lattice rotation vector $\vec{\boldsymbol{R}}(0)$ at the GC is fixed in the proportional loading case of RVE. At the same time, the relative lattice rotation angle $\left|\vec{\boldsymbol{R}}(0)\right|$ is proportional to the maximum principal plastic strain ε^{P} of RVE: $\left|\vec{\boldsymbol{R}}(0)\right| = C{\cdot}\varepsilon^{\mathrm{P}}$, and the resultant lattice rotation change rate C is the square root of the quadratic sum of three component lattice rotation change rates C_1, C_2 and C_3: $C = \sqrt{C_1^2 + C_2^2 + C_3^2}$. As a natural result, the resultant C also depends on the variable k and Euler angles ϕ_1, ϕ, ϕ_2: $C = C(k,\ \phi_1, \phi, \phi_2)$.

$$\vec{\boldsymbol{R}}(0) = \overbrace{C_1(k,\ \phi_1, \phi, \phi_2)\varepsilon^{\mathrm{P}}}^{R_1(0)}\vec{e}_1 + \overbrace{C_2(k,\ \phi_1, \phi, \phi_2)\varepsilon^{\mathrm{P}}}^{R_2(0)}\vec{e}_2 + \overbrace{C_3(k,\ \phi_1, \phi, \phi_2)\varepsilon^{\mathrm{P}}}^{R_3(0)}\vec{e}_3,\ \left|\vec{\boldsymbol{R}}(0)\right| = \overbrace{C(k,\ \phi_1, \phi, \phi_2)\varepsilon^{\mathrm{P}}}^{=\sqrt{C_1^2+C_2^2+C_3^2}} \tag{13}$$

Furthermore, the influence of variable k and Euler angles ϕ_1, ϕ, ϕ_2 on lattice rotation change rates $C_{1\sim3}(k,\ \phi_1, \phi, \phi_2)$ was investigated through numerical calculation of another five special cases. For each case, variable k and two of the Euler angles ϕ, ϕ_2 are fixed: (a) $k = 0.25,\ \phi = \phi_2 = \pi/4$; (b) $k = 0.75,\ \phi = \phi_2 = \pi/4$; (c) $k = 0.50,\ \phi = \phi_2 = \pi/4$; (d) $k = 0.50,\ \phi = \pi/5,\ \phi_2 = \pi/4$; (e) $k = 0.50,\ \phi = \pi/4,\ \phi_2 = \pi/5$, and another Euler angle ϕ_1 ranges from 0° to 360°, which describes the spatial rotation of crystallographic lattice and potential slip factors $\left(\vec{s}^{\alpha}, \vec{n}^{\alpha}\right)\Big|_{\alpha=1\sim12}$ along the z axis (the direction of maximum principal plastic strain ε^{p}). The calculation results are shown in Figure 11, from which we can see not only the dependence of $C_{1\sim3}$ on both variable k and Euler angles ϕ_1, ϕ, ϕ_2, but also an interesting phenomenon in $C_{1\sim3}$ evolution with the change of Euler angle ϕ_1.

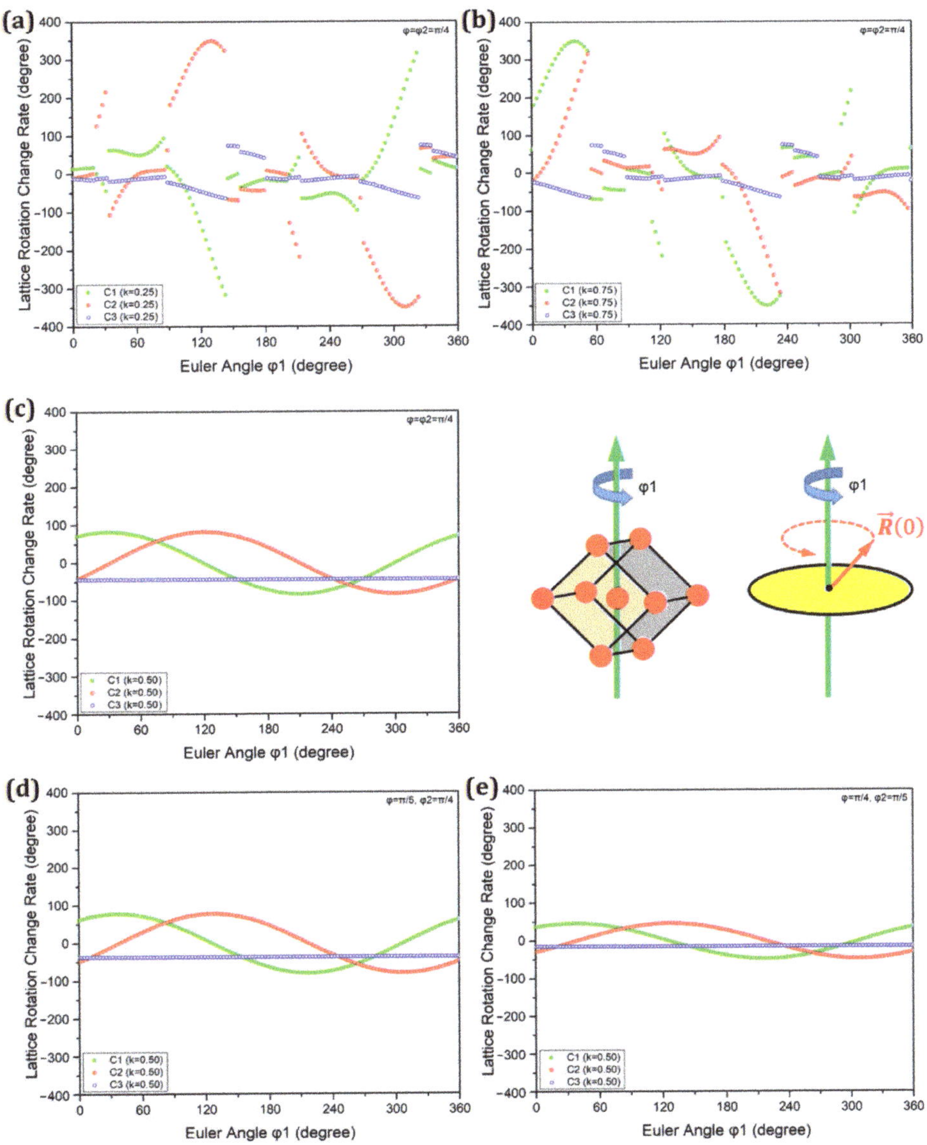

Figure 11. Influence of variable k and Euler angles ϕ_1, ϕ, ϕ_2 on lattice rotation change rates $C_{1\sim3}$ in cases of: (**a**) $k = 0.25$, $\phi = \phi_2 = \pi/4$; (**b**) $k = 0.75$, $\phi = \phi_2 = \pi/4$; (**c**) $k = 0.50$, $\phi = \phi_2 = \pi/4$; (**d**) $k = 0.50$, $\phi = \pi/5$, $\phi_2 = \pi/4$; (**e**) $k = 0.50$, $\phi = \pi/4$, $\phi_2 = \pi/5$ (ϕ_1 ranges from 0° to 360°).

The interesting phenomenon mentioned above is as follows. The $C_{1\sim3}$ values will suddenly "jump" with the change of Euler angle ϕ_1 if the variable $k \neq 0.5$ as shown in Figure 11a,b, while the $C_{1\sim2}$ values will smoothly change in a sinusoidal way and the C_3 value will keep constant with the change of Euler angle ϕ_1 if the variable $k = 0.5$ as shown in Figure 11c–e. This is because the two principal deviatoric stresses $-(1-k)\sigma$ along the x axis and $-k\sigma$ along the y axis of mesoscopic stress $\overline{\sigma}$ are unequal if $k \neq 0.5$, in which case the activated slip factors will naturally vary with the spatial rotation of crystallographic lattice along the z axis. On the contrary, if the mesoscopic stress $\overline{\sigma}$ is transversely isotropic in the x-O-y plane when $k = 0.5$, then the activated slip factors will be fixed during the spatial rotation of crystallographic lattice along the z axis, and the corresponding $\vec{R}(0)$ obtained from Equations (10) and (11) under the same mesoscopic plastic strain $\overline{\varepsilon}^p$ will rotate along the z axis synchronously with the change of Euler angle ϕ_1. Therefore, the $C_{1\sim2}$ values smoothly change in a sinusoidal way and the C_3 value keeps constant according to their roles in $\vec{R}(0)$ as shown in Equation (13).

4.4. Isotropic Evolution Law of Average $\overline{KAM}$ and $\overline{GROD}$

After clarifying the linear evolution law of intragranular misorientation $\vec{R}(0)$ at the GC as well as its distribution $\vec{R}(r)$ associated with the mesoscopic plastic strain $\overline{\varepsilon}^p$ of RVE, the intragranular geometrically necessary dislocation $\rho_{GND}(\psi)$ and its evolution with maximum principal plastic strain ε^p can then be determined by the continuum dislocation theory [41–43] as shown in Equation (14) and Figure 12a. The detailed derivation of Equation (14) can be referred to in Appendix A.

$$\rho_{GND}(\psi) = \tilde{\Omega}_{lattice}(r) \times \nabla = \frac{2\left|\vec{R}(0)\right|}{bD_{Grain}}\left(\vec{b}_n\vec{e}_\psi - \cos\psi\,\vec{b}_t\vec{e}_t\right) = \overbrace{\frac{2C(k,\ \phi_1,\phi,\phi_2)\cdot\varepsilon^p}{bD_{Grain}}}^{\rho_{GND}}\left(\overbrace{\vec{b}_n\vec{e}_\psi}^{\text{longitude}} - \overbrace{\cos\psi\,\vec{b}_t\vec{e}_t}^{\text{latitude}}\right) \tag{14}$$

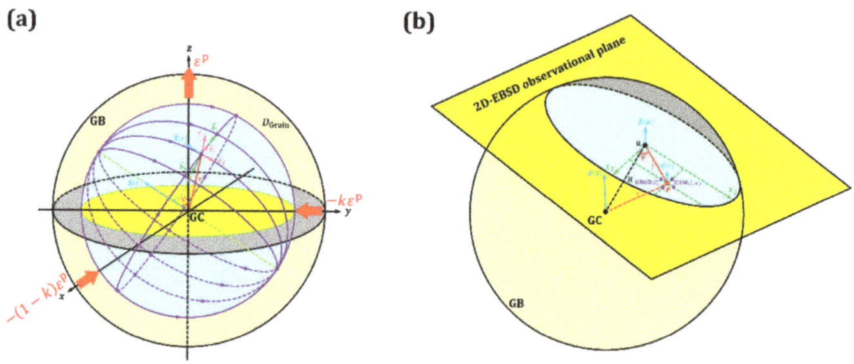

Figure 12. (**a**) Intragranular misorientation $\vec{R}(r)$ and geometrically necessary dislocation $\rho_{GND}(\psi)$, as well as (**b**) **KAM** and **GROD** values measured on arbitrary 2D-EBSD observational plane.

In the above expression, ρ_{GND} reflects the intragranular geometrically necessary dislocation density proportional to the maximum principal plastic strain ε^p but is inversely proportional to the grain diameter D_{Grain} [51], where the dependence of its change rate $C(k,\ \phi_1,\phi,\phi_2)$ on the grain orientation has been confirmed by Ref. [52]. ψ is the angle between the radius vector $\vec{r}$ and the relative lattice rotation vector $\vec{R}(0)$ at the GC. At the same time, $\vec{b}_t$ is the Burgers vector perpendicular to both $\vec{R}(0)$ and $\vec{r}$, and $\vec{b}_n$ is the

Burgers vector perpendicular to both $\vec{\boldsymbol{R}}(0)$ and $\vec{\boldsymbol{b}}_{\mathrm{t}}$, where $\left|\vec{\boldsymbol{b}}_{\mathrm{t}}\right| = \left|\vec{\boldsymbol{b}}_{\mathrm{n}}\right| = b$. In addition, $\vec{\boldsymbol{e}}_{t}$ is the unit vector parallel to $\vec{\boldsymbol{b}}_{\mathrm{t}}$, and $\vec{\boldsymbol{e}}_{\psi}$ is the unit vector perpendicular to both $\vec{\boldsymbol{b}}_{\mathrm{t}}$ and $\vec{\boldsymbol{r}}$.

From Equation (14) and Figure 12a, we can see that the structure of intragranular geometrically necessary dislocation $\rho_{\mathrm{GND}}(\psi)$ given by this 3D polycrystalline plasticity model is similar with the "longitude" and "latitude" mesh of the Earth, while the $\vec{\boldsymbol{R}}(0)$ is like the rotation axis of the Earth. Therein, $\vec{\boldsymbol{b}}_{\mathrm{n}}\vec{\boldsymbol{e}}_{\psi}$ represents the "longitude" dislocation, $\cos\psi\,\vec{\boldsymbol{b}}_{\mathrm{t}}\vec{\boldsymbol{e}}_{\mathrm{t}}$ represents the "latitude" dislocation, and the distribution of intragranular geometrically necessary dislocation density is uniform along the radius.

Once arbitrary 2D-EBSD observational plane with distance H away from the GC was determined as shown in Figure 12b, the 2D-EBSD scanning coordinate system (x_1, x_2) located on this plane can then be established. x_1 axis is along the scanning direction, x_2 axis is along the normal direction, and the origin R of this coordinate system is located at the center of grain cross profile. For each point P on this 2D-EBSD observational plane, the radial distance between GC and point P is $r = \sqrt{H^2 + l^2}$, where $l = \sqrt{x_1^2 + x_2^2}$ is the distance between origin R and point P on the 2D-EBSD observational plane, and φ is the angle between RP and x_1 axis. The intragranular misorientation parameters **KAM** and **GROD** measured on the above 2D-EBSD observational plane can be calculated from intragranular misorientation distribution $\vec{\boldsymbol{R}}(r)$ according to their definitions shown in Equations (3) and (4).

KAM is defined as the average absolute value of the $\vec{\boldsymbol{R}}(r)$ change within the EBSD scanning step size a along x_1 and x_1 axes, whose expression in partial derivation format is shown in Equation (15).

$$\mathbf{KAM}(l,\varphi) = \frac{a}{2}\left|\frac{\partial\vec{\boldsymbol{R}}(r)}{\partial r}\cdot\frac{\partial r}{\partial l}\cdot\frac{\partial l}{\partial x_1}\right| + \frac{a}{2}\left|\frac{\partial\vec{\boldsymbol{R}}(r)}{\partial r}\cdot\frac{\partial r}{\partial l}\cdot\frac{\partial l}{\partial x_2}\right| = \frac{l}{\sqrt{H^2+l^2}}(|\cos\varphi| + |\sin\varphi|)\cdot\frac{a\left|\vec{\boldsymbol{R}}(0)\right|}{D_{\mathrm{Grain}}} \quad (15)$$

KAM(l, φ) is influenced by the position metrics l and φ; thus, its local values were averaged over the cross profile of each grain by integral operation. The expression for $\mathbf{KAM}_{\mathrm{ave}}$ is shown in Equation (16).

$$\mathbf{KAM}_{\mathrm{ave}} = \frac{\int_0^{\sqrt{\left(\frac{D_{\mathrm{Grain}}}{2}\right)^2 - H^2}}\int_0^{2\pi}\mathbf{KAM}(l,\varphi)\cdot l d\varphi\cdot dl}{\pi\left[\left(\frac{D_{\mathrm{Grain}}}{2}\right)^2 - H^2\right]} = \overbrace{\left[\frac{1}{\sqrt{1-(2H/D_{\mathrm{Grain}})^2}} - \frac{(2H/D_{\mathrm{Grain}})^2}{1-(2H/D_{\mathrm{Grain}})^2}\ln\left(\frac{1+\sqrt{1-(2H/D_{\mathrm{Grain}})^2}}{2H/D_{\mathrm{Grain}}}\right)\right]}^{f(2H/D_{\mathrm{Grain}})}\cdot\frac{4a\left|\vec{\boldsymbol{R}}(0)\right|}{\pi D_{\mathrm{Grain}}} \quad (16)$$

In particular, the $f(2H/D_{\mathrm{Grain}})$ is a grain geometrical structure-related coefficient determined by the ratio $2H/D_{\mathrm{Grain}}$ ranging from 0 to 1. As two extreme cases, the expressions for $\mathbf{KAM}_{\mathrm{ave}}$ in $2H/D_{\mathrm{Grain}} \to 0$ and $2H/D_{\mathrm{Grain}} \to 1$ are shown in Equation (17) according to the limit operation on Equation (16). The case of $2H/D_{\mathrm{Grain}} \to 0$ means that the GC is located on the 2D-EBSD observational plane, where the expression for $\mathbf{KAM}_{\mathrm{ave}}$ degenerates into 2D polycrystalline plasticity model shown in Ref. [45].

$$\lim_{2H/D_{\mathrm{Grain}}\to 0}\mathbf{KAM}_{\mathrm{ave}} = \frac{4a\left|\vec{\boldsymbol{R}}(0)\right|}{\pi D_{\mathrm{Grain}}} = \frac{2ab}{\pi}\rho_{\mathrm{GND}} = \frac{4a\cdot C(k,\ \phi_1,\phi,\phi_2)}{\pi D_{\mathrm{Grain}}}\varepsilon^{\mathrm{P}},\quad \lim_{2H/D_{\mathrm{Grain}}\to 1}\mathbf{KAM}_{\mathrm{ave}} = 0 \quad (17)$$

$\mathbf{KAM}_{\mathrm{ave}}$ is still influenced by the grain geometrical structure-related coefficient $f(2H/D_{\mathrm{Grain}})$ and the grain-orientation-related coefficient $C(k,\ \phi_1,\phi,\phi_2)$. Therefore, these average values $\mathbf{KAM}^{\mathrm{A}}_{\mathrm{ave}}$, $\mathbf{KAM}^{\mathrm{B}}_{\mathrm{ave}}\cdots$ are further averaged over multiple grains by their cross profile areas S as shown in Equation (18).

$$\overline{\mathbf{KAM}} = \frac{S_A \cdot \mathbf{KAM}_{ave}^A + S_B \cdot \mathbf{KAM}_{ave}^B + \cdots}{S_A + S_B + \cdots} = \frac{2ab}{\pi} \overline{\rho}_{GND} = \overline{C}_f(k) \cdot \frac{4a\varepsilon^P}{\pi D_{Grain}} \tag{18}$$

The $\overline{\rho}_{GND}$ is an equivalent density of intragranular geometrically necessary dislocation density $\rho_{GND}^A, \rho_{GND}^B \cdots$ of multiple grains in the 2D-EBSD observational region, which can be calculated by Equation (19).

$$\overline{\rho}_{GND} = \frac{\left[1 - (2H_A/D_{Grain})^2\right] \cdot f(2H_A/D_{Grain}) \cdot \rho_{GND}^A + \left[1 - (2H_B/D_{Grain})^2\right] \cdot f(2H_B/D_{Grain}) \cdot \rho_{GND}^B + \cdots}{\left[1 - (2H_A/D_{Grain})^2\right] + \left[1 - (2H_B/D_{Grain})^2\right] + \cdots} \tag{19}$$

The $\overline{C}_f(k)$ is an equivalent factor of various grain orientation related coefficients $C(k, \phi_1^A, \phi^A, \phi_2^A)$, $C(k, \phi_1^B, \phi^B, \phi_2^B) \cdots$ of multiple grains in the 2D-EBSD observational region calculated by Equation (20).

$$\overline{C}_f(k) = \frac{\left[1 - (2H_A/D_{Grain})^2\right] \cdot f(2H_A/D_{Grain}) \cdot C(k, \phi_1^A, \phi^A, \phi_2^A) + \left[1 - (2H_B/D_{Grain})^2\right] \cdot f(2H_B/D_{Grain}) \cdot C(k, \phi_1^B, \phi^B, \phi_2^B) + \cdots}{\left[1 - (2H_A/D_{Grain})^2\right] + \left[1 - (2H_B/D_{Grain})^2\right] + \cdots} \tag{20}$$

Another intragranular misorientation parameter **GROD** is defined as the absolute value of the intragranular misorientation $\left|\vec{R}(r) - \vec{R}(H)\right|$ between the point P on 2D-EBSD observational plane and the origin R of 2D-EBSD scanning coordinate system, whose expression is shown in Equation (21).

$$\mathbf{GROD}(l) = \left|\vec{R}(r) - \vec{R}(H)\right| = \left(\sqrt{H^2 + l^2} - H\right) \cdot \frac{2\left|\vec{R}(0)\right|}{D_{Grain}} \tag{21}$$

$\mathbf{GROD}(l)$ is also influenced by the position metric l, thus its local values were averaged over the cross profile of each grain by integral operation. The expression for $\mathbf{GROD}_{ave}$ is shown in Equation (22).

$$\mathbf{GROD}_{ave} = \frac{\int_0^{\sqrt{\left(\frac{D_{Grain}}{2}\right)^2 - H^2}} \mathbf{GROD}(l) \cdot 2\pi l dl}{\pi\left[\left(\frac{D_{Grain}}{2}\right)^2 - H^2\right]} = \overbrace{\left[\frac{2 - 2H/D_{Grain} - (2H/D_{Grain})^2}{1 + 2H/D_{Grain}}\right]}^{g(2H/D_{Grain})} \cdot \frac{\left|\vec{R}(0)\right|}{3} \tag{22}$$

Similarly, the $g(2H/D_{Grain})$ is another grain geometrical structure related coefficient determined by the ratio $2H/D_{Grain}$ ranging from 0 to 1. As two extreme cases, the expressions for $\mathbf{GROD}_{ave}$ in $2H/D_{Grain} \to 0$ and $2H/D_{Grain} \to 1$ are shown in Equation (23) according to the limit operation on Equation (22). The case of $2H/D_{Grain} \to 0$ means that the GC is located on the 2D-EBSD observational plane, where the expression for $\mathbf{GROD}_{ave}$ degenerates into 2D polycrystalline plasticity model shown in Ref. [45].

$$\lim_{2H/D_{Grain} \to 0} \mathbf{GROD}_{ave} = \frac{2\left|\vec{R}(0)\right|}{3} = \frac{2 \cdot C(k, \phi_1, \phi, \phi_2)}{3}\varepsilon^P, \quad \lim_{2H/D_{Grain} \to 1} \mathbf{GROD}_{ave} = 0 \tag{23}$$

$\mathbf{GROD}_{ave}$ is still influenced by the grain geometrical structure-related coefficient $g(2H/D_{Grain})$ and the grain-orientation-related coefficient $C(k, \phi_1, \phi, \phi_2)$. Therefore, these average values $\mathbf{GROD}_{ave}^A$, $\mathbf{GROD}_{ave}^B \cdots$ are further averaged over multiple grains by their cross profile areas S as shown in Equation (24).

$$\overline{\mathbf{GROD}} = \frac{S_A \cdot \mathbf{GROD}_{ave}^A + S_B \cdot \mathbf{GROD}_{ave}^B + \cdots}{S_A + S_B + \cdots} = \overline{C}_g(k) \cdot \frac{\varepsilon^P}{3} \tag{24}$$

The $\overline{C}_g(k)$ is an equivalent factor of various grain-orientation-related coefficients $C(k,\ \phi_1^A, \phi^A,\ \phi_2^A)$, $C(k,\ \phi_1^B, \phi^B,\ \phi_2^B) \cdots$ of multiple grains in the 2D-EBSD observational region calculated by Equation (25).

$$\overline{C}_g(k) = \frac{\left[1 - (2H_A/D_{Grain})^2\right]\cdot g(2H_A/D_{Grain})\cdot C(k,\ \phi_1^A, \phi^A,\ \phi_2^A) + \left[1 - (2H_B/D_{Grain})^2\right]\cdot g(2H_B/D_{Grain})\cdot C(k,\ \phi_1^B, \phi^B,\ \phi_2^B) + \cdots}{\left[1 - (2H_A/D_{Grain})^2\right] + \left[1 - (2H_B/D_{Grain})^2\right] + \cdots} \quad (25)$$

Last but not least, the above equivalent metrics $\overline{\rho}_{GND}$, $\overline{C}_f(k)$ and $\overline{C}_g(k)$ shown in Equations (19), (20) and (25), respectively, are mesoscopic statistical parameters determined by the microscopic geometrical structure factor $2H/D_{Grain}$ and orientation factor ϕ_1, ϕ, ϕ_2 of each counted grain covered by the 2D-EBSD observational region. They will converge to constant values (assume that the variable k is fixed, because the $\overline{C}_f(k)$ and $\overline{C}_g(k)$ here are k-related) if the counted grains number is large enough as well as the possible values assigned to $2H/D_{Grain}$ and Euler angles ϕ_1, ϕ, ϕ_2 of each counted grain are completely random in their ranges. For polycrystalline metals with uniform equiaxial grains, the $2H/D_{Grain}$ and Euler angles ϕ_1, ϕ, ϕ_2 of multiple grains covered by arbitrary 2D-EBSD observational region will follow the similar statistical distribution law, whichever the plane we choose for the 2D-EBSD observation. The isotropy of $\overline{C}_f(k)$ and $\overline{C}_g(k)$ in the mesoscopic scale (the RVE scale, i.e., 2D-EBSD observational region scale) is the origin of isotropic evolution law of intragranular misorientation in deformed polycrystalline low alloy steel with uniform equiaxial grains. Based on the above discussions, both the average $\overline{\textbf{KAM}}$ and $\overline{\textbf{GROD}}$ follow the isotropic linear evolution with the maximum principal plastic strain ε^P, and are influenced by the ratio k between another two principal plastic strains of RVE at the same time according to Equations (18) and (24).

5. Conclusions

In this research, we focused on the evolution law of intragranular misorientation in deformed low alloy steel with uniform equiaxial grains. Intragranular misorientation parameters **KAM** and **GROD** were measured on 2D-EBSD observational planes with different angles to loading axis in both undeformed clamp sections and deformed gauge section of a interrupted tensile specimen. To explain the isotropic evolution law of average $\overline{\textbf{KAM}}$ and $\overline{\textbf{GROD}}$ with mesoscopic plastic strain observed in our experimental results and other literature studies, a modified 3D polycrystalline plasticity model was then developed in this research, based on our 2D polycrystalline plasticity model published elsewhere.

(1) The average $\overline{\textbf{KAM}}$ and $\overline{\textbf{GROD}}$ values in the deformed gauge section measured on 2D-EBSD observational planes with different angles to loading axis are almost the same, which reveals the isotropic evolution law of $\overline{\textbf{KAM}}$ and $\overline{\textbf{GROD}}$ during the deformation.

(2) Six fundamental assumptions including several necessary simplifications, such as spherical grain hypothesis and minimum activated slip factors number hypothesis, were made in this research to help us establish the modified 3D polycrystalline plasticity model based on our previous 2D model.

(3) The relative lattice rotation $\vec{R}(0)$ at the GC and the intragranular misorientation distribution $\vec{R}(r) = (1 - 2r/D_{Grain})\cdot\vec{\boldsymbol{R}}(0)$ were calculated in different cases based on the equations given by the 3D polycrystalline plasticity model. The linear relationship $\left|\vec{\boldsymbol{R}}(0)\right| = C(k,\ \phi_1, \phi, \phi_2)\cdot\varepsilon^P$ turned out to exist between the relative lattice rotation angle $\left|\vec{\boldsymbol{R}}(0)\right|$ at the GC and the maximum principal plastic strain ε^P of RVE, where the coefficient C was influenced by both the Euler angles ϕ_1, ϕ, ϕ_2 of any individual grain and the ratio k between another two principal plastic strains of RVE.

(4) The $\overline{\textbf{KAM}}$ and $\overline{\textbf{GROD}}$ were theoretically derived from the intragranular misorientation distribution $\vec{\boldsymbol{R}}(r)$ according to their definitions: $\overline{\textbf{KAM}} = \overline{C}_f(k)\cdot 4a\varepsilon^P/\pi D_{Grain}$

and $\overline{\textbf{GROD}} = \overline{C}_g(k) \cdot \varepsilon^{p}/3$. For polycrystalline metals with uniform equiaxial grains, $\overline{C}_f(k)$ and $\overline{C}_g(k)$ were turned out to be isotropic factors independent of 2D-EBSD observational plane selection. Therefore, both $\overline{\textbf{KAM}}$ and $\overline{\textbf{GROD}}$ follow the isotropic linear evolution law with the maximum principal plastic strain ε^{p} and are meanwhile influenced by the ratio k between another two principal plastic strains of RVE.

(5) Two laws given by this model were supported by experimental results: the linear evolution law of $\overline{\textbf{KAM}}$ and $\overline{\textbf{GROD}}$ has already been widely reported by previous studies, and the isotropic evolution law was verified by experimental result in this research.

Author Contributions: Conceptualization, S.-S.R.; methodology, S.-S.R.; software, Z.-H.S.; validation, Y.S.; formal analysis, Y.S.; investigation, S.-S.R.; resources, J.-M.Z.; data curation, S.-S.R.; writing—original draft preparation, S.-S.R.; writing—review and editing, S.-S.R.; visualization, Z.-H.S.; supervision, H.-J.S.; project administration, H.-J.S.; funding acquisition, H.-J.S. All authors have read and agreed to the published version of the manuscript.

Funding: This research work is financially supported by the National Natural Science Foundation of China (Nos. 12202446, 12172193, 12002155 and 12102348), the National Major Science and Technology Projects of China (No. J2019-VI-0002-0115), the Opening Fund of the Key Laboratory of Aero-engine Thermal Environment and Structure, Ministry of Industry and Information Technology (No. CEPE2022004), as well as the Opening Fund of the State Key Laboratory of Nonlinear Mechanics.

Data Availability Statement: Not applicable.

Acknowledgments: We thank Qi-Nan Han from the Nanjing University of Aeronautics and Astronautics for his invitation and kind suggestions during the submission of this work.

Conflicts of Interest: The authors declare no conflict of interest.

Appendix A

The derivation process of Equation (14) is attached as follows. Taking the lattice rotation vector at the grain center $\vec{R}(0)$ as the Earth axis, we establish the Earth-like longitude–latitude network and the spherical coordinate system $\left(\vec{e}_{\mathrm{r}}, \vec{e}_{\psi}, \vec{e}_{\mathrm{t}}\right)$ as shown in Figure 12a. Therein, ψ is the angle between position vector $\vec{r}$ and earth axis, $\vec{e}_{\mathrm{t}}$ is the tangent vector and $\vec{e}_{\mathrm{n}}$ is the normal vector of a latitude circle. The above vectors follow Equation (A1) below:

$$\vec{e}_{\mathrm{n}} \times \vec{e}_{\mathrm{r}} = \cos\psi\, \vec{e}_{\mathrm{t}},\ \vec{e}_{\mathrm{n}} \times \vec{e}_{\mathrm{r}} = \vec{e}_{\psi} \tag{A1}$$

The residual lattice rotation tensor $\tilde{\boldsymbol{\Omega}}_{\mathrm{lattice}}(r)$ distribution inside each grain can be given as following according to Assumption (6), under the orthogonal coordinate system $\left(\vec{e}_{\mathrm{t}}, \vec{e}_{\mathrm{n}}, \vec{e}_{\mathrm{t}} \times \vec{e}_{\mathrm{n}}\right)$, and the direction of lattice rotation vector $\vec{R}(r)$ is parallel to $\vec{e}_{\mathrm{t}} \times \vec{e}_{\mathrm{n}}$.

$$\tilde{\boldsymbol{\Omega}}_{\mathrm{lattice}}(r) = \left|\vec{R}(r)\right| \left(\vec{e}_{\mathrm{t}}\vec{e}_{\mathrm{n}} - \vec{e}_{\mathrm{n}}\vec{e}_{\mathrm{t}}\right) = \left|\vec{R}(0)\right| \left(1 - \frac{2r}{D_{\mathrm{Grain}}}\right) \left(\vec{e}_{\mathrm{t}}\vec{e}_{\mathrm{n}} - \vec{e}_{\mathrm{n}}\vec{e}_{\mathrm{t}}\right) \tag{A2}$$

Furthermore, $\boldsymbol{\nabla}$ is the Nabla operator for calculating gradient, divergence and curl, whose format under spherical coordinate system $\left(\vec{e}_{\mathrm{r}}, \vec{e}_{\psi}, \vec{e}_{\mathrm{t}}\right)$ can be expressed as Equation (A3):

$$\nabla = \vec{e}_{\mathrm{r}}\frac{\partial}{\partial r} + \vec{e}_{\psi} r \frac{\partial}{\partial \psi} + \vec{e}_{\mathrm{t}} r \sin\psi \frac{\partial}{\partial \mathrm{t}} \tag{A3}$$

Then, the geometrically necessary dislocation (GND) density tensor ρ_{GND} can be expressed as the right curl of residual lattice rotation tensor $\tilde{\boldsymbol{\Omega}}_{\mathrm{lattice}}(r)$, as shown in Equation (A4):

$$\rho_{\mathrm{GND}}(\psi) = \tilde{\boldsymbol{\Omega}}_{\mathrm{lattice}}(r) \times \nabla = \frac{\partial \left| \vec{\boldsymbol{R}}(r) \right|}{\partial r} \left(\vec{\boldsymbol{e}}_{\mathrm{t}} \vec{\boldsymbol{e}}_{\mathrm{n}} - \vec{\boldsymbol{e}}_{\mathrm{n}} \vec{\boldsymbol{e}}_{\mathrm{t}} \right) \times \vec{\boldsymbol{e}}_{\mathrm{r}} = \frac{2 \left| \vec{\boldsymbol{R}}(0) \right|}{D_{\mathrm{Grain}}} \left(\vec{\boldsymbol{e}}_{\mathrm{n}} \vec{\boldsymbol{e}}_{\psi} - \cos\psi \vec{\boldsymbol{e}}_{\mathrm{t}} \vec{\boldsymbol{e}}_{\mathrm{t}} \right) = \frac{2 \left| \vec{\boldsymbol{R}}(0) \right|}{b D_{\mathrm{Grain}}} \left(\vec{\boldsymbol{b}}_{\mathrm{n}} \vec{\boldsymbol{e}}_{\psi} - \cos\psi \vec{\boldsymbol{b}}_{\mathrm{t}} \vec{\boldsymbol{e}}_{\mathrm{t}} \right) \quad \text{(A4)}$$

References

1. Brewer, L.N.; Field, D.P.; Merriman, C.C. Mapping and Assessing Plastic Deformation Using EBSD. In *Electron Backscatter Diffraction in Materials Science*, 2nd ed.; Schwartz, A.J., Kumar, M., Adams, B.L., Field, D.P., Eds.; Springer: Boston, MA, USA, 2009; pp. 251–262.
2. Wright, S.I.; Nowell, M.M.; Field, D.P. A Review of Strain Analysis Using Electron Backscatter Diffraction. *Microsc. Microanal.* **2011**, *17*, 316–329. [CrossRef] [PubMed]
3. Wright, S.I.; Suzuki, S.; Nowell, M.M. In Situ EBSD Observations of the Evolution in Crystallographic Orientation with Deformation. *JOM* **2016**, *68*, 2730–2736. [CrossRef]
4. Kamaya, M. Measurement of local plastic strain distribution of stainless steel by electron backscatter diffraction. *Mater. Charact.* **2009**, *60*, 125–132. [CrossRef]
5. Kamaya, M.; Wilkinson, A.J.; Titchmarsh, J.M. Measurement of plastic strain of polycrystalline material by electron backscatter diffraction. *Nucl. Eng. Des.* **2005**, *235*, 713–725. [CrossRef]
6. Kamaya, M.; Wilkinson, A.J.; Titchmarsh, J.M. Quantification of plastic strain of stainless steel and nickel alloy by electron backscatter diffraction. *Acta Mater.* **2006**, *54*, 539–548. [CrossRef]
7. Kamaya, M. Assessment of local deformation using EBSD: Quantification of accuracy of measurement and definition of local gradient. *Ultramicroscopy* **2011**, *111*, 1189–1199. [CrossRef]
8. Kamaya, M.; da Fonseca, J.Q.; Li, L.M.; Preuss, M. Local Plastic Strain Measurement by EBSD. *Appl. Mech. Mater.* **2007**, *7–8*, 173–179. [CrossRef]
9. Rui, S.-S.; Shang, Y.-B.; Su, Y.; Qiu, W.; Niu, L.-S.; Shi, H.-J.; Matsumoto, S.; Chuman, Y. EBSD analysis of cyclic load effect on final misorientation distribution of post-mortem low alloy steel: A new method for fatigue crack tip driving force prediction. *Int. J. Fatigue* **2018**, *113*, 264–276. [CrossRef]
10. Rui, S.-S.; Shang, Y.-B.; Qiu, W.; Niu, L.-S.; Shi, H.-J.; Matsumoto, S.; Chuman, Y. Fracture mode identification of low alloy steels and cast irons by electron back-scattered diffraction misorientation analysis. *J. Mater. Sci. Technol.* **2017**, *33*, 1582–1595. [CrossRef]
11. Kobayashi, D.; Miyabe, M.; Kagiya, Y.; Sugiura, R.; Yokobori, A.T. An Assessment and Estimation of the Damage Progression Behavior of IN738LC under Various Applied Stress Conditions Based on EBSD Analysis. *Metall. Mater. Trans. A* **2013**, *44*, 3123–3135. [CrossRef]
12. Kobayashi, D.; Miyabe, M.; Achiwa, M. Failure Analysis and Life Assessment of Thermal Fatigue Crack Growth in a Nickel-Base Superalloy Based on EBSD Method. *ASME Turbo Expo* **2015**, *2015*, V006T21A004.
13. Kobayashi, D.; Ito, A.; Miyabe, M.; Kagiya, Y.; Yoshioka, Y. Crack Initiation Behavior and its Estimation for a Gas Turbine Rotor Based on the EBSD Analysis. *ASME Turbo Expo* **2012**, *2012*, 71–79.
14. Kobayashi, D.; Miyabe, M.; Achiwa, M. Failure Analysis Method of Ni-base Superalloy by EBSD Observation of the Cross Section. In Proceedings of the JSMS 13th Fractographic Conference, Wakayama, Japan, 14 November 2014.
15. Kamaya, M. Characterization of microstructural damage due to low-cycle fatigue by EBSD observation. *Mater. Charact.* **2009**, *60*, 1454–1462. [CrossRef]
16. Kamaya, M. Observation of Low-Cycle Fatigue Damage by EBSD(Microstructural Change in SUS316 and STS410). *Trans. Jpn. Soc. Mech. Eng.* **2011**, *77*, 154–169. [CrossRef]
17. Kamaya, M.; Kuroda, M. Fatigue Damage Evaluation Using Electron Backscatter Diffraction. *Mater. Trans.* **2011**, *52*, 1168–1176. [CrossRef]
18. Rui, S.-S.; Shang, Y.-B.; Fan, Y.-N.; Han, Q.-N.; Niu, L.-S.; Shi, H.-J.; Hashimoto, K.; Komai, N. EBSD analysis of creep deformation induced grain lattice distortion: A new method for creep damage evaluation of austenitic stainless steels. *Mater. Sci. Eng. A* **2018**, *733*, 329–337. [CrossRef]
19. Kobayashi, D.; Miyabe, M.; Kagiya, Y.; Nagumo, Y.; Sugiura, R.; Matsuzaki, T.; Yokobori, A.T., Jr. Creep damage evaluation of IN738LC based on the EBSD method by using a notched specimen. *Strength Fract. Complex.* **2011**, *7*, 157–167. [CrossRef]
20. Kobayashi, D.; Miyabe, M.; Kagiya, Y.; Nagumo, Y.; Sugiura, R.; Matsuzaki, T.; Yokobori, A.T. Geometrical influence for creep damage evaluation of IN738LC using electron backscatter diffraction. *Mater. High Temp.* **2012**, *29*, 301–307. [CrossRef]
21. Wei, S.; Kim, J.; Tasan, C.C. Boundary micro-cracking in metastable Fe45Mn35Co10Cr10 high-entropy alloys. *Acta Mater.* **2019**, *168*, 76–86. [CrossRef]
22. Han, Q.-N.; Rui, S.-S.; Qiu, W.; Ma, X.; Su, Y.; Cui, H.; Zhang, H.; Shi, H. Crystal orientation effect on fretting fatigue induced geometrically necessary dislocation distribution in Ni-based single-crystal superalloys. *Acta Mater.* **2019**, *179*, 129–141. [CrossRef]
23. Kobayashi, D.; Takeuchi, T.; Achiwa, M. Evaluation of Fatigue Crack Growth Rate by the EBSD Method. In Proceedings of the JSME Annual Conference 2015, Tokyo, Japan, 13–16 September 2015.
24. Jiang, J.; Britton, T.B.; Wilkinson, A.J. Measurement of geometrically necessary dislocation density with high resolution electron backscatter diffraction: Effects of detector binning and step size. *Ultramicroscopy* **2013**, *125*, 1–9. [CrossRef] [PubMed]

25. Littlewood, P.D.; Wilkinson, A.J. Geometrically necessary dislocation density distributions in cyclically deformed Ti–6Al–4V. *Acta Mater.* **2012**, *60*, 5516–5525. [CrossRef]
26. Wallis, D.; Hansen, L.N.; Britton, T.B.; Wilkinson, A.J. Geometrically necessary dislocation densities in olivine obtained using high-angular resolution electron backscatter diffraction. *Ultramicroscopy* **2016**, *168*, 34–45. [CrossRef] [PubMed]
27. Jiang, J.; Britton, T.B.; Wilkinson, A.J. Evolution of dislocation density distributions in copper during tensile deformation. *Acta Mater.* **2013**, *61*, 7227–7239. [CrossRef]
28. Wallis, D.; Hansen, L.N.; Britton, T.B.; Wilkinson, A.J. Dislocation Interactions in Olivine Revealed by HR-EBSD. *J. Geophys. Res. Solid Earth* **2017**, *122*, 7659–7678. [CrossRef]
29. Vilalta-Clemente, A.; Naresh-Kumar, G.; Nouf-Allehiani, M.; Gamarra, P.; di Forte-Poisson, M.A.; Trager-Cowan, C.; Wilkinson, A.J. Cross-correlation based high resolution electron backscatter diffraction and electron channelling contrast imaging for strain mapping and dislocation distributions in InAlN thin films. *Acta Mater.* **2017**, *125*, 125–135. [CrossRef]
30. Sarac, A.; Oztop, M.S.; Dahlberg, C.F.O.; Kysar, J.W. Spatial distribution of the net Burgers vector density in a deformed single crystal. *Int. J. Plast.* **2016**, *85*, 110–129. [CrossRef]
31. Kysar, J.W.; Gan, Y.X.; Morse, T.L.; Chen, X.; Jones, M.E. High strain gradient plasticity associated with wedge indentation into face-centered cubic single crystals: Geometrically necessary dislocation densities. *J. Mech. Phys. Solids* **2007**, *55*, 1554–1573. [CrossRef]
32. Dahlberg, C.F.O.; Saito, Y.; Öztop, M.S.; Kysar, J.W. Geometrically necessary dislocation density measurements at a grain boundary due to wedge indentation into an aluminum bicrystal. *J. Mech. Phys. Solids* **2017**, *105* (Suppl. C), 131–149. [CrossRef]
33. Dahlberg, C.F.O.; Saito, Y.; Öztop, M.S.; Kysar, J.W. Geometrically necessary dislocation density measurements associated with different angles of indentations. *Int. J. Plast.* **2014**, *54*, 81–95. [CrossRef]
34. Sarac, A.; Kysar, J.W. Experimental validation of plastic constitutive hardening relationship based upon the direction of the Net Burgers Density Vector. *J. Mech. Phys. Solids* **2018**, *111* (Suppl. C), 358–374. [CrossRef]
35. Kysar, J.W.; Saito, Y.; Oztop, M.S.; Lee, D.; Huh, W.T. Experimental lower bounds on geometrically necessary dislocation density. *Int. J. Plast.* **2010**, *26*, 1097–1123. [CrossRef]
36. Pantleon, W. Resolving the geometrically necessary dislocation content by conventional electron backscattering diffraction. *Scr. Mater.* **2008**, *58*, 994–997. [CrossRef]
37. Calcagnotto, M.; Ponge, D.; Demir, E.; Raabe, D. Orientation gradients and geometrically necessary dislocations in ultrafine grained dual-phase steels studied by 2D and 3D EBSD. *Mater. Sci. Eng. A* **2010**, *527*, 2738–2746. [CrossRef]
38. Konijnenberg, P.J.; Zaefferer, S.; Raabe, D. Assessment of geometrically necessary dislocation levels derived by 3D EBSD. *Acta Mater.* **2015**, *99*, 402–414. [CrossRef]
39. Gao, H.; Huang, Y. Geometrically necessary dislocation and size-dependent plasticity. *Scr. Mater.* **2003**, *48*, 113–118. [CrossRef]
40. Arsenlis, A.; Parks, D.M. Crystallographic aspects of geometrically-necessary and statistically-stored dislocation density. *Acta Mater.* **1999**, *47*, 1597–1611. [CrossRef]
41. Nye, J.F. Some geometrical relations in dislocated crystals. *Acta Metall.* **1953**, *1*, 153–162. [CrossRef]
42. Bilby, B.A.; Bullough, R.; Smith, E. Continuous distributions of dislocations: A new application of the methods of non-Riemannian geometry. *Proc. R. Soc. Lond. Ser. A. Math. Phys. Sci.* **1955**, *231*, 263.
43. Kröner, E. *Continuum Theory of Dislocation and Self-Stresses*; Springer: Berlin, Germany, 1958.
44. Harte, A.; Atkinson, M.; Preuss, M.; da Fonseca, J.Q. A statistical study of the relationship between plastic strain and lattice misorientation on the surface of a deformed Ni-based superalloy. *Acta Mater.* **2020**, *195*, 555–570. [CrossRef]
45. Rui, S.-S.; Niu, L.-S.; Shi, H.-J.; Wei, S.; Tasan, C.C. Diffraction-based misorientation mapping: A continuum mechanics description. *J. Mech. Phys. Solids* **2019**, *133*, 103709. [CrossRef]
46. Hutchinson, J.W.; Neale, K.W. Finite Strain J2 Deformation Theory. In *Proceedings of the IUTAM Symposium on Finite Elasticity*; Carlson, D.E., Shield, R.T., Eds.; Springer: Dordrecht, The Netherlands, 1982; pp. 237–247.
47. *Standard Test Methods for Tension Testing of Metallic Materials*; ASTM International: West Conshohocken, PA, USA, 2008.
48. Han, Q.-N.; Rui, S.-S.; Qiu, W.; Su, Y.; Ma, X.; Su, Z.; Cui, H.; Shi, H. Effect of crystal orientation on the indentation behaviour of Ni-based single crystal superalloy. *Mater. Sci. Eng. A* **2020**, *773*, 138893. [CrossRef]
49. Cho, J.-H.; Rollett, A.D.; Oh, K.H. Determination of a mean orientation in electron backscatter diffraction measurements. *Metall. Mater. Trans. A* **2005**, *36*, 3427–3438. [CrossRef]
50. Glez, J.C.; Driver, J. Orientation distribution analysis in deformed grains. *J. Appl. Crystallogr.* **2001**, *34*, 280–288. [CrossRef]
51. Ashby, M.F. The deformation of plastically non-homogeneous materials. *Philos. Mag.* **1970**, *21*, 399–424. [CrossRef]
52. Kundu, A.; Field, D.P. Influence of plastic deformation heterogeneity on development of geometrically necessary dislocation density in dual phase steel. *Mater. Sci. Eng. A* **2016**, *667*, 435–443. [CrossRef]

metals

MDPI

Article

Fatigue Crack Growth Behavior and Failure Mechanism of Nickel-Based Alloy GH4169 under Biaxial Load Based on Fatigue Test of Cruciform Specimen

Zhirong Wu [1,*], Ying Pan [1], Hang Lei [2], Shuaiqiang Wang [1] and Lei Fang [1]

[1] College of Energy and Power, Nanjing University of Aeronautics and Astronautics, Nanjing 210016, China
[2] AVIC Chengdu Aircraft Design and Research Institute, Chengdu 610041, China
* Correspondence: zrwu@nuaa.edu.cn

Abstract: Due to the complex geometry and various cyclic loads, aeroengine components are often in a multiaxial complex stress state during service. Multiaxial fatigue is a major cause of many air accidents. It is of great significance to study the fatigue failure mechanism of aeronautical materials. This paper carries out biaxial fatigue tests on cruciform specimens and uses the surface replication method to record the initiation and propagation process of crack. Based on these fatigue tests, this paper studies the multiaxial fatigue characteristics of nickel-based alloy GH4169 for aeroengines and analyzes the fatigue crack growth behavior and failure mechanism of nickel-based alloys under a complex multiaxial stress state.

Keywords: nickel-based alloy GH4169; complex stress state; biaxial fatigue; fatigue crack behavior

Citation: Wu, Z.; Pan, Y.; Lei, H.; Wang, S.; Fang, L. Fatigue Crack Growth Behavior and Failure Mechanism of Nickel-Based Alloy GH4169 under Biaxial Load Based on Fatigue Test of Cruciform Specimen. *Metals* **2023**, *13*, 588. https://doi.org/10.3390/met13030588

Academic Editor: George A. Pantazopoulos

Received: 17 February 2023
Revised: 9 March 2023
Accepted: 12 March 2023
Published: 14 March 2023

1. Introduction

1.1. Background

With the increase in thrust–weight ratio, an aeroengine's working temperature and rotational speed are also improved. The working conditions of aeroengine components are very harsh. Due to the complex geometry and various cyclic loads, these components are often under a multiaxial stress state, and the durability and reliability evaluation of aeroengine parts under a multiaxial stress condition is becoming more and more prominent. Therefore, it is very meaningful to study fatigue characteristic, investigate fatigue crack initiation and growth behavior and develop a fatigue crack life analysis method for aeroengine materials under a complex multiaxial stress state.

Biaxial tension fatigue is one of the main, typical multiaxial fatigues. Many pieces of research on biaxial tension fatigue have been carried out to study fatigue crack initiation and propagation behavior under a multiaxial stress state. Wolf et al. [1] studied the influence of different loading phase differences and stress ratios on the fatigue crack growth behavior with cruciform specimens made of aluminum alloy and austenitic stainless steel, respectively. The results showed that, under the same loading phase, the crack path of austenitic stainless steel specimens was zigzag, and that of aluminum alloy specimens was almost straight. When the biaxial fatigue load was not in the same phase, the crack path was deflected. Additionally, the crack growth paths of austenitic stainless steel and aluminum alloy specimens were basically the same under different stress ratios. Meng et al. [2] studied the fatigue crack growth direction of magnesium alloy under different loading phase differences and stress ratios with cruciform specimens made of magnesium alloy. The results showed that, under different loading phase differences, the initial crack growth direction angle of the magnesium alloy was different. When the loading phase difference was 0°, the prefabricated crack expanded along a straight line. When the loading phase difference was 90° or 180°, the prefabricated crack expanded into two bifurcated cracks. In addition, the crack growth rate increased with the increase in stress ratio under the same stress intensity factor amplitude. Abecassis et al. [3] studied the fatigue crack behavior of a titanium-base alloy under biaxial fatigue loading with cruciform specimens and studied the interaction of material microstructure, crack path and fatigue crack growth rate. The results showed that the deflection of crack growth path caused by the material microstructure reduced the crack growth rate. Misak et al. [4] studied the influence of different stress ratios on the fatigue crack growth path with cruciform specimens made of aluminum alloy. The results showed that the crack growth area was basically parallel to the notch direction of the specimen when the biaxial fatigue load ratios were 0, 0.5 and 1. When the biaxial fatigue load ratio was 1.5, the crack growth area was within the area with an angle of 45° to the notch direction of the specimen.

The research on biaxial fatigue crack has mainly focused on the influence of different loading phase differences, stress ratios and angles of pre-crack on fatigue crack growth behavior. However, pieces of research on crack initiation behavior under biaxial fatigue loading are relatively few. In addition, most of the materials used in these studies were aluminum alloys. Studies on other alloys are limited.

Nickel-based alloy GH4169 is a kind of precipitation-strengthening alloy in China. Similar to Inconel 718, GH4169 has the advantages of high-cost performance, good formability and fatigue resistance [5–7]. It is widely used in various key parts of aeroengines. Therefore, it is meaningful to carry out research on the damage tolerance mechanism of GH4169, which helps to master the internal evolution behavior, and the rules of the fatigue damage process of GH4169 and form a calculation method for fatigue crack initiation life based on damage evolution theory and crack propagation life based on fracture mechanics theory. It reflects the major needs of the aviation industry and helps to improve aeroengine research and development.

1.2. Purpose

This paper takes nickel-based alloy GH4169 as the research object and studies its fatigue crack behavior under a complex stress state, which has important theoretical significance and engineering application value.

The purpose is to develop a fatigue crack life analysis method and damage tolerance design method for key aeroengine components. The biaxial fatigue test is carried out on a cruciform specimen specially designed to simulate the hole in the casing of an aeroengine under a multiaxial fatigue condition during service.

2. Experimental Procedures

2.1. Materials and Mechanical Properties

The material used in this investigation was nickel-based alloy GH4169. Nickel-based alloy GH4169 has good fatigue resistance, oxidation resistance and corrosion resistance and is widely used in aeroengine turbine disks, casing and other structural parts. The main chemical composition of nickel-based alloy GH4169 is shown in Table 1. The Young's modulus, yield strength and ultimate strength of nickel-based alloy GH4169 at room temperature are 198 GPa, 1045 MPa and 1398 MPa, respectively. The microstructure of GH4169 is shown in Figure 1. The average grain size of GH4169 is about 16 μm, and it contains some inclusion defects.

Table 1. Main chemical composition of nickel-based alloy GH4169.

Element	Nb	Mo	Ti	Cr	Fe	Ni
Wt/%	5.67	3.84	1.40	19.14	17.20	52.76

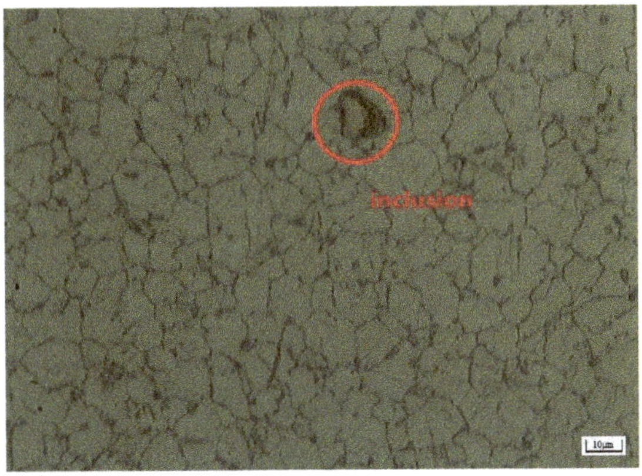

Figure 1. Microstructural result of GH4169.

2.2. *Preparation of Specimens*

Two kinds of cruciform specimens with different shapes of notch in the center area were designed in this paper. These specimens were center-thinned by equal thickness on both sides and had a circular notch or a 0° waist round notch in the central area. The total length of the test piece was 90 mm. The thickness of the central area was 0.8 mm. The four ends of the test piece were 10 mm thick and had circular holes with a diameter of 8 mm. The specific geometry and dimensions of the cruciform specimen are shown in Figure 2.

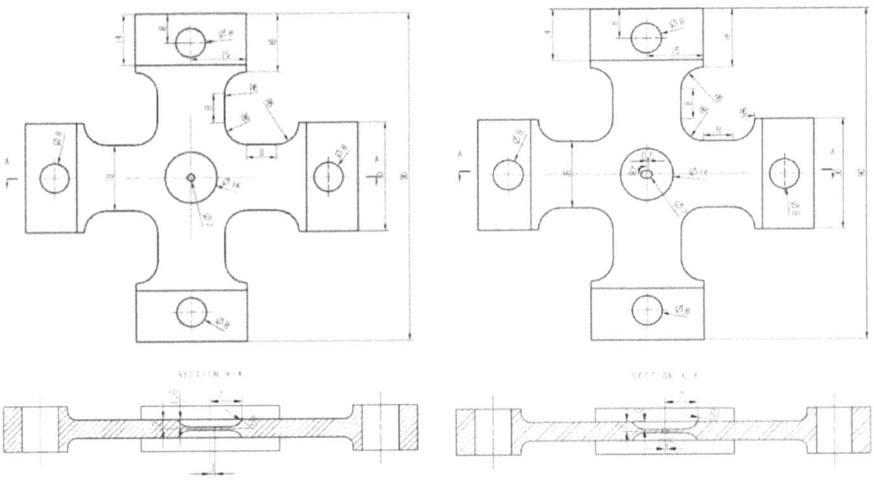

(a) (b)

Figure 2. Geometry and dimensions of cruciform specimen (mm). (**a**) Cruciform specimen with circular notch in the central area. (**b**) Cruciform specimen with 0° waist circle hole notch in the central area.

2.3. *Fatigue Tests and Replication Method*

The biaxial fatigue test was carried out on a uniaxial tensile testing machine in combination with a biaxial fixture device. The test temperature was room temperature. The loading waveform was sine wave. The loading frequency was 10 Hz. The loading amplitudes on both axes were the same. The stress ratio was 0.1. The maximum test load F_A was 30 kN and 33 kN, respectively. A total of 9 specimens were tested, of which, 5 specimens were under intermittent loading with the surface replication (SR) method to record the initiation and propagation process of the crack in the central area. The other 4 specimens were under continuous loading without SR method. The detailed biaxial fatigue test conditions are shown in Table 2. The biaxial fatigue test device is shown in Figure 3.

Table 2. Test conditions of cruciform specimen.

Specimen	Notch Shape	Stress Radio	$F_{A,\,max}$ (kN)	Test Condition
1	circular notch	0.1	30	Non-SR
2	circular notch	0.1	30	SR
3	circular notch	0.1	30	SR
4	circular notch	0.1	33	Non-SR
5	circular notch	0.1	33	SR
6	waist round notch	0.1	30	Non-SR
7	waist round notch	0.1	30	SR
8	waist round notch	0.1	33	Non-SR
9	waist round notch	0.1	33	SR

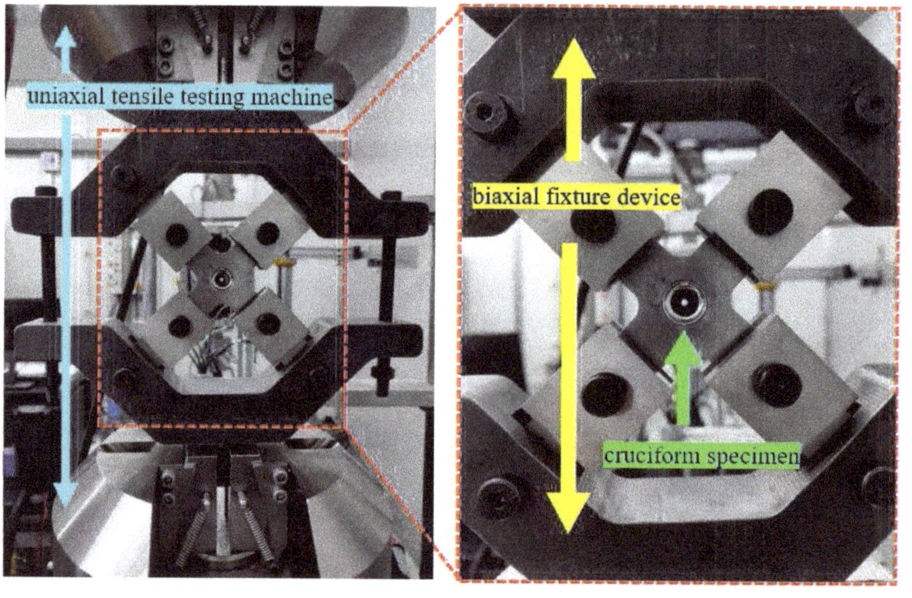

Figure 3. Biaxial fatigue test device.

Surface replication is one common crack monitoring method. Compared with other methods, such as the scanning electron microscope method and electric potential method, the SR method has the advantage of repeatable observation of a crack, which is helpful to determine the location of crack initiation and observe the behavior of crack initiation and early growth.

The replication material used in this paper was RepliSet, produced by Struers. This material is a fast-curing binary silicone rubber which can be replicated with a minimum length of 10~20 μm crack. The replication system used for spraying RepliSet is shown in Figure 4. The components of the replication system are manual applicator, RepliSet cylinder box, static mixing nozzle and nozzle head. The replication material is stored in the RepliSet cylinder box; it was sprayed on the specimen through the mixing nozzle and nozzle head. The cylinder box and nozzle are connected to the manual applicator to facilitate the operation.

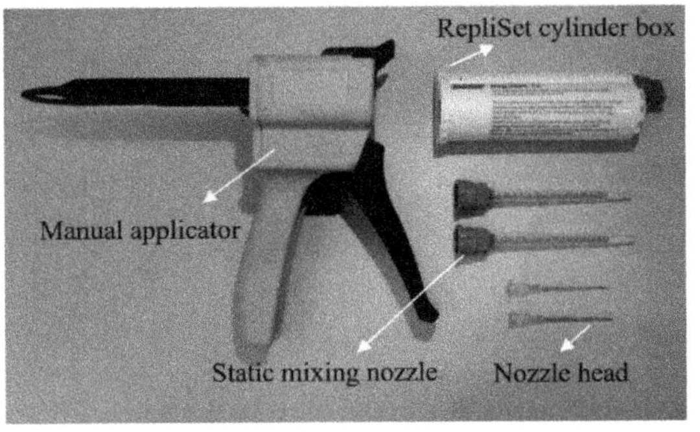

Figure 4. Replication system for surface replication.

The procedure of the test using the SR method was as follows [8]. Fatigue load was applied firstly. Then, the static tensile load was applied at 80% of the maximum fatigue load. After the static tensile load, the replication material was sprayed on the central area of the cruciform specimen and then a piece of paper with a size of about 35 × 25 mm^2 was pasted on the replication material to facilitate its removal. After the material was completely dry, the paper with the material recording the path of the fatigue crack was removed from the cruciform specimen. The interval of the spraying operations was reasonably selected according to the results of the fatigue test without SR method to ensure that there were about 20 effective copies of each cruciform specimen during its whole fatigue life. As shown in Figure 5, the biaxial fatigue test was terminated when the fatigue crack in the central area of the cruciform specimen extended to the edge of the central area. The microscope shown in Figure 6 successively measured the crack length on each replica from back to front and recorded the relative position of the crack to obtain the biaxial fatigue crack initiation position and the biaxial fatigue crack propagation path. The fatigue fracture morphology of the cruciform specimen was analyzed by scanning electron microscope. The crack projection length obtained from the replica measurement was α, and the definition of crack projection length is shown in Figure 7.

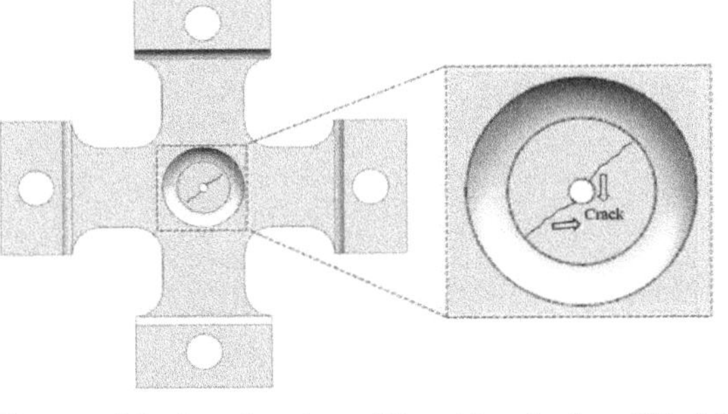

Figure 5. Criterion of crack condition at termination of biaxial fatigue test.

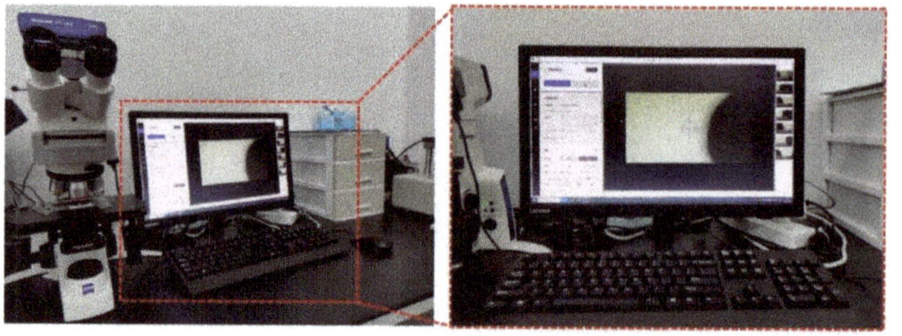

Figure 6. Microscope for observing fatigue crack remodeling.

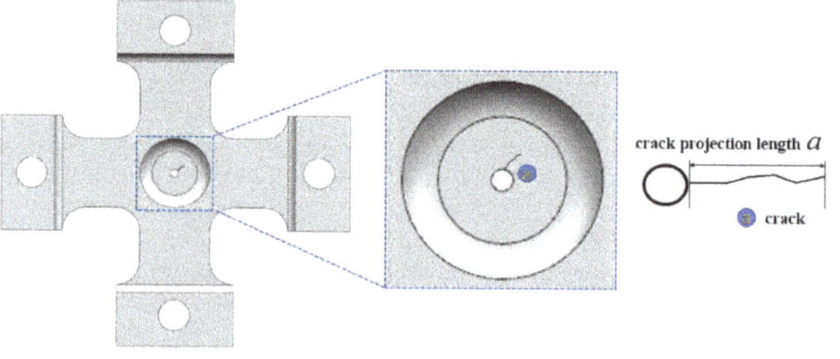

(a)

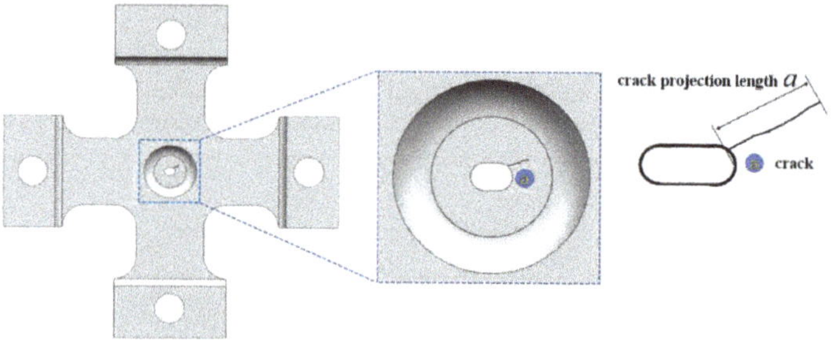

(b)

Figure 7. Crack projection length definition. (**a**) Crack projection length of circular notch specimen in central region. (**b**) Crack projection length of 0° waist circular hole notch specimen in central region.

3. Experimental Results and Discussion

3.1. Result of Fatigue Tests

The biaxial fatigue test results are shown in Table 3, where N_f is the full fatigue life of the cruciform specimen. N_{ini} is the fatigue crack initiation life of the cruciform specimen (fatigue crack initiation life is defined as the cycle when the crack is first detected [9]). $a_{crack\text{-}ini}$ is the crack length when the first crack is first observed, and N_{ini}/N_f is the ratio of fatigue crack initiation life to full fatigue life.

From the test results, it can be seen that the replication operation had no obvious effect on the biaxial fatigue life, and the fatigue life of the cruciform specimen with a circular notch in the central area was similar to that of the cruciform specimen with a 0° waist hole notch in the central area.

Table 3. Biaxial fatigue test results of cruciform specimens.

Specimen	N_f (Cycles)	N_{ini} (Cycles)	$a_{crack\text{-}ini}$ (μm)	N_{ini}/N_f	Number of Cracks
1	37,784	/	/	/	2
2	42,100	9000	193.9	21.38%	2
3	34,500	7700	193	22.32%	2
4	31,500	/	/	/	2
5	23,700	7000	42.4	29.53%	2
6	34,000	/	/	/	2
7	34,500	12,500	37.4	36.23%	2
8	21,500	/	/	/	2
9	24,900	8500	58.5	34.14%	2

The test data relating to when the second crack appeared on the cruciform specimen are shown in Table 4, where $N_{crack\text{-}2\text{-}ini}$ is the fatigue life of the cruciform specimen when the second crack appeared. $N_{crack\text{-}2\text{-}ini}/N_f$ is the ratio of fatigue life when the second crack appeared to full fatigue life, and $a_{crack\text{-}2\text{-}ini}$ is the length of the second crack when it was observed for the first time.

Table 4. Test data of cruciform specimen with the second crack.

Specimen Number	N_f (Cycles)	$N_{crack\text{-}2\text{-}ini}$ (Cycles)	$N_{crack\text{-}2\text{-}ini}/N_f$	$a_{crack\text{-}2\text{-}ini}$ (μm)
2	42,100	22,200	52.7%	100.7
3	34,500	18,500	53.6%	100.6
5	23,700	12,800	54.0%	26.70
7	34,500	23,000	66.6%	64.4
9	24,900	13,000	52.2%	36.3

From Table 4, it can be seen that the ratio of fatigue life when the second crack appeared to the full fatigue life of specimen 2 was about 52.7%, and the length of the second crack was about 100.7 μm. From Table 4, it can be seen that the ratio of fatigue life when the second crack appeared to the full fatigue life of specimen 7 was about 66.6%, and the length of the second crack was about 64.4 μm. The results of the other specimens were very close. To sum up, the ratio of the fatigue life when the second crack appeared to the full fatigue life was between 52% and 67%, which was similar for specimens with a circular notch in the central area or a 0° waist hole notch in the central area.

The final failure diagram of the cruciform specimen is shown in Figure 8, where (a) is the final failure diagram of the cruciform specimen with a circular notch in the central area, and (b) is the final failure diagram of the cruciform specimen with a 0° waist hole notch in the central area. From Figure 8, it can be seen that two cracks appeared in the central area of the cruciform specimen, and the propagation paths of the two cracks were basically symmetrical.

The crack naming method of the cruciform specimen is shown in Figure 9, where (a) is the crack naming method of the cruciform specimen with a circular notch in the central area, and (b) is the crack naming method of the cruciform specimen with a 0° waist round hole notch in the central area. In these methods, m and n are the specimen number. The black dot is the starting point of the fatigue crack propagation path. The angle between the crack path and the x-axis was measured with the use of ImageJ.

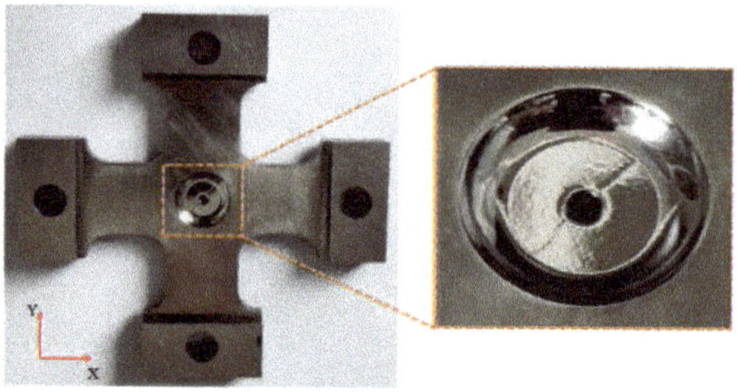

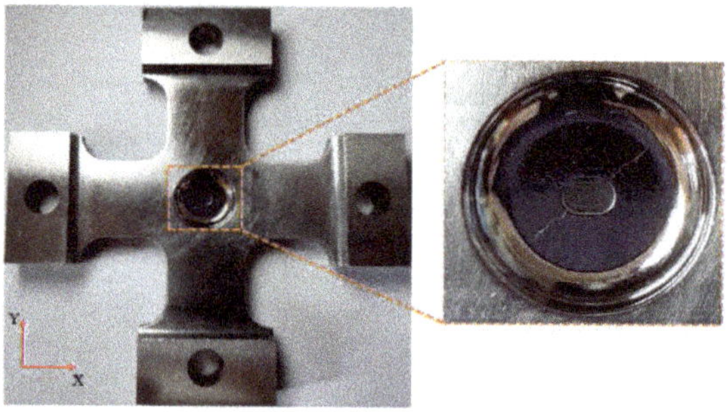

Figure 8. Final failure diagram of cruciform specimen. (**a**) Cruciform specimen with circular notch in the central area. (**b**) Cruciform specimen with 0° waist circle hole notch in the central area.

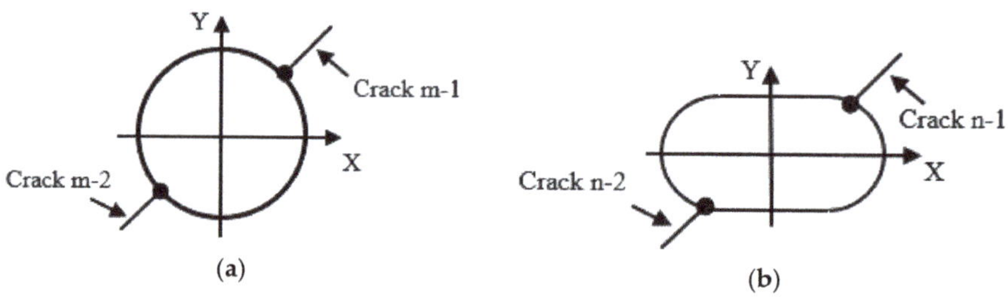

Figure 9. Crack naming method of cruciform specimen. (**a**) Cruciform specimen with circular notch in the central area. (**b**) Cruciform specimen with 0° waist circle hole notch in the central area.

The details of the angle are shown in Table 5. From Table 5, it can be seen that the angle between the crack propagation path and the x-axis on the cruciform specimen with a circular notch in the central area was 42–46°. The angle between the crack propagation path and x-axis on the cruciform specimen with a 0° waist round hole notch in the central area was 40–48°. The starting point of the fatigue crack propagation path is the fatigue crack initiation point, which is also the fatigue dangerous point in terms of the physical mechanism of fatigue damage. The fatigue dangerous point of the cruciform specimen under biaxial equal proportion fatigue load was basically consistent with the position of the maximum principal stress point on the notch edge, which is consistent with Chaves et al.'s conclusions relating to the crack initiation point under torsional and proportional loading [10].

Table 5. Angle between crack propagation path and *x*-axis in the cruciform specimen.

Specimen Number	Crack Number	Angle
1	Crack 1-1	45.2°
	Crack 1-2	44.3°
2	Crack 2-1	45.3°
	Crack 2-2	43.3°
3	Crack 3-1	43.2°
	Crack 3-2	45.5°
4	Crack 4-1	44.1°
	Crack 4-2	42.3°
5	Crack 5-1	43.5°
	Crack 5-2	44.3°
6	Crack 6-1	40.7°
	Crack 6-2	42.9°
7	Crack 7-1	40.5°
	Crack 7-2	41.4°
8	Crack 8-1	44.4°
	Crack 8-2	47.5°
9	Crack 9-1	40.1°
	Crack 9-2	42.6°

Figure 10 is a replica photo of crack 2-1's crack initiation and propagation process; Figure 11 is the replica photos of crack 2-2's crack initiation and propagation process. This group of photos shows the crack initiation and propagation path process of cruciform specimen 2 with a circular notch in the central area under biaxial fatigue load.

Figure 12 is a replica photo of crack 7-1's crack initiation and propagation process; Figure 13 is the replica photos of crack 7-2's crack initiation and propagation process. This group of photos shows the crack initiation and propagation path process of cruciform specimen 7 with a 0° waist hole notch in the central area under biaxial fatigue load.

From Figure 10a, it can be seen that crack 2-1 originated from the circular notch. When n = 9000 cycles, crack 2-1 was monitored for the first time, and the crack length was about 193.9 μm. From Figure 10b–h, it can be seen that, after crack 2-1 sprouted from the circular notch, the crack gradually expanded, and the crack path deflected many times during the crack propagation process. From Figure 11a, it can be seen that crack 2-2 originated from the circular notch. When n = 22,200 cycles, crack 2-2 was monitored for the first time. At this time, the crack length of crack 2-2 was about 100 μm. From Figure 11b–h, it can be seen that, after crack 2-2 sprouted from the circular notch, the crack gradually expanded, and the crack path deflected many times during the crack propagation process. From Figure 12a, it can be seen that crack 7-1 originated from the notch of the 0° waist hole. When n = 12,500 cycles, crack 7-1 was monitored for the first time, and the crack length was about 37.4 μm. From Figure 12b–h, it can be seen that, after crack 7-1 initiated from the notch of the 0° waist hole, the crack gradually expanded, and the crack path deflected many times during the crack propagation process. From Figure 13a, it can be seen that crack 7-2 originated from the notch of the 0° waist hole. When n = 23,000 cycles, crack 7-2 was monitored for the first time. At this time, the crack length of crack 7-2 was about 64.4 μm. From Figure 13b–h, it can be seen that, after crack 7-2 sprouted from the notch of the 0° waist hole, the crack gradually expanded, and the crack path deflected many times during the crack propagation process.

To sum up, the crack of the cruciform specimen initiated from the notch. After the crack initiation, the crack gradually expanded, and the crack propagation path deflected many times during the propagation process. There were different orientations between adjacent grains or small defects somewhere on the crack growth path, which may have led to the deflection of the crack growth path, which is consistent with the research on the fatigue crack growth mechanism of nickel-based alloy GH4169 under uniaxial tensile fatigue load by Zhu et al. [11].

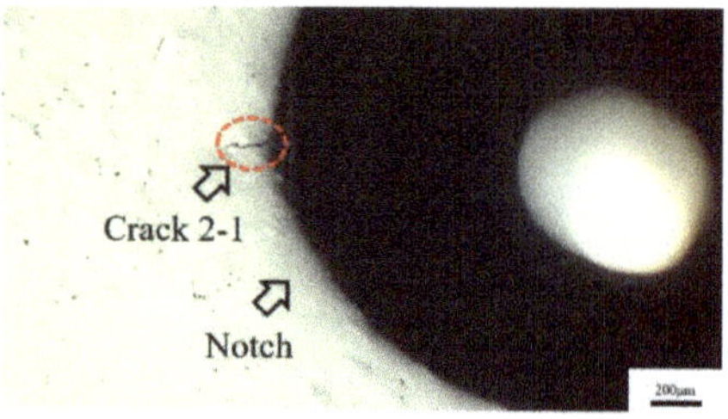

(**a**) $N = 9000$

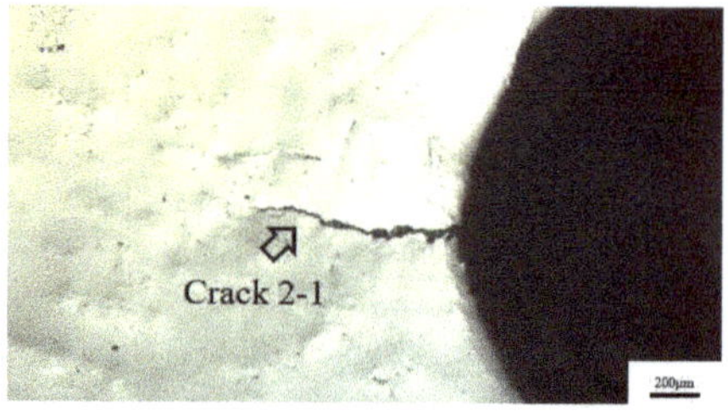

(**b**) $N = 16{,}000$

(**c**) $N = 19{,}800$

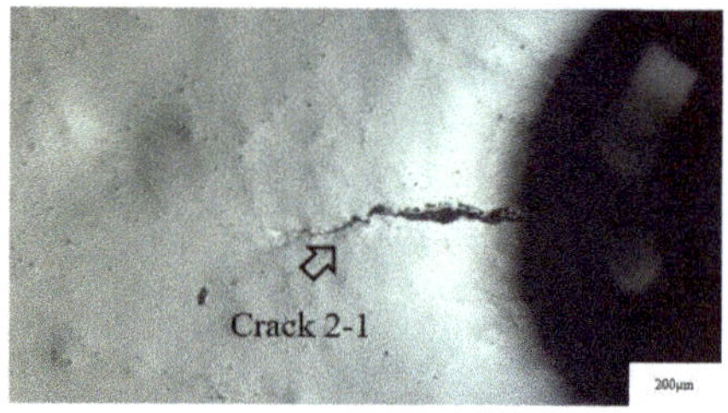

(**d**) $N = 22{,}200$

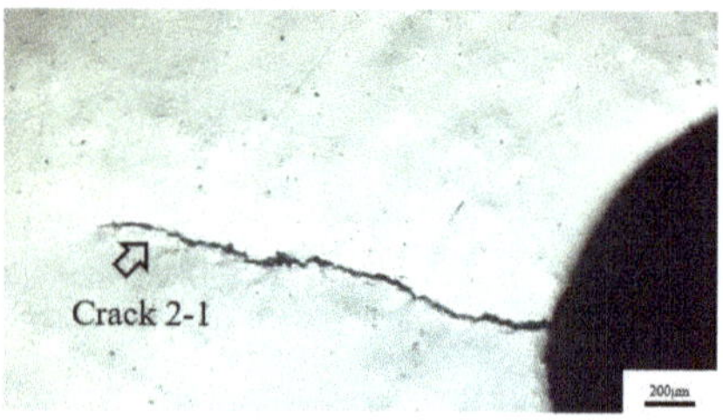

(**e**) $N = 25{,}600$

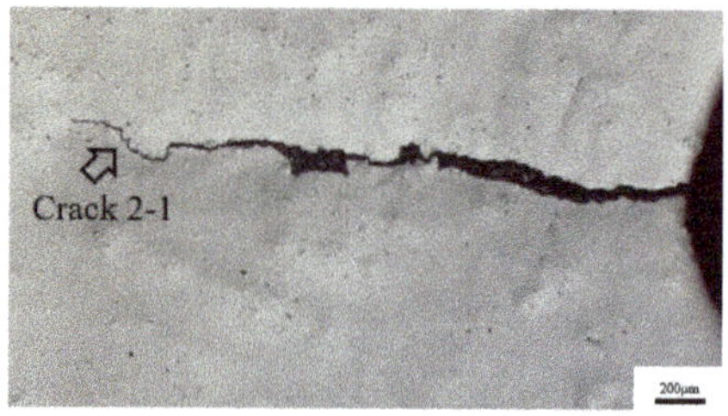

(**f**) $N = 28{,}600$

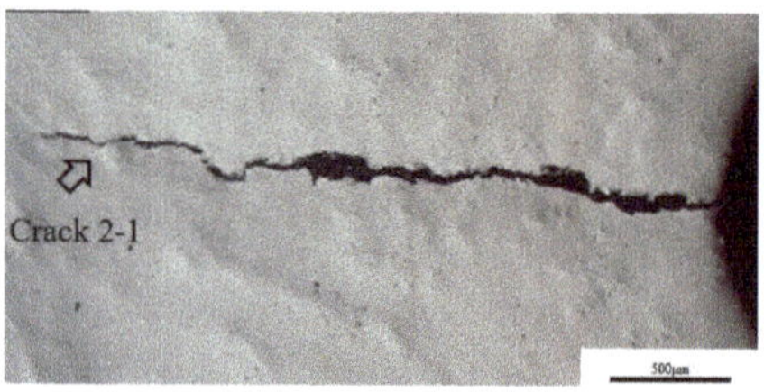

(**g**) $N = 31{,}800$

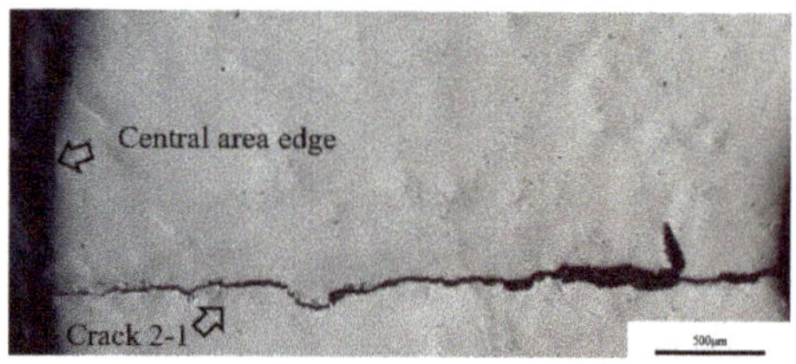

(**h**) $N = 34{,}200$

Figure 10. Crack 2-1 initiation and propagation process.

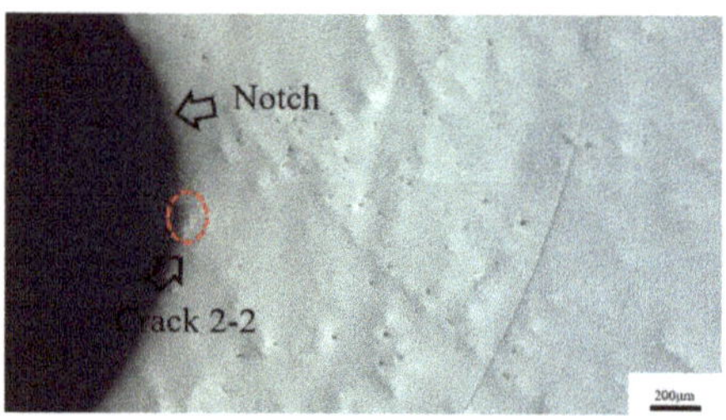

(**a**) $N = 22200$

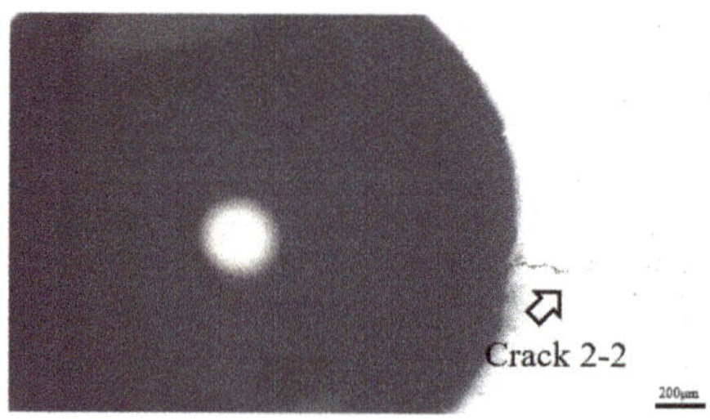

(**b**) $N = 25{,}600$

Figure 11. *Cont.*

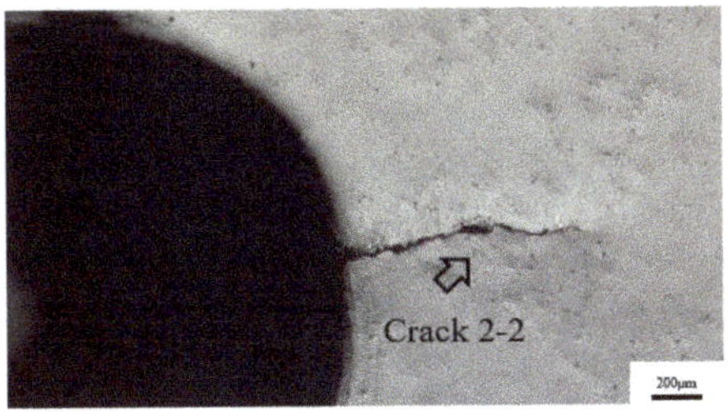

(c) $N = 29{,}400$

(d) $N = 32{,}600$

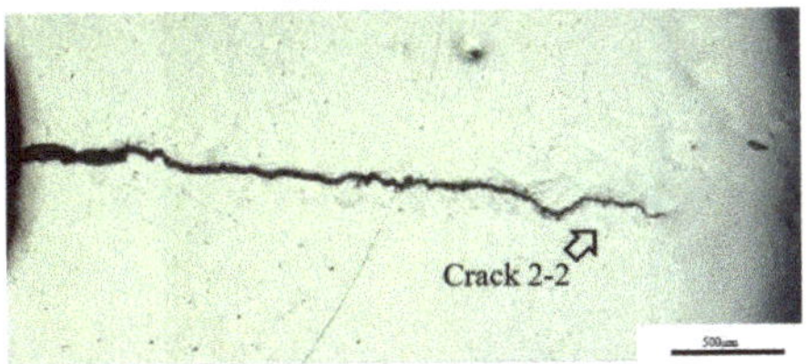

(e) $N = 38{,}700$

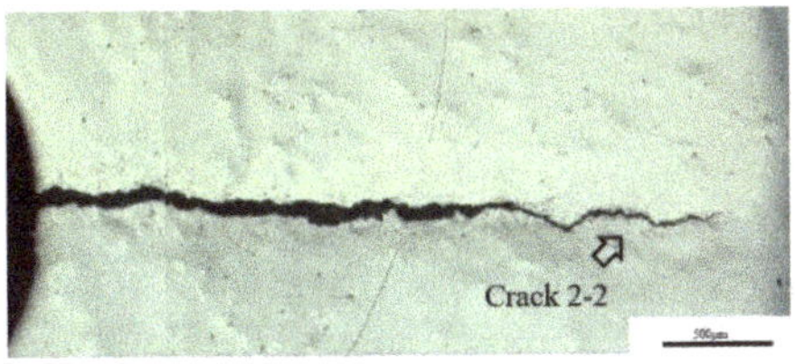

(f) $N = 40{,}000$

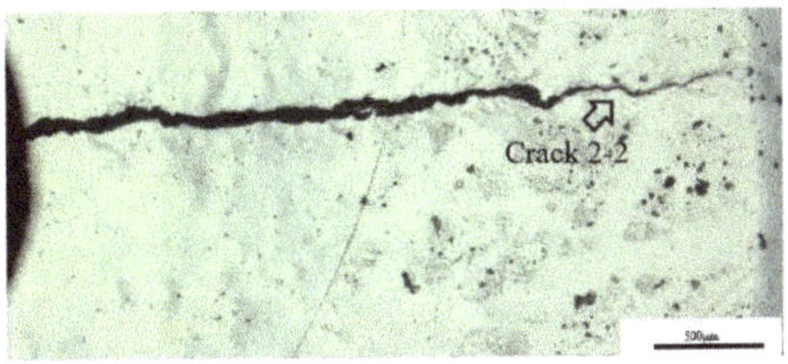

(g) $N = 40{,}800$

(h) $N = 42{,}100$

Figure 11. Crack 2-2 initiation and propagation process.

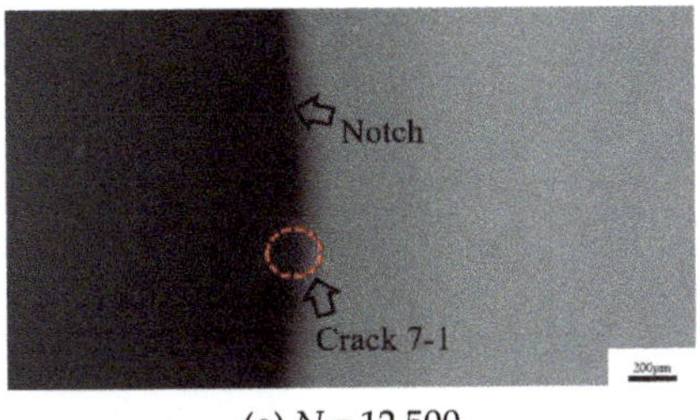

(a) $N = 12{,}500$

(b) $N = 17{,}000$

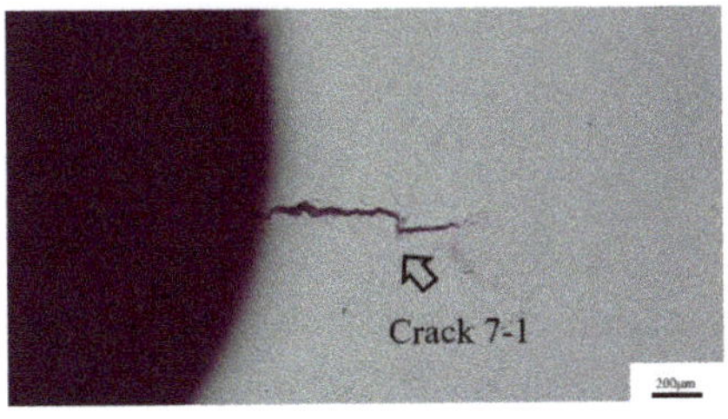

(c) $N = 23{,}000$

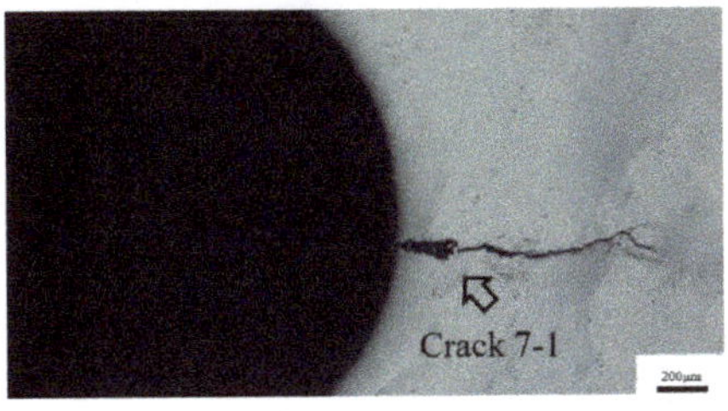

(d) $N = 26{,}000$

Figure 12. *Cont.*

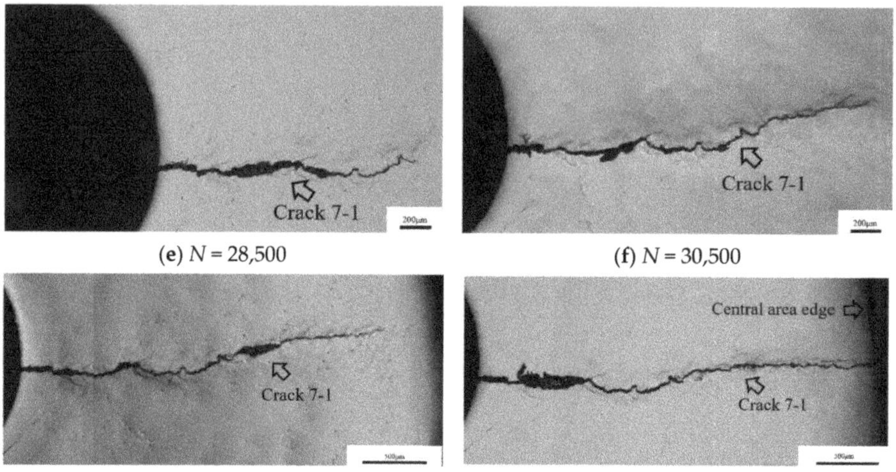

(**e**) $N = 28{,}500$ (**f**) $N = 30{,}500$

(**g**) $N = 32{,}100$ (**h**) $N = 34{,}500$

Figure 12. Crack 7-1 initiation and propagation process.

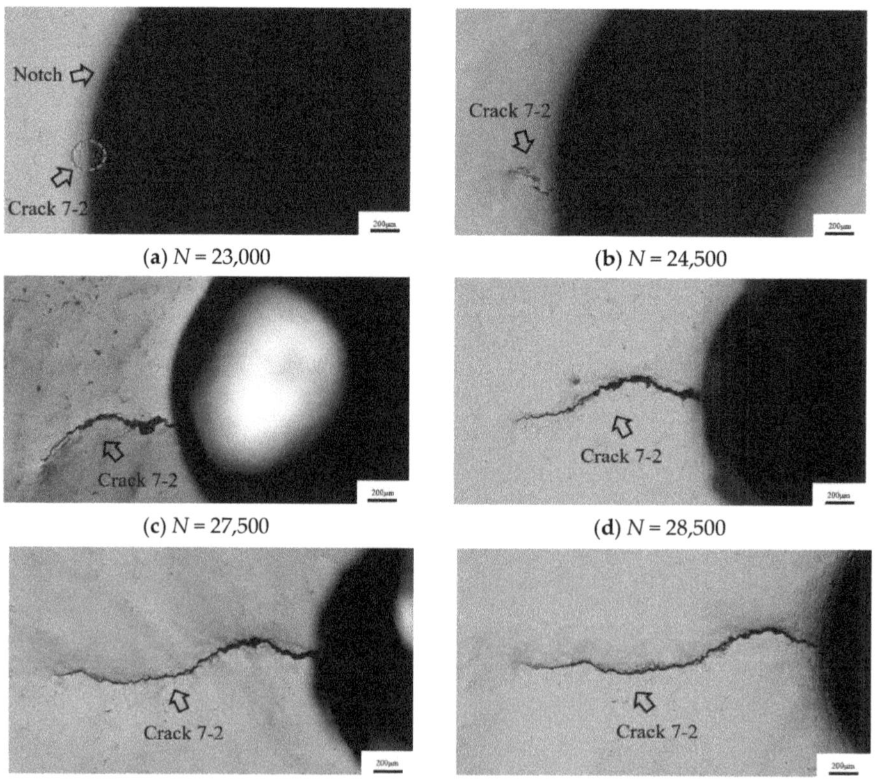

(**a**) $N = 23{,}000$ (**b**) $N = 24{,}500$

(**c**) $N = 27{,}500$ (**d**) $N = 28{,}500$

(**e**) $N = 30{,}500$ (**f**) $N = 31{,}300$

Figure 13. *Cont.*

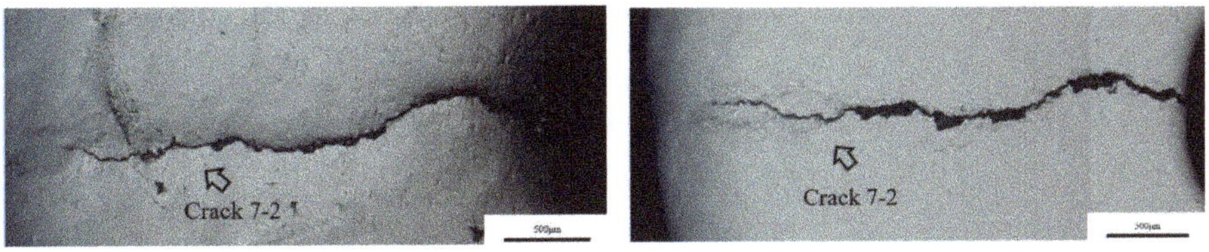

(**g**) $N = 33{,}700$ (**h**) $N = 34{,}500$

Figure 13. Crack 7-2 initiation and propagation process.

3.2. *Biaxial Fatigue Crack Growth Behavior*

The variation of the fatigue crack length of the cruciform specimen with a circular notch in the central area is shown in Figure 14, where (a) is the variation diagram of fatigue crack length with the number of cycles, and (b) is the variation diagram of fatigue crack length with cycle ratio. The variation of the crack length of the cruciform specimen with a 0° waist round hole notch in the central area is shown in Figure 15, where (a) is the variation diagram of the fatigue crack length with the number of cycles, and (b) is the variation diagram of the fatigue crack length with cycle ratio.

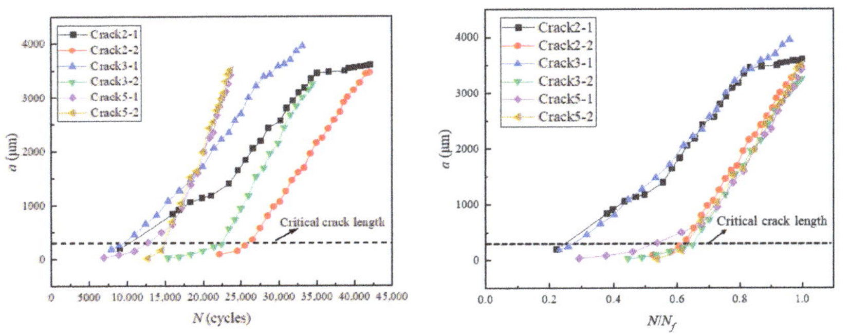

(**a**) (**b**)

Figure 14. Variation of crack length of cruciform specimen with circular notch in the central area. (**a**) Variation of crack length with cycle number. (**b**) Variation of crack length with cycle ratio.

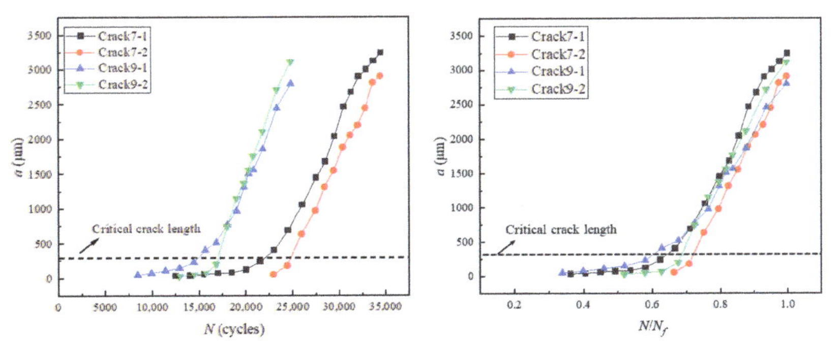

(**a**) (**b**)

Figure 15. Variation of crack length of cruciform specimen with 0° waist circle hole notch in the central area. (**a**) Variation of crack length with cycle number. (**b**) Variation of crack length with cycle ratio.

From Figures 14a and 15a, it can be seen that there is a critical crack length. When the crack length is less than this critical crack length, the crack propagation is very slow. When the crack length exceeds the critical crack length, the crack expands rapidly and leads to the failure of the cruciform specimen. Zhu et al. [12] considered that the critical fatigue crack length of nickel-based alloy GH4169 under uniaxial loading is about 500 μm when studying the small crack initiation and propagation mechanism of nickel-based alloy GH4169 using unilateral notch tensile specimens. Deng et al. [13] considered that the critical fatigue crack length of GH4169 under uniaxial tensile load is about 200 μm when studying the effect of grain size on the small crack initiation and propagation mechanism of nickel-based alloy GH4169. According to Figures 14a and 15a, it can be seen that, under the biaxial fatigue load conditions adopted in this paper, the critical crack length of nickel-based alloy GH4169 under biaxial proportional fatigue load is about 300 μm.

The variation of the biaxial fatigue crack length with cycle ratio (N/N_f) of the cruciform specimen with a circular notch in the central area is shown in Figure 14b, and the variation of the biaxial fatigue crack length with cycle ratio (N/N_f) of the cruciform specimen with a 0° waist circle hole notch in the central area is shown in Figure 15b, where N is the number of fatigue loading cycles, and N_f is the full fatigue life.

From Figure 14b, it can be seen that the variation curves of the two crack lengths of specimen 2 and specimen 3 with the cycle ratio are almost coincident. The cycle times of the first crack of specimen 2 and specimen 3 when it changed from slow growth to rapid growth were about 25% of the whole fatigue life, and the cycle times of the first crack of specimen 5 when it changed from slow growth to rapid growth were about 55% of the whole fatigue life. The difference between them was caused by the magnitude of load. When the load is larger, the crack growth rate in rapid growth is faster, and the rapid growth's proportion of the whole fatigue life is reduced. The cycle times of the second crack of specimen 2, specimen 3 and specimen 5 when it changed from slow propagation to rapid propagation were about 65% of the whole fatigue life. The difference between the first and second crack of specimen 2 and specimen 3 was because the existing crack could accelerate the growth rate of the second crack in rapid growth. From Figure 15b, it can be seen that the cycle times when the first crack of specimen 7 and specimen 9 changed from slow growth to rapid growth were about 65% of the whole fatigue life, and the cycle times when the second crack of specimen 7 and specimen 9 changed from slow growth to rapid growth were about 70% of the whole fatigue life. This difference was not that big because the shape of a waist circle hole notch is not as sensitive to load and crack sequence as that of a circular notch.

The secant method was used to calculate the crack growth rate:

$$\left(\frac{da}{dN}\right) = \frac{\Delta a}{\Delta N} = \frac{a_{i+1} - a_i}{N_{i+1} - N_i} \tag{1}$$

where Δa is the difference of crack length, ΔN is the interval of cycle times, a_i is the crack length at the number of N_i cycles, which is the average crack length, and $\overline{a} = (a_{i+1} + a_i)/2$.

The variation of the biaxial fatigue crack growth rate of the cruciform specimen with a circular notch in the central area is shown in Figure 16, where (a) is the change of crack growth rate with crack length, and (b) is the variation of crack growth rate with cycle ratio. The variation of the biaxial fatigue crack growth rate of the cruciform specimen with a waist hole notch in the central area is shown in Figure 17, where (a) is the variation of crack growth rate with crack length, and (b) is the variation of crack growth rate with cycle ratio.

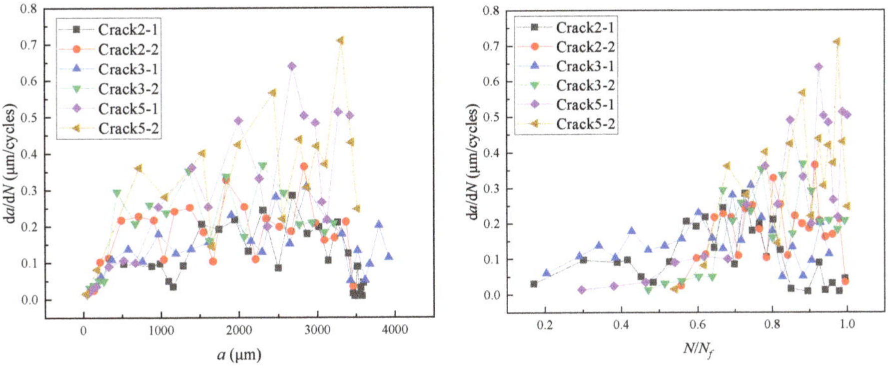

(a) (b)

Figure 16. Variation of crack growth rate of cruciform specimen with circular notch in the central area. (**a**) Variation of crack growth rate with crack length. (**b**) Variation of crack growth rate with cycle ratio.

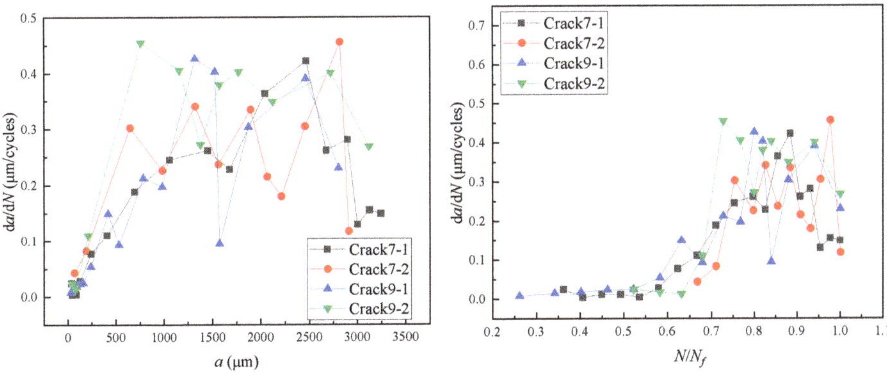

(a) (b)

Figure 17. Variation of crack growth rate of cruciform specimen with 0° waist hole notch in the central area. (**a**) Variation of crack growth rate with crack length. (**b**) Variation of crack growth rate with cycle ratio.

Suresh [14] classified small cracks into microstructure small cracks, mechanical small cracks, physical small cracks and chemical small cracks. The length of physical small cracks is generally less than 0.5–1 mm. In this paper, the size of small cracks is defined as less than 1 mm. From Figures 16a and 17a, it can be seen that the biaxial fatigue crack growth rate of the cruciform specimen fluctuated. In the early stage of small fatigue crack propagation, when the crack length was close to the grain size of the material, the crack propagation was affected by the microstructure properties such as grain orientation and grain boundary morphology. When the angle between the fatigue crack tip and grain boundary was small, the inhibitory effect on the fatigue crack propagation caused by grain boundary was relatively poor. When the angle was large, the inhibitory effect was intense and could even make the crack propagation stop. This is consistent with the conclusion reached by Zhu [11] and Wu [15] when studying the fatigue crack growth behavior of nickel-based alloy GH4169.

The fluctuation of fatigue long crack growth rate is related to the specimen structure, crack closure effect and crack deflection. The central area of the cruciform specimen selected in this paper had equal thickness thinning on both sides, and the thinning transition area of the cruciform specimen had a circular chamfer transition, that is, the thickness of the edge of the central area was larger than the thickness of the central area, which led to the phenomenon that the crack propagation rate decreased during the propagation of the long fatigue crack. The crack closure effect also led to the reduction or retardation of fatigue crack growth rate. Neerukatti et al. [16] found that crack closure reduced the fatigue crack growth rate when studying the fatigue crack growth of an Al7075-T651 cruciform specimen under biaxial tensile load. Elber [17] found through experimental research that, even if the far-field load is a tensile load, the fatigue crack is closed. Under the action of crack closure, the crack surfaces contact in advance, resulting in the arrest of fatigue crack growth.

3.3. Biaxial Fatigue Crack Deflection Behavior

Figure 18 is the relationship between the crack propagation path and the crack propagation rate of cruciform specimen 3 with a circular notch in the central area. The white dot is the crack deflection point, *a* is the crack length and da/dN is the crack propagation rate. Based on Figure 18a, the decrease in crack growth rate between points 1 and 2, 3 and 4, 5 and 7 and 10 and 13 was related to the deflection of the crack growth path. Based on Figure 18b, the decrease in crack growth rate between points 1 and 2, 3 and 4, 5 and 6, 7 and 8 and 9 and 12 was related to the deflection of the crack growth path. The reduction in crack growth rate under biaxial fatigue loading was related to the deflection of the crack growth path. The effective stress intensity factor of the deflection crack was less than that of a straight crack with the same projection length. When the loading amplitude is the same, the deflection behavior of the fatigue crack path significantly reduces the growth rate of the fatigue crack [18]. The conclusion that the deflection of crack propagation path leads to the reduction of crack propagation rate is consistent with the research results of Bui et al. [19,20] on crack bifurcation and crack deflection.

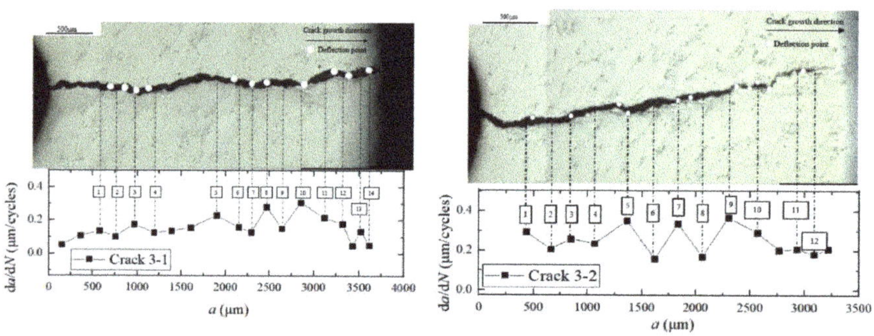

(**a**) Crack 3-1. (**b**) Crack 3-2.

Figure 18. Relationship between crack growth path and crack growth rate.

3.4. Fracture Morphology Analysis

The fatigue fracture morphology of the cruciform specimen with a circular notch in the central area was observed and analyzed by scanning electron microscope (SEM). The fatigue fracture morphology is shown in Figures 19–22. Figure 19 is the overall morphology near the root of the circular notch in the central area of the cruciform specimen. Figure 20 shows the source area of fatigue crack initiation. From Figure 20, it can be seen that there were many crack initiation sources at the notch root of the cruciform specimen. Fatigue crack propagation can be divided into two stages: the first stage of fatigue crack propagation and the second stage of fatigue crack propagation. Figure 21 shows the characteristics of the first stage of fatigue crack growth. From Figure 21, it can be seen that the first stage of fatigue crack propagation of nickel-based alloy GH4169 presented a parallel, serrated section and a cleavage-like fracture plane. Figure 22 shows the characteristics of the second stage of fatigue crack growth. Fatigue bands can be found in Figure 22, which are perpendicular to the crack propagation direction. It is generally believed that one load cycle corresponds to one fatigue band. Therefore, some scholars [21] measure the fatigue band spacing to calculate the crack growth rate. The results show that this method can also effectively estimate the fatigue crack growth rate.

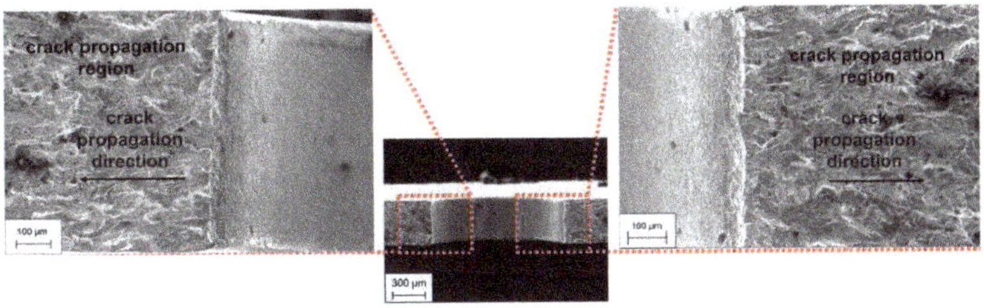

Figure 19. Overall morphology near circular notch.

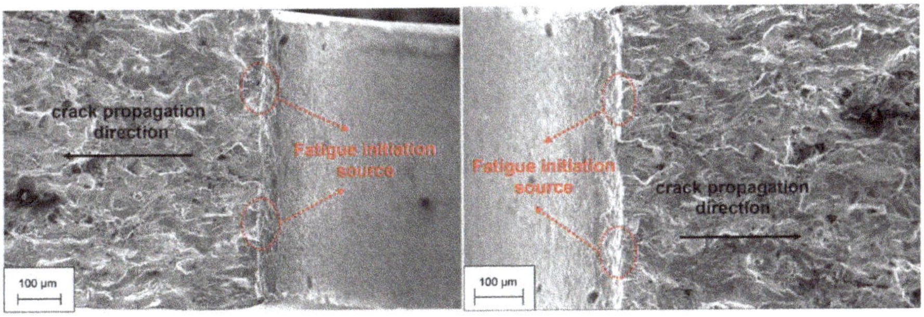

Figure 20. Fatigue crack initiation source region.

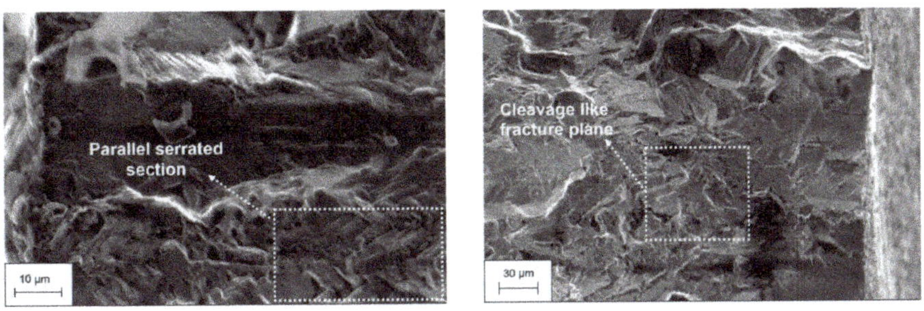

Figure 21. Characteristics of the first stage of fatigue crack growth.

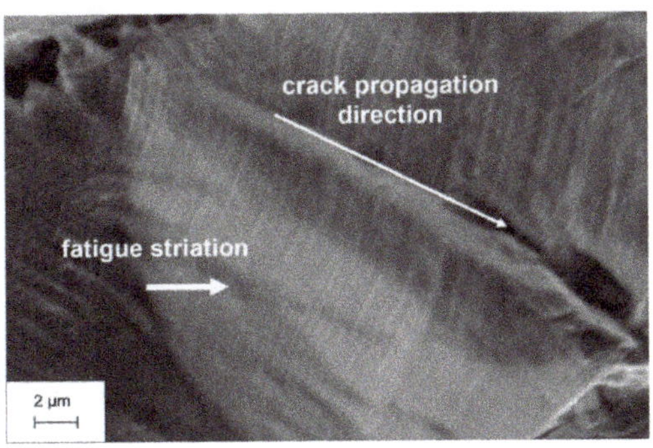

Figure 22. Characteristics of the second stage of fatigue crack growth.

4. Conclusions

Based on the biaxial fatigue test results and the fatigue crack initiation and propagation process in the central area of the cruciform specimen monitored and recorded by the replica method, the crack propagation path, propagation behavior and deflection behavior of nickel-based alloy GH4169 under biaxial fatigue load were studied, and the fatigue fracture morphology was observed and analyzed by scanning electron microscope. The result can help to develop a fatigue crack life analysis method and damage tolerance design method for key aeroengine components. The conclusions are as follows:

(1) The crack of the cruciform specimen initiated from the notch. In the process of propagation, the crack propagation path deflected many times. In the early stage of small crack propagation, the crack propagation rate fluctuated greatly due to the influence of microstructure properties such as grain orientation and grain boundary morphology. At the stage of long crack growth, the crack growth rate fluctuated due to the specimen structure and crack deflection. The deflection of the crack propagation path reduced the crack propagation rate;

(2) Under biaxial proportional fatigue loading, the critical crack length of nickel-based alloy GH4169 during crack propagation is about 300 μm. When the crack length is less than 300 μm, the crack grows slowly, and when the crack length is larger than 300 μm, the crack propagates rapidly;

(3) There were many crack initiation sources at the notch root of the cruciform specimen. The propagation of a biaxial fatigue crack is divided into the first stage and the second stage. The first stage of fatigue crack propagation presents a parallel, serrated section and cleavage-like fracture small plane; the second stage of fatigue crack propagation presents fatigue bands, and the fatigue bands are perpendicular to the crack propagation direction;

(4) The method of combining a uniaxial tensile testing machine and biaxial fixture device has the advantage of low cost and easy access.

Author Contributions: Conceptualization, Z.W.; methodology, L.F.; software, H.L.; validation, Y.P. and H.L.; formal analysis, S.W. and Y.P.; investigation, S.W.; resources, H.L. and L.F.; data curation, S.W.; writing—original draft preparation, Y.P. and S.W.; writing—review and editing, Y.P.; visualization, S.W. and Y.P.; supervision, L.F.; project administration, Z.W.; funding acquisition, Z.W. All authors have read and agreed to the published version of the manuscript.

Funding: This research was funded by the Major National Science and Technology Project grant number J2019-IV-0008-0076.

Data Availability Statement: Data not available due to ethical restrictions.

Conflicts of Interest: The authors declare that they have no conflict of interest.

References

1. Wolf, C.H.; Henkel, S.; Burgold, A.; Qiu, Y.X.; Kuna, M.; Biermann, H. Investigation of fatigue crack growth under in-phase loading as well as phase-shifted loading using cruciform specimens. *Int. J. Fatigue* **2019**, *124*, 595–617. [CrossRef]
2. Meng, Y.X.; Gao, H.; Yan, Y.Z.; Gao, L.L. Effects of phase difference and stress ratio on biaxial tension–tension fatigue crack propagation behavior of rolled ZK60 magnesium alloy. *Mater. Today Commun.* **2020**, *24*, 101159. [CrossRef]
3. Abecassis, M.; Köster, A.; Esin, V.A.; Chiaruttini, V.; Maurel, V. Crack Growth Behavior in Dissimilar Welded Ti Based Alloys under Biaxial Fatigue Loading. *Int. J. Fatigue* **2018**, *118*, 209–224. [CrossRef]
4. Misak, H.E.; Perel, V.Y.; Sabelkin, V.; Mall, S. Biaxial tension–tension fatigue crack growth behavior of 2024-T3 under ambient air and salt water environments. *Eng. Fract. Mech.* **2014**, *118*, 83–97. [CrossRef]
5. Limin, S.; Chao, Z. Experimental Research on Turning of Superalloy GH4169 under High Pressure Cooling Condition. *Integr. Ferroelectr.* **2020**, *207*, 75–85. [CrossRef]
6. Xu, R.F.; Zhou, Y.X.; Li, X.; Yang, S.L.; Han, K.N.; Wang, S.J. The Effect of Milling Cooling Conditions on the Surface Integrity and Fatigue Behavior of the GH4169 Superalloy. *Metals* **2019**, *9*, 1179. [CrossRef]
7. Wu, D.X.; Zhang, D.H.; Yao, C.F. Effect of Turning and Surface Polishing Treatments on Surface Integrity and Fatigue Performance of Nickel-Based Alloy GH4169. *Metals* **2018**, *8*, 549. [CrossRef]
8. Zhu, L.; Wu, Z.R.; Hu, X.T.; Song, Y.D. Small Fatigue Crack Growth Behavior of Nickel-Based Alloy GH4169. *J. Aeronaut. Power* **2017**, *32*, 1984–1991. Available online: https://www.webofscience.com/wos/alldb/full-record/CSCD:6079438 (accessed on 23 August 2017).

9. Zhu, L.; Hu, X.T.; Jiang, R.; Song, Y.D.; Qu, S.D. An investigation of small fatigue crack behavior in titanium alloy TC4 under different stress levels. *Proc. Inst. Mech. Eng. Part G J. Aerosp. Eng.* **2019**, *233*, 5567–5578. [CrossRef]
10. Chaves, V.; Beretta, G.; Balbín, J.A.; Navarro, A. Fatigue life and crack growth direction in 7075-T6 aluminium alloy specimens with a circular hole under biaxial loading. *Int. J. Fatigue* **2019**, *125*, 222–236. [CrossRef]
11. Zhu, L.; Wu, Z.R.; Hu, X.T.; Song, Y.D. Investigation of small fatigue crack initiation and growth behaviour of nickel base superalloy GH4169. *Fatigue Fract. Eng. Mater. Struct.* **2016**, *39*, 1150–1160. [CrossRef]
12. Zhu, L.; Wu, Z.; Hu, X.T.; Song, Y.D. Comparative study of small crack growth behavior between specimens with and without machining-induced residual stress of alloy GH4169. *J. Mech. Sci. Technol.* **2018**, *32*, 5251–5261. [CrossRef]
13. Deng, G.J.; Tu, S.T.; Zhang, X.C.; Wang, Q.Q.; Qin, C.H. Grain size effect on the small fatigue crack initiation and growth mechanisms of nickel-based superalloy GH4169. *Eng. Fract. Mech.* **2015**, *134*, 433–450. [CrossRef]
14. Suresh, S. Crack deflection: Implications for the growth of long and short fatigue cracks. *Metall. Trans. A* **1983**, *14*, 2375–2385. [CrossRef]
15. Wu, Z.R.; Liu, F.L.; Li, X.; Fang, L.; Song, Y.D. Multistage fatigue modeling of single-edge-notch tension specimens for Ni-based superalloy GH4169. *Adv. Mech. Eng.* **2017**, *9*, 168781401772994. [CrossRef]
16. Neerukatti, R.K.; Datta, S.; Chattopadhyay, A.; Iyyer, N.; Phan, N. Fatigue crack propagation under in-phase and out-of-phase biaxial loading. *Fatigue Fract. Eng. Mater. Struct.* **2018**, *41*, 387–399. [CrossRef]
17. Elber, W. *The Significance of Fatigue Crack Closure*; ASTM International: West Conshohocken, PA, USA, 1971; p. 486. [CrossRef]
18. Wang, Y.; Huang, H.; Gao, K.Y.; Wen, S.P.; Zhang, P.P.; Nie, Z.R. Fatigue Fracture Behavior of Cold Rolled 5E06 and 5E83 Aluminum Alloy Plate. *Chin. J. Rare Met.* **2011**, *35*, 791–798. Available online: https://www.webofscience.com/wos/alldb/full-record/CSCD:4369885 (accessed on 1 June 2011).
19. Bui, T.P.; Miyashita, Y.; Mutoh, Y.; Morikage, Y.; Tagawa, T.; Handa, T.; Otsuka, Y. Fatigue crack deflection and branching behavior of low carbon steel under mechanically large grain condition. *Int. J. Fatigue* **2021**, *148*, 106217. [CrossRef]
20. Bui, T.P.; Miyashita, Y.; Morikage, Y.; Tagawa, T.; Handa, T.; Mutoh, Y.; Otsuka, Y. Contributions of Grain Size and Crystal Orientation to Fatigue Crack Deflection and Branching Behavior in Low Carbon Steel Plates. *ISIJ Int.* **2020**, *61*, 424–433. [CrossRef]
21. Caton, M.J.; Jhon, R.; Porter, W.J.; Burba, M.E. Stress ratio effects on small fatigue crack growth in Ti-6Al-4V. *Int. J. Fatigue* **2011**, *38*, 36–45. [CrossRef]

metals MDPI

Article

Change in the Properties of Rail Steels during Operation and Reutilization of Rails

Kassym Yelemessov [1], Dinara Baskanbayeva [1], Nikita V. Martyushev [2,*], Vadim Y. Skeeba [3], Valeriy E. Gozbenko [4,5] and Antonina I. Karlina [6]

1 Institute of Energy and Mechanical Engineering, Satbayev University, Almaty 050013, Kazakhstan; k.yelemessov@satbayev.university (K.Y.); d.baskanbayeva@satbayev.university (D.B.)
2 Department of Advanced Technologies, Tomsk Polytechnic University, 634050 Tomsk, Russia
3 Department of Industrial Machinery Design, Novosibirsk State Technical University, 630073 Novosibirsk, Russia; skeeba_vadim@mail.ru
4 Department of Organization of Transportation and Management on Motor Transport, Angarsk State Technical University, 60, Chaykovskogo str., 665835 Angarsk, Russia
5 Department of Mathematics, Irkutsk State Transport University, 15, Chernyshevskiy str., 664074 Irkutsk, Russia
6 Stroytest Research and Testing Center, Moscow State University of Civil Engineering, 26, Yaroslavskoye Shosse, 129337 Moscow, Russia
* Correspondence: martjushev@tpu.ru

Abstract: The paper considers the possibility of reusing previously used railway rails. The analysis is conducted using the standards and operating conditions of the rails of one of the Central Asian states, Kazakhstan, as an example. The operation of these rails causes significant stresses, while the surface layers are strengthened as a result of cold hammering. These phenomena significantly change the physical and mechanical characteristics of rails. As a result, they may not be suitable in terms of parameters for basic use but can be suitable for installation on other tracks. The conducted studies have shown that when the standard service life of the RP65 rail expires, the surface layer is deformed to a depth of up to 300 microns, hardness increases, and internal residual stresses are formed. These changes lead to an increase in the strength properties of the rails. However, at the same time, cracks originate in the surface layer of the rail, thus worsening operational characteristics. The RP65 rails are used under a cyclic load of 700 kN (which is determined by the national standard), withstanding 790,000 cycles. When the load is reduced to 510 kN, these rails can withstand the 2,000,000 cycles required by the standard without failure. Thus, these rails can be reutilized only on non-loaded and non-critical sections.

Keywords: rails; rail steel; reutilization; fatigue life; rail reliability

Citation: Yelemessov, K.; Baskanbayeva, D.; Martyushev, N.V.; Skeeba, V.Y.; Gozbenko, V.E.; Karlina, A.I. Change in the Properties of Rail Steels during Operation and Reutilization of Rails. *Metals* **2023**, *13*, 1043. https://doi.org/10.3390/met13061043

Academic Editor: Haitao Cui

Received: 20 April 2023
Revised: 20 May 2023
Accepted: 27 May 2023
Published: 30 May 2023

1. Introduction

Rail transport plays a crucial role in the economies of the majority of countries, providing a significant share of all cargo turnovers and the transportation of most export and transit goods. This share in Central Asian countries may reach 50% or more. Railway rails perform an important function: the elastic transformation and transfer of loads proceeding from wheels to tracks, directing the running gears of locomotives and cars [1]. High requirements are imposed on the quality of rails during the operation of the railway track [2,3]. R65 rails are used in the manufacture of rail tracks in the most massive sections of traffic (main tracks) in Central Asia, such as the sections for cargo and passenger traffic. The speed of passenger trains in these sections reaches its maximum. In addition, freight trains are also involved in operations [4]. In such areas, steel with sufficiently good plastic properties and relatively low hardness (HB = 360–380 MPa) is used as the rail material. These rails are made of low-alloy carbon steel with a ferrite-pearlite structure [5]. There are many other sections in addition to the main railway, including industrial sections, safety dead ends, crossovers, turnouts, marshalling, engine tracks, and station tracks, that have completely different requirements for rails [6,7]. Of the listed types of tracks, industrial ones make up a considerable proportion of their total number (ranking second as compared to the main tracks). Rails of the RP65 type are used on these tracks. Their difference from R65 rails consists of the type of material used in their production. The chemical composition of the RP65 material differs by having a higher content of sulfur and phosphorus as compared to that of R65. Owing to this fact, R65 has a higher plasticity and a lower tendency to crack. The same low-alloy carbon steel with a ferrite-pearlite structure is used to produce RP65 rails. The proportion of doped components in such steel is slightly lower, and the number of sulfur and phosphorus impurities is slightly higher, though such steel also has a slightly higher hardness and lower ductility compared to R65 steel [8]. The same material is often used for RP65 rails and R65 rails when the controlled parameters of the rail bed technological process decrease, thus changing the mechanical properties of the steel. In this case, stresses are formed in the surface layer during the long-term operation of the main track rail. The studies show that these are compressive stresses that will result in the rail surface hardening. The studies revealed that the stresses in each of the rail elements (head, wall, and base) are distributed individually [9]. These stresses depend not only on the operating conditions but also on the rail manufacturing technology [10].

Another phenomenon observed during the operation of rails that leads to a change in the properties of the surface layer is cold hammering. The authors of the work [11] show that in the subsurface layer of the rail head (located at a depth of 2–10 mm), the most significant physical strengthening mechanism is dislocation. This mechanism is due to the interaction of moving dislocations with stationary ones. In the surface layer of the rail head, however, the substructural mechanism is quite different. It is caused by the interaction between dislocations and small-angle boundaries of fragments and nanometer-range subgrains. The authors noted that tonnage ranging from 691.8 to 1411 million tons passes along the rails, increasing their strength 1.5–2.0 times. The authors of the work [12] also state that the main influence on the strengthening of rail steel is the dislocation substructure, which is formed during the rail operation. This work also notes that there is a change in the chemical composition of the rail surface during operation, which also contributes to the service life of the rails.

Both cold hammering and compression sets form in the surface layer of the rail and lead to its strengthening. As a result of these phenomena, after a certain period of rail operation, the strength properties of the rail surface will be higher than those in the initial state, i.e., before the operation. However, at the same time, cracks will form in the surface of the rail, leading to a decrease in its performance. The authors of the study [13] show that cracks often originate in the spots of various inclusions and defects in the surface layer of rails. Often, such cracks begin to develop on the side surface of the rail head. Grinding can prolong the life of the rails by crushing the head, removing damaged material, or moving the contact tape away from the stress concentrator [14,15]. There are very few research articles devoted to a comprehensive study of the properties of rails at various stages of their operation due to the significant labor input required. In addition, rail manufacturing standards and test equipment in different countries differ significantly from each other [16]. Although mass elements are similar, they may have different materials and slightly different geometric parameters. Therefore, the rails of each standard require a particular study. Regional operating conditions are also important. In [17], the authors study the influence of climatic changes on the track twist and rail joints under the conditions of the Spanish climate. The studies showed that temperature fluctuations strongly affected the durability of rails. Having analyzed the existing experimental data, the authors of [18] come to similar conclusions. In addition, based on the data obtained from the Swedish Portal for Climate Change Adaptation [19], they show that in climate conditions in some countries characterized by frequent variations in negative and positive temperatures during the year, the problem of the rail twist is critical.

Hence, it can be concluded that each region must conduct a specific study to accurately determine the changes in the properties of rails during operation. The results obtained through such studies allow us to think only about the possibility of reusing rails.

The aim of the work is to study comprehensively the properties of the high-duty rails of the R65 type after the warranty period of their operation and obtain information about changes that take place in their structure and properties. Obtaining comprehensive knowledge about such changes fills in the blanks in this knowledge area and helps determine the correct area for reusing these rails.

2. Methods and Materials

The R65 rails that were studied in this work had an actual operating time for the handled tonnage that amounted to 494 million t-km gross. The samples made from the RP65 rails manufactured by ARBZ LLP (Aktobe rail plant LLP, Aktobe, Republic of Kazakhstan) were used for comparison. The R65 rails were used on the main tracks. The choice of the RP65 rails for comparison is dictated by the fact that the rails are used to travel along industrial tracks for less critical purposes. The R65 rails with a hardened surface will be considered as an option to replace the RP65 rails that are less crucial.

The chemical composition of the materials used for the rails is shown in Table 1. Table 1 reflects the chemical composition of the used rails according to the industry standard. The chemical composition of the R65- and RP65-grade rail steels is almost identical, with the content of alloying elements (Mn, Si, and V) differing by no more than 0.1%. Such a difference does not significantly and noticeably influence the mechanical properties. The main difference is observed in the content of harmful impurities such as sulfur and phosphorus. Even minor changes in the concentration of these substances can significantly worsen the plastic properties of the steel.

Table 1. Chemical composition of the rail steel.

Steel Grade	Content, %					
	C	Mn	Si	V	P	S
					Not more than	
R65	0.71–0.82	0.75–1.15	0.25–0.60	0.03–0.15	0.025	0.025
RP65	0.72–0.84	0.75–1.25	0.18–0.55	0.03–0.10	0.045	0.035

The metallographic specimens were made from the samples obtained using conventional technology to perform structural studies. To identify the carbon steel structure, a 3% solution of nitric acid in ethyl alcohol was used. The surfacing layer structure was detected by means of a solution of HNO_3 and HCl acids in a ratio of 1:3. Structural studies were carried out using a Carl Zeiss AxioObserver Z1m light microscope and a Carl Zeiss EVO 50 XVP scanning electron microscope (Jena, Germany). The phase composition was studied using an ARL X'TRA (Thermo Fisher Scientific, Waltham, MA, USA) X-ray diffractometer in the CuKα radiation.

The following methods were used to identify defects in the welded joints: a visual-optical method using a Carl Zeiss AxioObserver A1m (Carl Zeiss Microscopy Deutschland GmbH, Oberkochen, Germany) microscope; a capillary method; and an eddy current method using a VD–70 eddy current flaw detector.

2.1. Mechanical Testing of Samples

The hardness of the rolling head surface was determined as follows: The rail's hardness was monitored using a Brinell device. The hardness of the surface and cross section of the rail was determined by a WILSON BH3000 (ITW Test & Measurement GmbH, Duesseldorf, Germany) hardness tester according to GOST 9012-59 (Figure 1). The spot for determining the hardness of the rail's rolling surface was cleaned to remove scale and a decarbonized metal layer to a depth of no more than 0.5 mm. The roughness of the cleaned surface was no more than 25 μm.

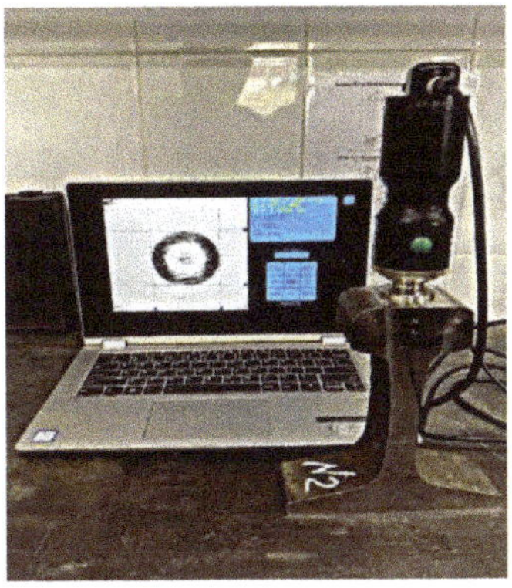

Figure 1. Measuring the hardness of the samples with the WILSON BH3000 hardness tester.

The mechanical tension properties of the rail were determined using the Instron HDX-1000 universal rapture test machine. The blanks of samples for the tensile test were cut along the rolling direction taken from the upper part of the head in the fillet zone as close as possible to the surface at a distance of at least 150 mm from the end of the rail (Figure 2). The main dimensions of the samples are also shown in Figure 1. The mechanical characteristics were recorded automatically and calculated by the control and calculation software of the test machine. The measurement error of this machine is ±0.5%. Tensile strength mechanical characteristics were determined according to the uniaxial tension scheme on the cylindrical samples (Figure 2), which had a pick-up movement rate of 10 mm/min.

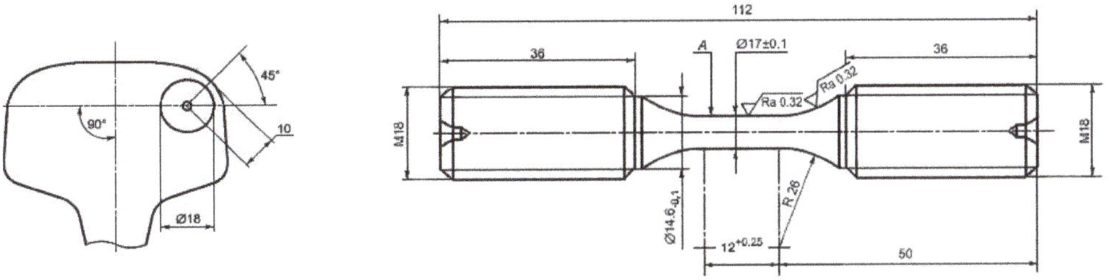

(a) (b)

Figure 2. The spot where the samples for tensile testing were cut and the sample dimensions. (**a**)—the place where the sample was cut from the rail, (**b**)—test specimen drawing.

The experimental results were statistically processed in Statistica (StatSoft Inc., Tulsa, OK, USA), Table Curve 2D, and Table Curve 3D software. For each value (each point in the diagram), the tests were conducted on at least 5 samples.

2.2. Cyclic Life

The cyclic life during fatigue tests is determined according to GOST 25.502 under hard loading (strain control) of the samples with a constant amplitude of the full strain (longitudinal) equal to 0.00135. Cyclic durability tests were performed using the Instron 8801 (Instron Engineering Corporation, Norwood, MA, USA) universal machine (Figure 3). This machine has built-in sensors for measuring the deformation value. Sample blanks for the test were cut along the rolling direction from the upper part of the head in the fillet area as close as possible to the surface at a distance of at least 150 mm from the end of the rail (Figure 2). The basic dimensions of the samples are also shown in Figure 1 (Loading scheme—cyclic tension-compression). Cyclic durability tests were conducted indoors at an ambient temperature of 22 °C and a relative humidity of 65%. The samples had a temperature of 22 °C before the tests. A longitudinal uniaxial cyclic load was applied to the sample, which had a loading cycle asymmetry coefficient of minus 1 and a loading frequency of 40 Hz. The test base is 5 million loading cycles. The testing was terminated when a crack or fracture in the sample formed or when the test base was reached.

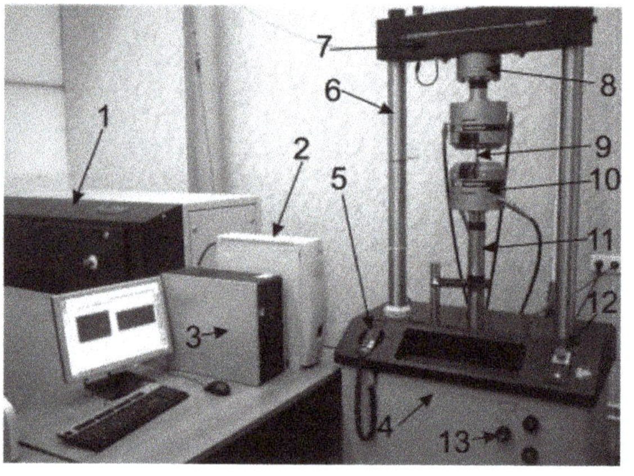

Figure 3. Appearance of the test complex, Instron 8801. 1—hydroelectric power station, 2—controller, 3—computer, 4—power frame, 5—control panel, 6—columns, 7—rigid horizontal traverse, 8—force action sensor, 9—sample, 10—hydraulic grips, 11—hydraulic servo drive column, 12—panel for controlling the hydroelectric station, and 13—valves for controlling the stiffening beam position.

2.3. *Rail Endurance Limit (Fatigue Tests)*

The tests were carried out on full-profile samples (1200 + 10 mm long), cut from the rails by cold mechanical cutting methods. The loading scheme is a flat three-point symmetric bending (Figure 4). The distance between the lower supports is 1000 ± 5 mm. The upper punch is installed in the middle between the supports within 500 ± 5 mm. The samples are tested under soft loading (force control) in the position "head down" of the rail when the asymmetry of the loading cycle is plus 0.1. The test base is 2 million cycles.

Figure 4. Experimental plant and a general view of tests for determining the endurance limit of the rails.

Cyclic life tests are performed indoors at an air temperature of 22 °C and a relative air humidity of 65%. The temperature of the samples before testing was 22 °C.

Measuring tools or equipment must provide a cyclic load of at least 1000 kN and have a maximum permissible relative measurement error of ±3%. The maximum relative error of the sensor must be ±2%. The loading frequency is 10 Hz, with a maximum relative error of ±2%.

2.4. Residual Stresses

Residual stresses in the rail neck are determined by the divergence of the groove as a rail height difference (H2-H1) along the axis at the end of the sample before and after cutting the groove according to the scheme shown in Figure 5. The samples intended for studying residual stresses by the mechanical method were prepared on an electrochemical die-sinking and hole-contouring machine, SFE-12000M (STANKOFINEXPO, Kirov, Russia). Such equipment was chosen because no additional technological residual stresses were introduced when using this method of preparing experimental samples. Initially, a piece of the rail having a length of 600 ± 3 mm was cut off, and then a groove 6 ± 1-mm wide was cut through, and the accuracy of this size was ensured by the accuracy of manufacturing with an electrode tool.

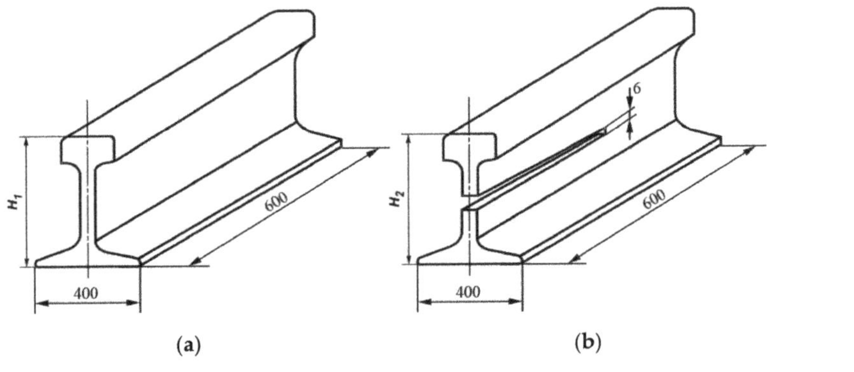

Figure 5. Schematic of residual stress control in a track neck. (**a**)—rail before testing, (**b**)—rail after cutting for residual stresses testing.

The control of residual stresses in the rail bottom is referred to as "periodic tests" and, according to item 6.5 "Periodic tests" of GOST 51685-2013, is carried out at least once every three years. The level of residual stress in the rail bottom is determined from the full-profile samples. The samples, 1.0 ± 0.1-m long, were cut from six rails by cold mechanical cutting. It is necessary to perform abrading to a depth of 0.3 to 0.5 mm on the bearing surface of the rail bottom in the middle part of the sample, after which a strain gauge is attached to the cleaned area in the longitudinal direction according to the recommendations of the manufacturer of the sensor. The permissible relative error of the sensor shall not exceed ±1%.

3. Results and Discussion

3.1. Study of Rail Microstructure

First, the microstructures of already-used and new rails were studied. The photos of the etched microstructure of the used R65 rails and the new R65 rails are shown in Figure 6.

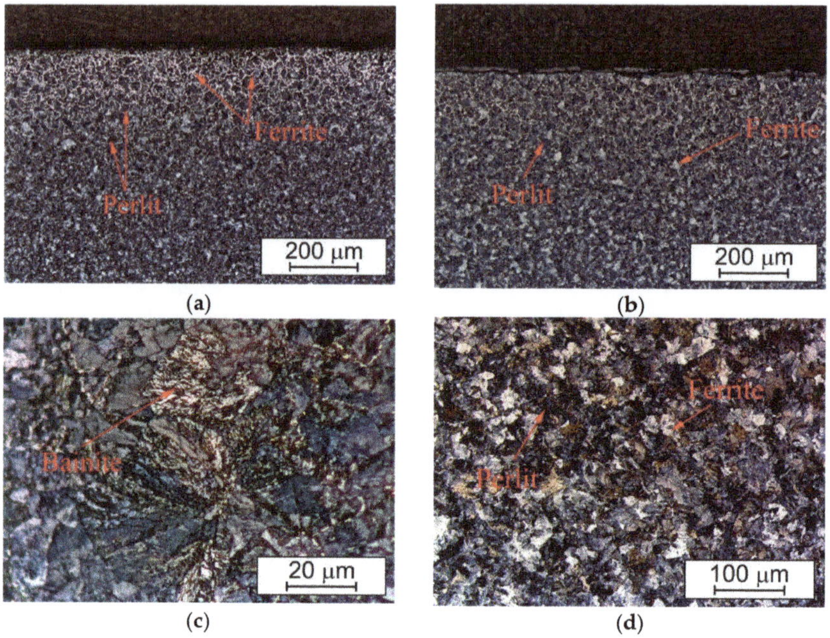

Figure 6. Microstructure of the studied R65 rails: (**a**) surface of the rail after the standard service life of 494 million t-km gross; (**b**) new (unused before) rail in a state of delivery; (**c**) pearlite grain on the rail surface after the standard service life; and (**d**) ferrite-pearlite structure section of the rail surface after the standard service life (at a depth of ~3 mm from the surface).

The metal microstructure of the rails after operation is represented by a lamellar pearlite estimated at 2-3 points (0.3–0.4 microns) with scattered ferrite sections along the grain boundaries. The number of ferrite grains present in the used R65 rails (Figure 6d) is less than 5%, and the average size is about 0.24–0.26 μm. There is practically no perlite in the new R65 rails (Figure 6b), and bainite was not detected in the microstructure of the studied rail. When moving away from the surface of the rail rolling head, perlite acquires a more coarse-grained structure. This is related to the specificity of rail manufacturing technology. The rails taken for this study were made by hot rolling. Since the rail cools unevenly after rolling, a finer-grained structure is formed on the surface; however, when moving away from the surface of the rail head, the pearlite grain size becomes larger. Characteristic images of this structure with a larger grain are shown in Figure 6d.

A deformed structure is observed to a depth of up to 300 microns on the etched cuts from the surface of the working fillet. The amount of the decarbonized layer on the surface identified by the solid ferrite grid does not exceed 320 μm (Table 2). Local light-etched areas of a work-hardened metal are observed on the rolling surface, along which cracking develops (Figure 6a).

Table 2. Analysis of microstructure test results.

Rail Type	Actual Value			
	Structure Type	Point	Scale	Decarbonized Layer Depth, mm
Used R65	Pearlite Ferrite	3 1	1 7	Impossible to detect due to surface wear
RP65	Pearlite	3	1	0.318

According to the classical definition, pearlite is a structural component of steel and cast iron, a eutectoid mixture of ferrite and cementite [20]. When crystallized under normal conditions, pearlite has a lamellar structure consisting of alternating plates of pearlite and cementite [21]. The studies performed in the present work revealed numerous imperfections in the structure of lamellar pearlite, namely, an alternating "comb" type structure (Figure 7a) and breaks of cementite plates (ferrite bridges) (Figure 7b). Quite often, there are curved plates of cementite of varying thickness in the pearlite colony.

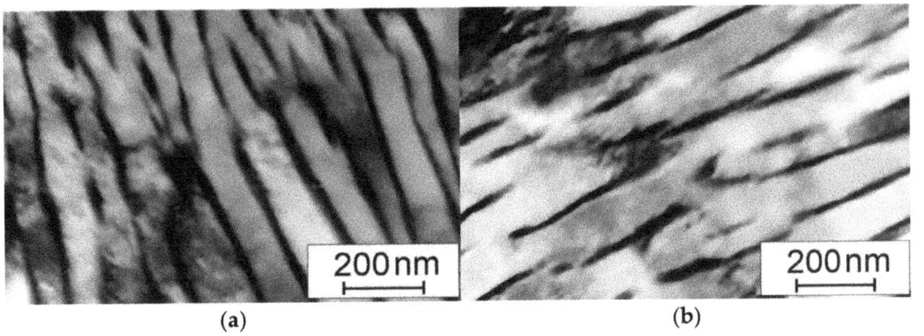

(a) (b)

Figure 7. Pearlite structure in the surface layer: (**a**) alternating pearlite structure and (**b**) broken structure of cementite plates.

The ferrite plates of pearlite colonies quite often have an alternating light and gray contrast. This diffraction contrast indicates that the ferrite plates are obviously divided into weakly misaligned areas as a result of elastic stresses.

The metallography of etched cuttings (Figure 6d) revealed that structurally free ferrite grains are present in the steel structure of the used rails, i.e., ferrite grains, in the volume of which there are no carbide phase particles. The relative content of such grains is small and does not exceed 5% of the steel structure. Ferrite grains, as a rule, are located in the form of interlayers along the borders of pearlite grains (Figure 6a) or at the joints of the borders of pearlite grains (Figure 6d).

Since the rails are obtained by means of rolling, the absence of line defects is important. Therefore, we conducted a rail study for the parameters of individual globular defects (ED) and line defects (EB). To identify the possibility of reutilizing rails, it is necessary to study their structure for the presence of non-metallic inclusions since such inclusions weaken the properties of steel. In general, only the shape and size of inclusions may change during operation, and the number of inclusions may only vary in the surface layer. To determine the average sizes of the inclusions, 8 samples for the new RP 65 rail and the already exploited R65 rail were studied. Eight samples were taken from different rails of each type to reproduce the results. A non-etched microstructure study showed that the average diameter of the inclusions for the new and already used rails is almost the same (Table 3). However, the length of inclusions for the latter is much greater, and at the same time, the total coefficient Ka for the used rails is still lower than the standard value.

3.2. Study of the Mechanical Properties of Rails

The studies of the mechanical properties of the rails showed that there was a slight difference in the values of the strength limits and the plastic characteristics of the new RP65 rails and the previously exploited R65 rails (Table 4, Figure 8). The obtained results demonstrated that the strength properties (σ_B) for the material of the previously used R65 rails are 40 MPa lower than those of the material of the new RP65 rails. However, at the same time, the plastic properties of R65 are higher by about 8.5%. This can be explained by an appropriate combination of factors. On the one hand, the previously used rails have larger stitch defects, which should lead to a decrease in plastic properties. On the other hand, the initial chemical composition of R65 contains less sulfur and phosphorus.

In addition, the surface layer undergoes plastic deformation during operation and has a finer-grained structure.

Table 3. Analysis of steel contamination with non-metallic inclusions.

Sample	ED Group Diameter Pd, μm *	EB Group Length Pl, μm *	Normalized Aggregate Coefficient, Not more than	Aggregate Coefficient, Ka, $\mu m^2/mm^2$
Actual values of used R65 rail				
1	11.0	353.0	–	–
2	11.0	353.0	–	–
3	11.0	353.0	–	–
4	11.0	353.0	–	–
5	11.0	353.0	30	33.5
6	11.0	353.0	–	–
7	11.0	353.0	–	–
8	11.0	353.0	–	–
Av.	11.0	353.0	–	–
Actual values of RP65 rail				
1	11.0	353.0	–	–
2	11.0	44.0	–	–
3	11.0	88.0	–	–
4	22.0	88.0	–	–
5	11.0	44.0	30	3.3
6	11.0	88.0	–	–
7	11.0	88.0	–	–
8	11.0	88.0	–	–
Av.	12.4	110.1	–	–

* Note: non-metallic inclusions: ED—individual globular inclusions; EB—line globular defects.

Table 4. Tensile mechanical properties.

Rail Type	Breaking Stress, σB, N/mm^2	$\sigma_{0.2}$, N/mm^2	Elongation, δ, %	Contraction, ψ, %
Used R65	1240	800	13	33
RP65	1280	840	12	30

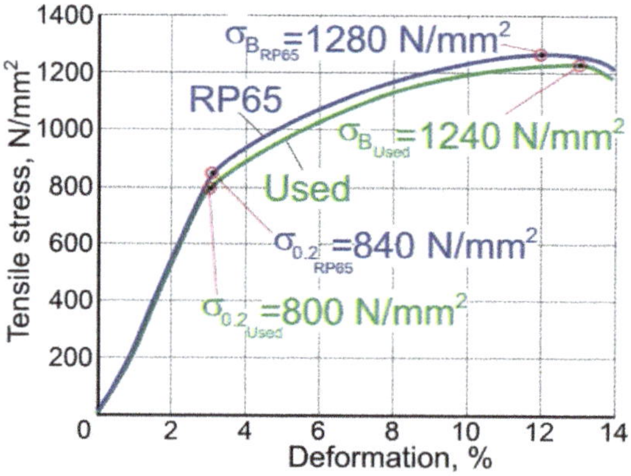

Figure 8. Tensile diagram for the studied samples cut from R65 and RP65 rails.

The difference in the hardness of the rolling surface along the length of the rails and samples was determined by three measurements on the middle line of the rolling surface.

An interval of at least 25 mm was taken for each of the three samples, which were taken from the ends and the middle part of the rail or on the rail surface (Tables 5 and 6).

Table 5. Hardness of the head rolling surface.

Brinell Hardness Scale	
	kg/mm²
Previously used R65	401
RP65	352

Table 6. Hardness along the rail cross-section.

Measurement Depth on the Rail Surface, mm	5	10	15	20	25	30	35
				kg/mm²			
Previously used R65	363	363	363	336	331	341	341
RP65	357	349	346	352	311	311	311

The diagram of hardness measurements along the rail section is shown in Figure 9.

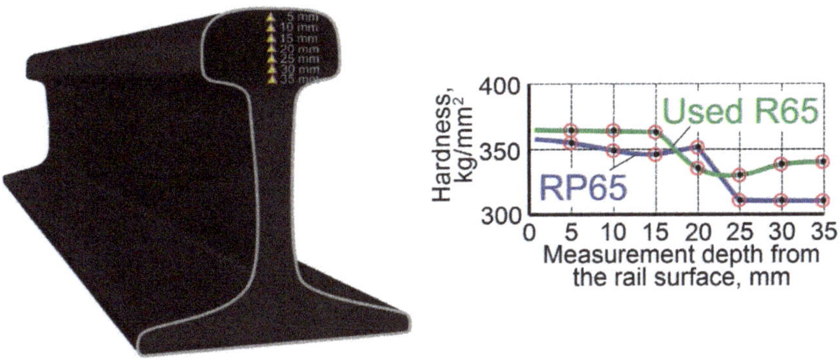

(a) (b)

Figure 9. Hardness measurement: (**a**) schematic of hardness measurement and (**b**) dependence of hardness on the distance from the rail surface.

Long-term operation of rails is usually accompanied by deformation and transformation of the material structure [22]. This can be justified by the hardness values obtained for the operating rails. The upper hardness limit is HB401So for the R65 rails. The hardness at a distance of 10 and 22 mm from the rolling surface must be 345–360 ($HB_{10\ mm}$) and 325–350 ($HB_{25\ mm}$). The hardness of the studied rails is within the upper boundary of these values or exceeds them, and the property changes are consistent with the changes in the rail structure. The quantitative analysis of the morphological state of the steel structure performed in this work showed that the operation of rails is accompanied by the transformation of the state of lamellar pearlite grains, namely, the destruction of cementite plates. According to the photographs in Figure 6, regardless of the position of the material's analyzed volume (rolling surface or fillet surface), the destruction of the lamellar pearlite structure is at its maximum in the surface layer of the rails, with a thickness of less than 2 mm. However, the destruction degree of the lamellar pearlite structure depends substantially on the position of the volume to be analyzed; namely, on the rolling surface, the relative content of the broken pearlite grains is more than 2 times higher than the content in the surface layer of the working fillet.

As noted above, pearlite grains become broken during rail operation. One of the main mechanisms of such destruction during plastic deformation of the steel is cutting cementite plates by gliding dislocations [23]. The operation of rails is accompanied by an increase in the level of elastoplastic stresses in the steel. The value of elastic plastic stresses in the steel, in accordance with [24], is characterized by an excessive dislocation density and curvature-torsion amplitude of the material crystal lattice. Both of these characteristics of the steel are noted to be determined in the analysis of the bending extinction contours of the material.

The above results of the study of the rail metal after long-term operation indicate the structural transformation of the lamellar pearlite.

The change in the cementite's elemental composition during crushing is minimal. At the initial stage of the transformation, the cementite plates of the pearlite colony are covered with gliding dislocations. This is accompanied by breaking the cementite plates into separate, weakly oriented fragments. Then, the structure of carbide changes as the plastic deformation degree of the material increases due to the stripping of carbon atoms from the cementite crystal lattice. It is worth remembering that this process is possible due to a noticeable difference in the average bond energy of carbon atoms with dislocations (0.6 eV) and iron atoms in the cementite crystal lattice (0.4 eV) [25].

The considered deformation transformations of the rail steel structure during operation on the railway should be noted as not adversely influencing the product's cyclic life. Table 7 shows the results of testing the cyclic life of the rails. The tests were carried out on the rolling surface and, for comparison, on the cross section at a distance of at least 10 mm from the rolling surface. The tests were stopped when the sample became cracked or fractured or when the test base was reached.

Table 7. Analysis of the cyclic life test.

Rail Type	Indicator	Sample	Sample Temperature, °C	Rated Value, Not less than	Actual Value
Previously used R65 RP65	Cyclic life at a constant amplitude of total deformation 0.00135	1	22	5,000,000 cycles	5,000,000 cycles
		1	22	5,000,000 cycles	5,000,000 cycles

The test results are considered positive if there are no fractures or cracks in all the tested samples upon reaching the test base. The test results are considered negative if the formation of a crack or fracture in at least one sample occurred within a number of loading cycles less than the test base.

The studies of the hardness along the cross section of the used rail show that the changes in properties can be considerable. Hence, the rolling of the wheel on the surface of the rail not only causes changes in hardness but can also form residual stresses at a considerable depth. The testing of the samples to determine residual stresses showed that there are considerable differences in values for the new and already used rails (Table 8). Residual stresses in the neck of the used rails are 1.42 times higher than those in the new rails (Figure 10). However, the stress values are in the upper range of values allowed for 2 mm rail use. A similar situation is observed for the middle part of the rail base. The residual stress level for the used rails is 1.25 times higher than that for the new ones (Table 9). In addition, the residual stresses of the used rails lie in the upper range of permitted values for use. Such differences in voltage values are certainly related to the operating conditions of the used R65 rails. Rolling the wheels of railway trains under a significant load deforms not only the surface layer but also the rail. This, in turn, determines the appropriate level and nature of the distribution of residual stresses. In the new rails, small residual stresses are formed at the manufacturing stage. Since the rails are made by hot rolling, internal stresses are formed in them [22].

Table 8. Residual stress indicators in a rail neck.

Rail Type	Indicator	Profile Height H1	Profile Height H2	Difference H2-H1
Already used R65	Slot deviation, mm	176.7	178.7	2.0
RP65		180.2	181.6	1.4

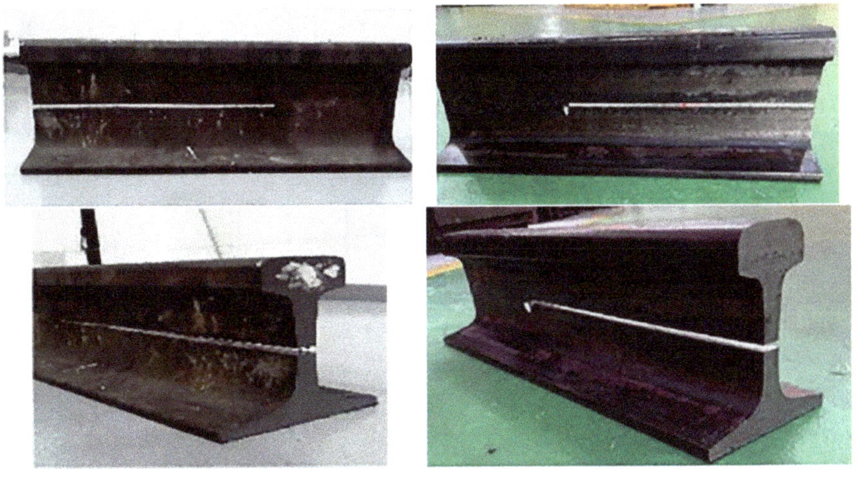

Figure 10. Samples after residual stress control in a rail neck.

Table 9. Analysis of test results for determining residual stresses in the middle part of the rail base.

Rail Type	Indicator	Ambient Temperature, °C	Relative Humidity, %	Normalized Value, Not More than, MPa	Actual Value, MPa
Already used R65	Stress, N/mm^2 (MPa)	22	65	250	242
RP65		22	65	250	194

The tests show that during operation, there are changes in the structure of the surface layer. Hardness increases along almost the entire rail section, and residual stresses are formed, but the defective surface layer of the rail also increases. The number and size of individual defects do not change, but the length of horizontally extended defects increases. In addition, cracks will form on the surface of the rail head. In general, such changes significantly affect one of the main operational properties of rails, i.e., the endurance limit.

The normalized values for using the rails in Central Asia are 2,000,000 cycles under a load of 700 kN for rails R65 and RP65. The new rails certainly meet these requirements (Table 10). In the case of the previously used rails, the maximum load under which they could operate for 2,000,000 cycles was 510 kN. Under a standard load of 700 kN, the used rails can only withstand 790,000 cycles (Figure 11a).

Table 10. Analysis of test results for the determination of the rail endurance limit.

Rail Type	Sample	Rated Value of Endurance Limit, Not Less than, MPa	Relative Humidity,%	Maximum Load, kN	Number of Cycles	Endurance Limit, MPa
Already used R65	1	370	60	700	790,056	489
	2		60	600	1,111,966	419
	3		60	510	2,000,000	355
RP65	1	370	69	700	2,000,000	489

(a) (b)

Figure 11. Rail failure after cyclic tests: (**a**) under a load of 700 kN, number of cycles: 790,056; (**b**) fracture under a load of 600 kN, number of cycles: 1,111,966.

This result is largely due to the formation of cracks in the surface layer of the rails. A photograph of the fracture surface (Figure 11a) shows that the fracture originated on the side surface of the rail. The reason for the crack development was the zone of lateral collapse of the rail head formed under operating conditions (Figure 12a). This crack nucleus led to its fatigue development. Figure 12a clearly shows the zone of stable fracture development. It is characterized by fatigue lines representing approximate concentric contours. The focus of these contours is at the origin of the fatigue crack (Figure 12b). The surface of this zone is smooth and level. This effect is the result of cold hammering caused by repeated presses of two surfaces of the crack on each other.

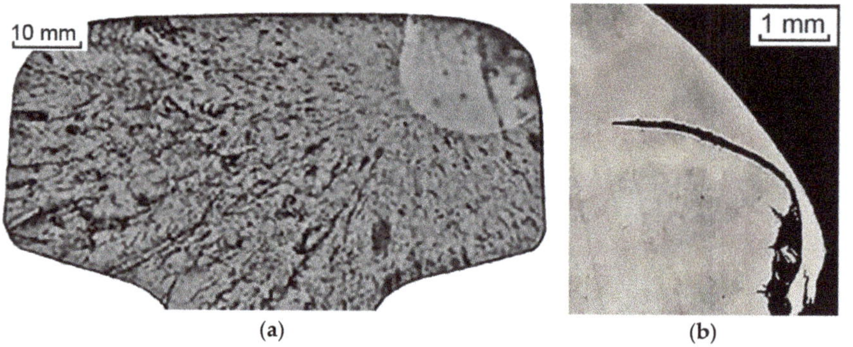

(a) (b)

Figure 12. Failure of a used rail after endurance limit tests. (**a**) A general view of the rail head and (**b**) the place where the crack originated on the rail fluting.

A set of experimental studies showed considerable changes in the structure of the R65 rail during its operation. The hardness of the surface layer and the rail itself increases to a sufficiently significant depth (up to 35 mm). In addition, considerable stresses are formed in the rail. It should be noted that both hardness and stress increase to the upper limit of the values allowed by the state standard (sometimes slightly exceeding it). Despite the increase in hardness and the formation of stresses, the plastic and strength properties of the rails remain almost at the same level and do not fall outside the tolerance field of the values determined by the state standard. The change in hardness is conditioned by changes in the structure of the rail. When the structure is deformed, the condition of the grains of lamellar pearlite is transformed, with the maximum amount of structure damage occurring in the surface layer of rails with a thickness of less than 2 mm. The lengths of the line defects also change, which then affects the cyclic life of the rails. New RP65 rails under a standard load of 700 kN can withstand 2,000,000 cycles. The previously used R65 rails under the same load can withstand significantly less than 790,000 cycles. They can withstand the 2,000,000 cycles determined by the state standard only under a load of 510 kN.

The changes that took place in the mechanical properties, especially in the cyclic properties, are related to both the operation intensity, structure, and mechanical composition of the R65 and RP65 rails. The analysis of the chemical composition of the analyzed steels is shown in Table 1. It shows that the chemical composition of the main alloying elements differs slightly, by no more than 0.1%. Such changes in the composition vary the properties of the steel by no more than a percentage [26]. Moreover, manganese in the RP65 steel may increase by 0.1%, and vanadium may increase by less than 0.05%. Such changes in the chemical composition will not noticeably change the mechanical properties [27,28]. The main difference between these steels is the content of sulfur and phosphorus. In RP65 steel, the content of sulfur and phosphorus can be higher by 0.01–0.02%, which somewhat reduces its plastic properties [27]. Based on this particular chemical composition, the R65 rails are designed for the main, working tracks [28], while the RP65 rails are intended for industrial routes, where the intensity of the movement is lower and there is not as much frequent deforming effect. The initial mechanical properties of the new R65 rail are slightly higher than those of the R65 rail. However, our tests have shown that both brands can withstand a certain standard load. Our studies have shown that during operation, microcracks form in the surface layer of the R65 rail, and the surface layer is decarbonized. These processes run in parallel with the processes of riveting the surface layer and the formation of internal stresses in the rails. The result is that the mechanical properties of the R65 rail (hardness and tensile strength) practically do not change or even increase by a small amount (5–10%). The formation of elongated defects in the surface layer and the formation of microcracks during operation lead to a decrease in the cyclic properties (rail endurance limit) of the rail. As a result, while the main mechanical properties of the rail material remain at the required level, the value of the rail endurance limit indicator for the used R65 rails becomes lower than that of the RP65 and no longer meets the requirements of the standard.

Based on this fact, the R65 rails can be said to have been industrially used for a standard period, which is not suitable for reutilization in the Republic of Kazakhstan. Their use is possible only in low-congested areas or areas without intensive movement (storage areas, dead ends, and other inactive routes). The number of passageways per day determines the possibility of using these rails. The regulatory framework for testing, amounting to 2,000,000 cycles that the rail must withstand, is determined by a guaranteed 20-year period of operation [29]. At the same time, they assume that up to 90 trains should pass per day. Based on the obtained results, the maximum number of cycles that the previously used R65 rail can withstand is 790 thousand, which is 2.53 times less. Therefore, it is possible to use these rails only within the sections where the traffic intensity is lower than 35 trains per day on average or various sections with non-intensive traffic, i.e., stops, dead ends, and other inactive ways. At the same time, the already-used R65 rails will provide the operation for the required period of 20 years.

At the same time, the standards and requirements for rails employed in different countries may differ significantly, as determined by rolling stock weight standards, climatic conditions, etc. Standards and requirements for the number of cycles of suitable rails can be different. Therefore, the suitability of rails for different countries can vary, necessitating meeting the standards of the country to determine the possibility of rail reutilization.

4. Conclusions

1. The studies revealed that the microstructure of the rail R65 metal after operation is represented by a 2–3-point lamellar pearlite (0.3–0.4 μm) with isolated ferrite sections along the grain boundaries, while the number of ferrite grains in the previously used rails is less than 5% (average size is 0.24–0.26 μm). The depth of the deformed layer is up to 300 μm, and the value of the decarbonized layer of the surface, identified by a solid ferrite network, does not exceed 320 μm. The average diameter of non-metallic inclusions for the new RP65 and already-used R65 rails is almost the same. However, the length of inclusions for R65 is much higher. After the warranty period of R65

operation, the length of line-extended defects increases. The long-term operation of the rails is accompanied by deformation and transformation of the material structure. The hardness values of the previously used rail during the warranty period were at the level of HB 350–360.

2. The quantitative analysis of the morphological state of the steel structure showed that the state of the grains of lamellar pearlite is transformed during operation. Its structure destruction is at its maximum in the upper layer of the rails, with a thickness of less than 2 mm regardless of the position of the analyzed volume. It was experimentally confirmed that the relative content of broken pearlite grains on the rolling surface was more than 2 times higher than the content in the surface layer of the working fillet.
3. During the operation, a change in the structure of the upper layer was revealed. At the same time, the hardness increases along the entire rail section, and residual stresses are formed. Its level in the neck of the already used rails is 1.42 times higher than that of the new ones, and the depth of the maximum residual stresses is 2 mm, which is in the upper range of the permitted values. The values of residual stresses in the middle part of the base of the used rails are 1.25 times higher than those for the new ones, but the total level of the stress-strain state of the rail is within the permissible interval.
4. Deformation transformations of the rail steel structure during their exploitation on the railway did not negatively influence the cyclic durability of the product. However, at the same time, the results of the endurance limit tests showed that the maximum load under which the used rails could run 2,000,000 cycles was 510 kN, and under a standard load of 700 kN, the already used rails could only withstand 790,000 cycles.
5. Hence, if an overall level of reduction in the operational indicators of the used rails is provided, it is possible to recommend their reutilization in less critical areas at a lower level than normal to prevent sudden and difficult-to-predict failures.

Author Contributions: Conceptualization, K.Y. and D.B.; methodology, D.B.; validation, V.E.G. and A.I.K.; formal analysis, V.E.G. and A.I.K.; investigation, V.E.G.; data curation, K.Y. and D.B.; writing—original draft preparation, K.Y., N.V.M. and V.Y.S.; writing—review and editing, N.V.M. and V.Y.S.; supervision, N.V.M. and V.Y.S.; project administration, N.V.M. and V.Y.S.; funding acquisition, A.I.K. and V.E.G. All authors have read and agreed to the published version of the manuscript.

Funding: The work was supported by the Ministry of Science and Higher Education of the Republic of Kazakhstan within the framework of the BR18574141 project Theme of the project: "Comprehensive multi-purpose program for improving energy efficiency and resource saving in the energy and mechanical engineering for the industry of Kazakhstan", for which the authors express their deep gratitude to them.

Institutional Review Board Statement: Not applicable.

Informed Consent Statement: Not applicable.

Data Availability Statement: The data presented in this study are available from the corresponding authors upon reasonable request.

Conflicts of Interest: The authors declare no conflict of interest.

References

1. Dementiev, V.P.; Korneva, L.V. *Railway Rails for Siberia*; Irkutsk State University, University of Communications: Irkutsk, Russia, 2010.
2. Kozyrev, N.A.; Dementiev, V.P. *Production of Railway Rails from Electric Steel*; Novokuznetsk Institute for Advanced Studies: Novokuznetsk, Russia, 2000.
3. Nikolin, A.I. *Improving the Processes of Welding and Heat Treatment of Rails of Mainline Railways*; VNIIZhT Ministry of Railways of Russia: Moscow, Russia, 2004.
4. Markov, A.A.; Kuznetsova, E.A. *Rail Flaw Detection. Formation and Analysis of Signals*; Ultra Print: St. Petersburg, FL, USA, 2014.
5. Smirnov, L.A. Improving the quality of domestic railway rails. *Bull. Sci. Tech. Econ. Inf. Ferr. Metall.* **2005**, *6*, 43–49.
6. Panda, B.; Balasubramaniam, R.; Moon, A. Microstructure and mechanical properties of novel rail steels. *Mater. Sci. Technol.* **2009**, *25*, 1375–1382. [CrossRef]

7. Vitez, I.; Krumes, D.; Kladarić, I. The comparison of the properties of railway steels by codex UIC 860V and draft new European standard. *Tehnički Vjesnik* **2003**, *10*, 35–41.
8. Tokmakova, E.N.; Perkov, I.E.; Ivanov, P.V.; Zagranichek, K.L. Development of rails of a new category for use in particularly difficult operating conditions. *Bull. Sci. Res. Inst. Railw. Transp. (Bull. VNIIZhT)* **2022**, *81*, 339–346. [CrossRef]
9. Muravev, V.; Volkova, L.; Buldakova, I. Analysis of Stress Distribution in a Rail by Electromagnetic Acoustic Method. *AIP Conf. Proc.* **2016**, *1785*, 030016. [CrossRef]
10. Lu, W.Y.; Dike, J.; Modjtahedzadeh, A. Finite element modeling of waves in a rail, in Quantitative Nondestructive Evaluation. *AIP Conf. Proc.* **2002**, *620*, 1835–1843.
11. Yur'ev, A.A.; Kormyshev, V.E.; Gromov, V.E.; Ivanov, Y.F.; Shlyarova, Y.A. The physical nature of the hardening of the surface of the rails during long-term operation. *Izv. Vuzov Ferr. Metall.* **2021**, *64*, 886–894. [CrossRef]
12. Ivanov, Y.F.; Gromov, V.E.; Yur'ev, A.A. Contributions of Various Mechanisms to the Hardening of Differentially Quenched Rails during Long-Term Operation. *Russ. Metall.* **2018**, *2018*, 985–989. [CrossRef]
13. Office of Railroad Policy and Development, *Rolling Contact Fatigue: A Comprehensive Review*; U.S. Department of Transportation, Federal Railroad Administration: Washington, DC, USA, 2011. Available online: https://railroads.dot.gov/sites/fra.dot.gov/files/fra_net/89/TR_Rolling_Contact_Fatigue_Comprehensive_Review_final.pdf (accessed on 6 April 2023).
14. Magel, E.; Roney, M.; Kalousek, J.; Sroba, P. The blending of theory and practice in modern rail grinding. *Fatigue Fract. Eng. Mater. Struct.* **2003**, *26*, 921–929. [CrossRef]
15. De Vries, R.; Sroba, P.; Magel, E. Preventive grinding moves into the 21st century on Canadian Pacific Railroad. In Proceedings of the AREMA Annual Conference, Chicago, IL, USA, 9–12 September 2001.
16. Vaičiūnas, G.; Bureika, G.; Steišūnas, S. Measurement Repeatability of Rail Wheel Loads Caused by Rolling Surface Damages. *Appl. Sci.* **2023**, *13*, 4474. [CrossRef]
17. Sanchis, I.V.; Franco, R.I.; Zuriaga, P.S.; Fernández, P.M. Risk of increasing temperature due to climate change on operation of the Spanish rail network. *Transp. Res. Procedia* **2020**, *45*, 5–12. [CrossRef]
18. Thaduri, A.; Garmabaki, A.; Kumar, U. Impact of climate change on railway operation and maintenance in Sweden: A State-of-the-art review. *Maint. Reliab. Cond. Monit.* **2021**, *1*, 52–70. [CrossRef]
19. Klimatanpassning. "Climate Change in Sweden". Available online: http://www.klimatanpassning.se/en/climate-changein-sweden (accessed on 6 April 2023).
20. Schastlivtseva, V.M.; Mirzaev, D.A.; Yakovleva, I.L.; Okishev, K.Y.; Tabatchikova, T.I.; Khlebnikova, Y.V. *Perlit in Carbon Steels*; Ural Branch of the Russian Academy of Sciences: Yekaterinburg, Russia, 2006.
21. Gridnev, V.N.; Gavrilyuk, V.G.; Nadutov, V.M.; Polushkin, Y.A. Redistribution of carbon and alloying elements during plastic deformation and subsequent heating of steel. *FMM* **1980**, *50*, 582–587.
22. Ivanov, Y.F.; Morozov, K.V.; Peregudov, O.A.; Gromov, V.E.; Popova, N.A.; Nikonenko, E.L. Structural-phase gradients in rails during long-term operation. *Probl. Ferr. Metall. Mater. Sci.* **2015**, *3*, 59–65.
23. Gavrilyuk, V.G.; Herzriken, D.S.; Polushkin, Y.A.; Filchenko, V.M. Mechanisms of cementite decay during plastic deformation. *FMM* **1981**, *51*, 147–151.
24. Ivanov, Y.F.; Gromov, V.E.; Nikitina, E.N. *Bainitic Structural Steel: Structure and Strengthening Mechanisms*; SibSIU Publishing Center: Novokuznetsk, Russia, 2015.
25. Gromov, V.E.; Peregudov, O.A.; Ivanov, Y.F. *Evolution of Structural-Phase States of Rail Metal during Long-Term Operation*; Publishing House of the SB RAS: Novosibirsk, Russia, 2017.
26. Rezanov, V.A.; Martyushev, N.V.; Kukartsev, V.V.; Tynchenko, V.S.; Kukartsev, V.A.; Grinek, A.V.; Skeeba, V.Y.; Lyosin, A.V.; Karlina, A.I. Study of Melting Methods by Electric Resistance Welding of Rails. *Metals* **2022**, *12*, 2135. [CrossRef]
27. Jena, B.K.; Gupta, N.; Singh, B.; Ahoo, G.S. Mechanical properties of low alloy high phosphorus weathering steel. *J. Min. Metall. Sect. B Metall.* **2015**, *51*, 5. [CrossRef]
28. Skeeba, V.Y.; Ivancivsky, V.V.; Martyushev, N.V. Peculiarities of High-Energy Induction Heating during Surface Hardening in Hybrid Processing Conditions. *Metals* **2021**, *11*, 1354. [CrossRef]
29. Ministry of Investment and Development of the Republic Kazakhstan. ST RK 2432-2013 Differentially Hardened and Non-Heat-Strengthened Railway Rails. Available online: https://pdf.standartgost.ru/catalog/Data2/1/4293741/4293741095.pdf (accessed on 10 January 2023).

metals

MDPI

Article

Design for the Vent Holes of Gas Turbine Flow Guide Disks Based on the Shape Optimization Method

Changlong Wen [1,*], Yanbing Zheng [1], Dong Mi [1,*], Zhengming Qian [1] and Hongjian Zhang [2]

1 AECC Hunan Aviation Powerplant Research Institute, Zhuzhou 412002, China
2 College of Energy and Power Engineering, Nanjing University of Aeronautics and Astronautics, Nanjing 210016, China
* Correspondence: changlongwen91@163.com (C.W.); midong2023@126.com (D.M.)

Abstract: Two shape optimization methods, based on non-parametric and geometric parameters, were developed to address stress concentrations in the vent holes of gas turbine flow guide disks. The design optimization focused on reducing the maximum equivalent stress at the hole edge in an aero-engine gas turbine flow guide disk. The effectiveness of both methods in achieving this objective was studied. The results indicated that the non-parametric-based optimization method reduced the maximum equivalent stress at the hole edge by 24.5% compared to the initial design, while the geometric parameter-based optimization method achieved a reduction of 20.2%. Both shape optimization methods proved effective in reducing stress concentrations and improving fatigue life. However, the non-parametric shape optimization method resulted in a better design for the vent holes based on the study's findings.

Keywords: shape optimization; flow guide disk; vent hole; design optimization

Citation: Wen, C.; Zheng, Y.; Mi, D.; Qian, Z.; Zhang, H. Design for the Vent Holes of Gas Turbine Flow Guide Disks Based on the Shape Optimization Method. *Metals* **2023**, *13*, 1151. https://doi.org/10.3390/met13071151

Academic Editor: Alireza Akhavan-Safar

Received: 29 May 2023
Revised: 19 June 2023
Accepted: 19 June 2023
Published: 21 June 2023

1. Introduction

Aero-engine components feature a significant number of hole features that serve various purposes, including connection, weight reduction, air system functionality, and heat transfer. These holes play a crucial role in the overall performance and efficiency of the engine. However, it is essential to carefully consider the location and design of these holes to prevent the occurrence of severe stress concentrations at their edges. Stress concentrations can lead to structural fatigue and a reduction in a component's overall lifespan [1–4], and the stress concentration problem has received great attention from aero-engine strength researchers [5,6]. Among the various components in an aero-engine, the gas turbine flow guide disk vent serves as a prime example, highlighting the challenges faced in the design of such component structures. The design of these structures is often constrained by the relative positions of the holes and the required cross-sectional areas. As a result, the design space for these components becomes limited, posing significant difficulties for engineers and designers.

To address these challenges and optimize the design of aero-engine components with hole features, researchers have undertaken extensive studies in the field of design optimization. These studies have aimed to enhance the performance and longevity of these components by reducing stress concentrations and improving fatigue life. Chen et al. [7,8] focused their research on a specific component—the high-pressure turbine disk-mounting side bolt hole. They approached the optimization problem by describing the hole boundary using a biaxially symmetric-shaped hole. By modeling and optimizing the geometric dimension parameters of this shaped hole, they successfully reduced the maximum equivalent stress value at the hole edge, thereby enhancing the component's performance and fatigue life, and they also expanded the biaxial symmetry non-circular hole to the uniaxial symmetry non-circular hole and obtained different results. Sun et al. [9] pursued a different approach by conducting shape optimization based on the hyper elliptic equation and the sequential response surface method for the open hole characteristics of a rotary casing. Their research demonstrated that shape optimization techniques could effectively improve the dynamic characteristics of the structure, leading to further enhancements in performance and reliability. In another study, Yan et al. [10] employed a multi-island genetic algorithm and introduced a non-circular-shaped hole design to optimize the gas turbine flow guide disk vent hole. By combining this approach with finite element analysis sub-model technology, they successfully reduced the maximum equivalent stress value at the hole edge and improved the component's fatigue life. In another article by Yan et al. [11], surrogate-based optimization with improved support vector regression was used for the non-circular vent hole on an aero-engine turbine disk. Hossein et al. [12–14] discussed the stress concentration factors (SFCs) in circular hollow-section X-connections retrofitted with a fiber-reinforced polymer under different load conditions, and the SCFs were greatly reduced. Zhu et al. [15] focused their efforts on optimizing the shape of a bolt hole in an engine sealing disk using multi-segment curves. The results of their research were impressive as they achieved a remarkable 32.8% reduction in the maximum equivalent stress at the hole edge compared to the original single round hole design. These optimization strategies have exhibited significant potential for enhancing a component's performance and durability.

In the current research landscape, the optimization of aero-engine component designs with hole features has gained significant attention. Researchers worldwide are striving to improve the structural integrity and performance of these components through innovative approaches and advanced optimization techniques. One prominent area of research has focused on advanced optimization algorithms and methodologies. Genetic algorithms, particle swarm optimization, response surface methods, and surrogate modeling techniques are some of the widely employed optimization strategies [16–21]. These algorithms have aimed to explore the design space efficiently and identify optimal solutions that reduce stress concentrations and enhance fatigue life.

Researchers have primarily relied on traditional geometric parameter-based optimization methods and finite element analysis to enhance the design of aero-engine components featuring hole features. The studies mentioned above have demonstrated the effectiveness of various optimization techniques in reducing stress concentrations and improving the overall performance and fatigue life of these structures. However, certain limitations, such as vent area optimization, still require further exploration. By addressing these challenges, researchers can continue to advance the field and contribute to the development of more efficient and reliable aero-engine components. The current approach employed by researchers for optimizing the design of hole structural features involves using the traditional geometric parameter-based shape optimization method. This method relies heavily on the engineering design experience of researchers and utilizes specific geometric parameters such as elliptical, multi-circular arc, or shaped holes to define the design boundaries. The optimization outcomes are highly dependent on the selection of these design variables. However, an alternative approach known as non-parametric shape optimization offers a promising solution [22–27]. Unlike the traditional method, non-parametric shape optimization does not rely on geometric curves to describe the design boundaries. This method holds the potential to overcome the limitations of poor design optimization results caused by the lack of relevant structural design experience among designers. It also offers a theoretical avenue to enhance the design optimization process.

To explore the effectiveness of the non-parametric shape optimization method, this study focused on the design optimization of a gas turbine flow guide disk vent hole. By utilizing the non-parametric approach, we aimed to obtain optimal design solutions that could outperform those achieved through the traditional geometric parameter-based shape optimization method. Furthermore, a comparative study was conducted between these two methods, providing valuable insights and serving as a reference for future design optimization endeavors involving similar structural features. By adopting the non-parametric shape optimization method, researchers can aspire to enhance the efficiency and accuracy of design optimization for hole structural features, ultimately contributing to advancements in the field of engineering design and promoting innovative solutions for various applications.

2. Shape Optimization Method and Process

2.1. Geometric Parameter-Based Shape Optimization Method and Process

To ensure minimal and independent design variables, as well as facilitate ease of processing and manufacturing, researchers employing the geometrical parameters of the shape optimization method must carefully choose a suitable curve to construct the parametric model for the hole boundary. Common approaches for constructing the hole scheme include selecting straight lines, multi-segment arcs, and ellipses.

In this paper, we present the flowchart of the geometric parameter-based shape optimization method (illustrated in Figure 1). Initially, a CAD parametric model is established, where the geometric parameters are modified and linked to the CAD model. This model undergoes preprocessing through finite element software, and it is subsequently solved. The optimizer then verifies the constraints and determines if the optimization results are optimal. If the design fails to meet convergence criteria, the process returns to updating the design by modifying the CAD geometric parameter model, and the aforementioned steps are repeated. It is important to note that this optimization method requires re-meshing after each iteration, and the selection of design variables significantly impacts the resulting optimization outcomes.

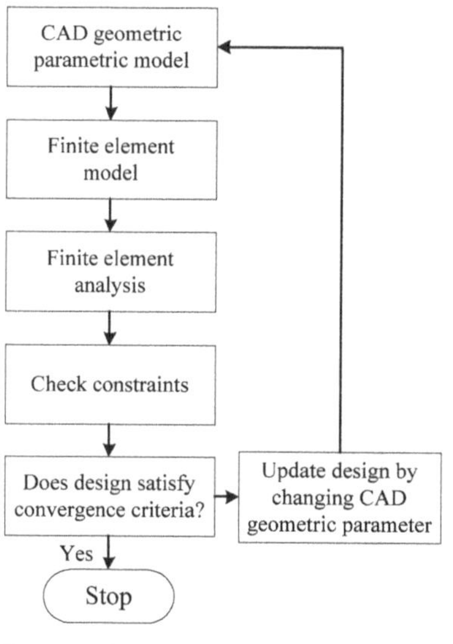

Figure 1. Flowchart of geometric parameter-based shape optimization.

2.2. Non-Parametric-Based Shape Optimization Method and Process

Non-parametric shape optimization is a method of optimizing designs that does not rely on specific geometric parameters to describe the shape. Instead, it offers flexibility by allowing the shape to be modified freely within the design space. This method is particularly useful in the finite element analysis (FEA) framework, where mesh deformation techniques can be employed to alter the mesh geometry.

In non-parametric shape optimization, the shape is typically represented using a set of control points or a mesh. The position of these control points or nodes can be adjusted to deform the shape. Mesh deformation techniques, such as the Free-Form Deformation (FFD) or Radial Basis Function (RBF) [28,29], can be utilized to smoothly modify the shape while maintaining mesh quality.

The optimization process involves defining an objective function that quantifies the desired performance of the design. This function can be based on various criteria, such as minimizing stress concentration, maximizing stiffness, or optimizing fluid flow characteristics. Additionally, constraints may be included to satisfy design requirements or limitations.

To optimize the shape, mathematical optimization algorithms, such as gradient-based or evolutionary algorithms, are employed. These algorithms iteratively adjust the positions of the control points or nodes to improve the objective function while satisfying the constraints. The optimization process continues until a satisfactory design solution is obtained [22].

Using the non-parametric shape optimization method, we can consider a hole-type structure as an example. In this approach, the coordinates of finite element nodes on the hole boundary are selected as design variables for optimization. During the optimization process, each node can be individually moved from its adjacent nodes. A schematic representation of coordinate movements for the design variable nodes is shown in Figure 2. Additionally, the nodes near the design variables can be adaptively moved, and to prevent jagged boundaries in the optimized structure, a mesh-smoothing algorithm is employed to ensure mesh quality.

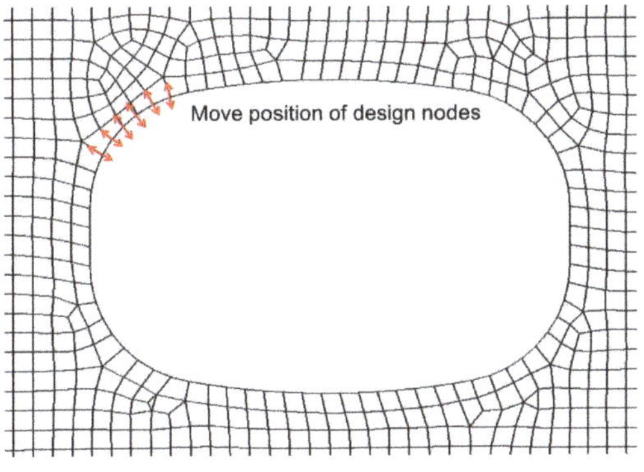

Figure 2. Schematic of design node movement for a hole-type structure.

In this study, we illustrated the non-parametric-based shape optimization process, as depicted in Figure 3. A finite element model of a vent hole was set up using the commercial FEA tool, ANSYS 19.2(ANSYS Inc., Pittsburgh, PA, USA). The non-parametric optimization process was performed by Tosca Structure 2020 (Dassault Systèmes, Vélizy-Villacoublay, France) and the process of minimization of the deviation from a reference stress value was based on the following hypothesis by Neuber: the optimum form of a component is achieved when the stresses running along the considered surface zone is fully constant (stress homogenization). Unlike conventional optimization iterations, our approach eliminated the need for modifying the CAD parametric model in each iteration. Instead, we directly adjusted the coordinates of the design variable nodes, thereby eliminating the time-consuming CAD parametric model modification, and then we repeated finite element pre-processing and the interpolation process for the temperature field.

Furthermore, after optimization by Tosca Structure' shape optimization, we obtained the mesh nodes file of a vent hole, and then the coordinates of the finite element nodes were obtained by using the user-developing languages of the ANSYS 19.2, namely, the APDL scripts. Then, we reconstructed the vent hole boundary in CAD software by combining spline curves with the finite element nodes coordinates from the previous step. In this process, this integration formed the final outcome of our optimization methodology.

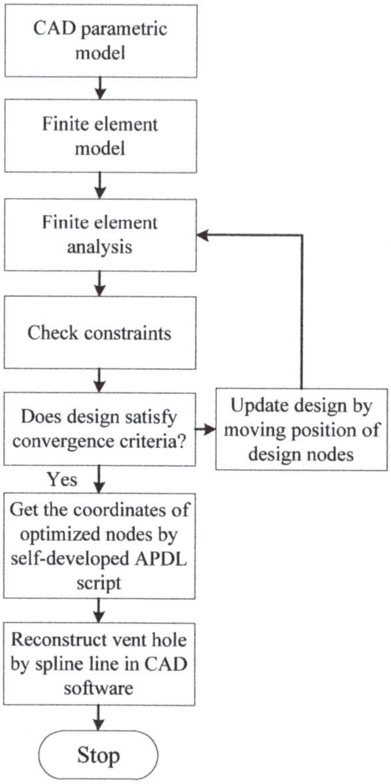

Figure 3. Flowchart of non-parametric-based shape optimization.

3. Vent Hole of a Gas Turbine Flow Guide Disk

3.1. Introduction of a Vent Hole

In this study, we focused on optimizing a gas turbine flow guide disk vent hole, which served as a representative example. Figure 4 illustrates the schematic diagram of a flow guide disk, wherein the vent holes play a crucial role in supplying cooling air to the turbine rotor. It is important to note that the location and configuration of these vent holes significantly impact the performance of the air system, as well as the fatigue life of the disk.

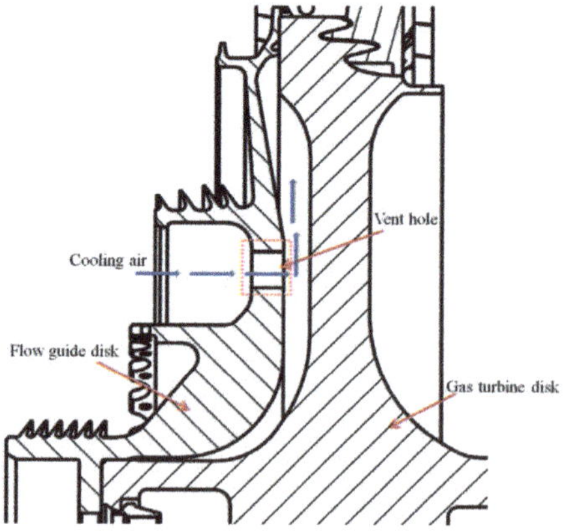

Figure 4. Schematic diagram of the gas turbine flow guide disk.

Considering that the positioning and size of the vent holes often result in considerable circumferential and radial stresses, the structural designer devised a solution to mitigate the stress concentration at the hole edges. As depicted in Figure 5, the designer initially employed a design approach featuring a straight line superimposed on multiple circular arcs for the hole, effectively reducing the stress concentration. By addressing these design considerations and optimizing the vent hole configuration, we aimed to enhance the overall performance and durability of the gas turbine system.

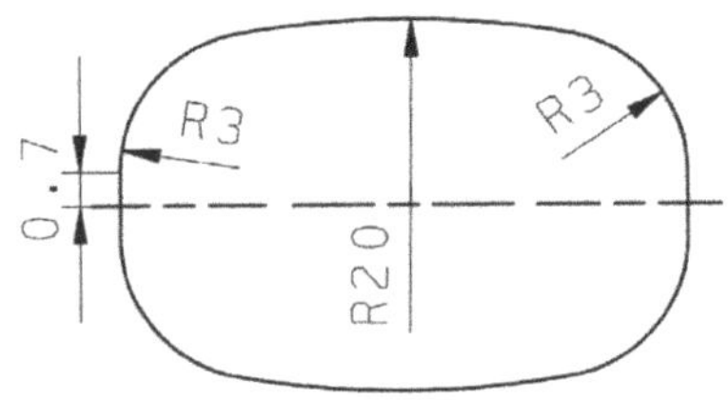

Figure 5. Schematic diagram of the vent hole (initial design).

3.2. *Strength Evaluation of the Initial Design*

The material used for the flow guide disk was the FGH95 Ni-based superalloy, which is commonly used in aerospace and gas turbine applications, especially for aero engines in China. Some of the mechanical properties of the FGH95 superalloy are shown in Table 1.

Table 1. Mechanical properties of FGH95.

Temperature (°C)	Young's Modulus E (GPa)	Poisson's Ratio μ	Yield Strength $\sigma_{0.2}$ (MPa)	Ultimate Strength σ_b (MPa)	Density (kg/m^3)
25	214.4	0.305	1200	1600	8270
500	197.2	0.337	1140	1520	

The flow guide disk had N holes evenly distributed in the circumferential direction, and the structure and loads exhibited cyclic symmetry. To optimize the design process and save on computational time, the calculation model considered a 1/N cyclic symmetric section, which included a vent hole.

Since the vicinity of a hole was the focus of this study, in the finite element model, the mesh near the hole edge was refined based on a mesh sensitivity analysis, as shown in Figure 6. It was observed that the maximum equivalent stress near the hole edge converged when the mesh size was approximately 0.2 mm. Then, the local mesh size for strength evaluation around the hole edge was determined to be 0.2 mm, while a mesh size of 2 mm was used in the remaining regions to reduce the computational analysis time. The finite element analysis model consisted of a total of 182,490 elements, and 289,011 nodes, as depicted in Figure 7. In Figure 7, the x-axis represents the radial direction, the y-axis represents the circumferential direction, and the z-axis represents the axial direction.

The finite element model was subjected to centrifugal and temperature loads. The centrifugal loads were applied to the model in the form of a rotational speed of 20,868 r/min, while the temperature loads were applied to the nodes, and the temperature distribution was as shown in Figure 8. The axial and circumferential displacements of all the nodes on face A, as well as the axial displacements of all the nodes on face B, were constrained. Additionally, the cyclic symmetry constraint was applied to both the lower and upper boundaries.

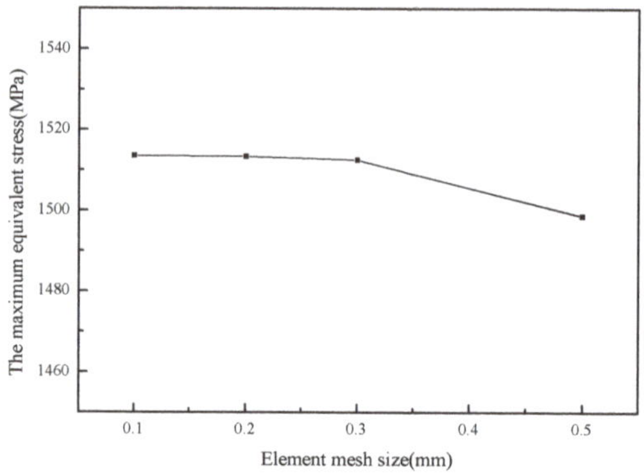

Figure 6. Mesh sensitivity analysis of the finite element model.

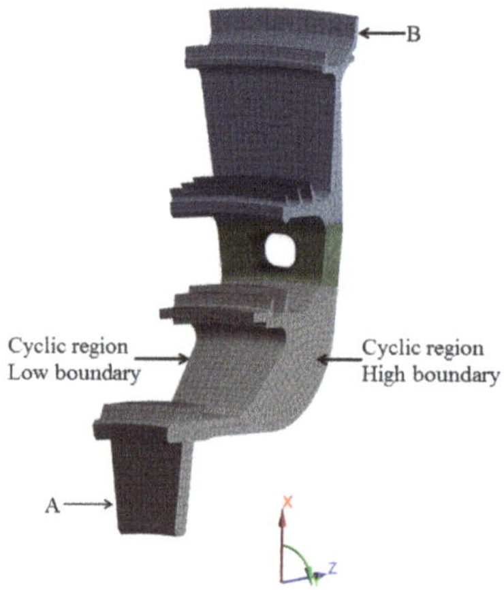

Figure 7. Finite element mesh of the gas turbine flow guide disk.

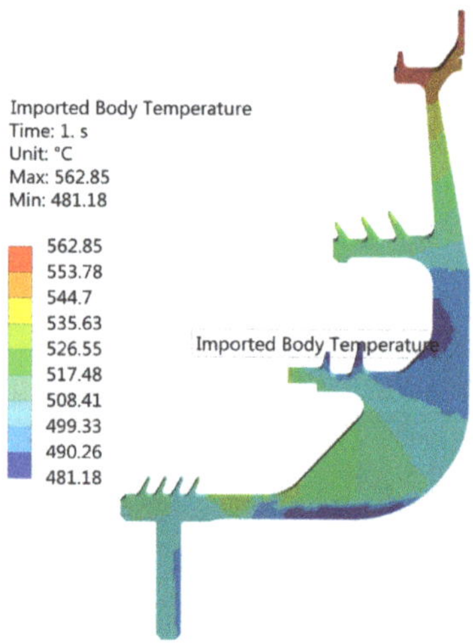

Figure 8. Temperature distribution of the gas turbine flow guide disk.

Under the given load conditions, the linear elastic finite element simulation analysis was performed with a calculation time of approximately 5 min on an Intel Xeon E5-1620 V4 Core 32 G memory computing platform. The equivalent stress distribution of the flow guide disk was as shown in Figure 9. The maximum equivalent stress was 1513.3 MPa, and it was located at the R3 circular arc section of the hole edge, as shown in Figure 5. This indicated that the edge of the vent hole served as a weak point in the fatigue life of the flow guide disk, and the stress distributions along the upper and lower edges of the hole were closely symmetric with respect to the middle line of the enlarged position shown in Figure 9.

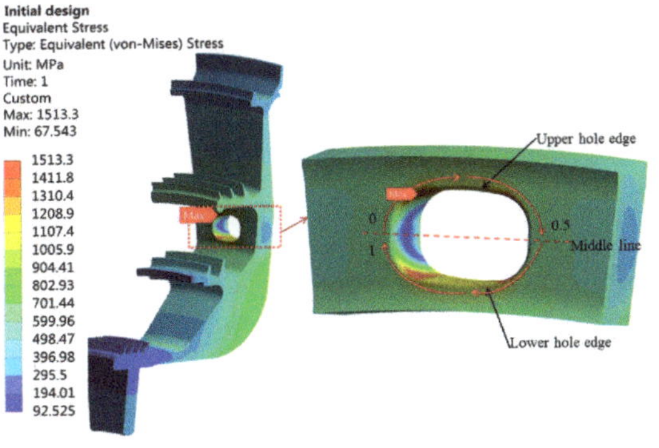

Figure 9. Equivalent stress distributions in the gas turbine flow guide disk's initial design.

In order to facilitate the subsequent design optimization, the detailed stress distribution and composition of the hole's front edge were extracted according to the normalized path of 0~0.5~1, as shown in Figure 9. Figure 10 illustrates the stress distribution along the vent hole edge of the initial design, encompassing both the equivalent stress and normal stress. An analysis of the figure revealed the following: the maximum circumferential stress (normal stress on the Y-axis) measured 1392 MPa, while the maximum radial stress (normal stress on the X-axis) amounted to 791 MPa. The circumferential stress distribution closely resembled the equivalent stress and surpassed the radial stress, implying that the circumferential stress served as the predominant stress component. Consequently, to mitigate the maximum stress value at the hole edge, the overall optimization approach proposed in this paper focused on reducing the curvature of the curve at the R3 edge of the hole and employing the "compress the upper and lower edge, stretch the left and right edge" strategy. This study incorporated methods based on geometric parameters and non-parametric techniques.

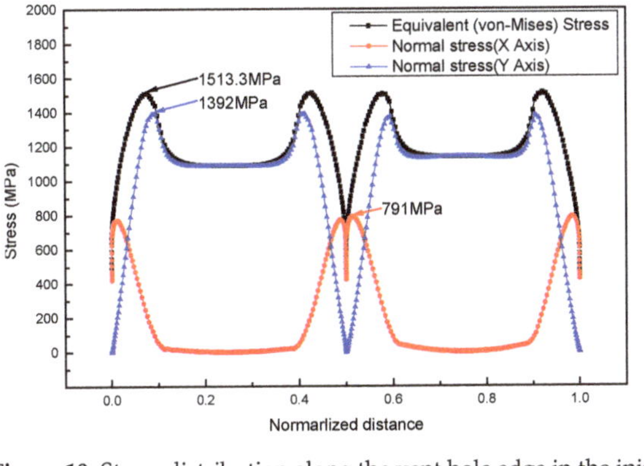

Figure 10. Stress distribution along the vent hole edge in the initial design.

4. Design and Optimization

4.1. Geometric Parameter-Based Optimization

After conducting the aforementioned strength analysis, the vent hole scheme was redesigned in this study with the aim of reducing the stress value at the hole edge. As illustrated in Figure 11, the vent hole was modified to an elliptical shape based on our engineering experience, with the semi major axis denoted as '*a*' and the semi minor axis denoted as '*b*'. The major axis of the ellipse defined the circumferential direction of the vent hole, while the minor axis determined the radial direction. Since the air system imposed restrictions on the vent hole cross-sectional area, the cross-sectional area of the vent hole remained constant at '*S*'. This modification addressed the stress-related concerns and ensured compliance with the vent hole cross-sectional area requirements.

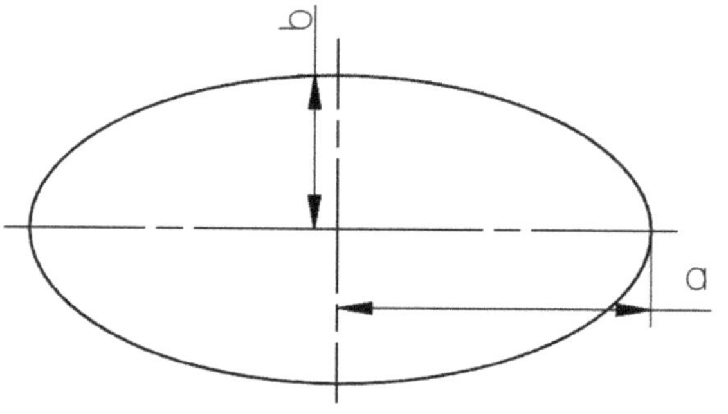

Figure 11. Schematic diagram of the vent hole based on the geometric parameters for vent holes.

In this study, the mathematical model for geometric parameter-based shape optimization can be represented by Equation (1), as follows:

$$\begin{aligned} &\min\ \sigma_{eqv,\max} \\ &find\ a \\ &s.t.\ S = S_0 \end{aligned} \tag{1}$$

In Equation (1), $\sigma_{eqv,\max}$ is the maximum equivalent stress at the hole edge, a is the semi-major axis of the ellipse, and S represents the cross-sectional area of vent hole, which is equal to a specific value, denoted as 'S_0', according to the requirements of the air system on the vent hole. In this case, S_0 was equal to 84.198 mm^2. The shape optimization method based on the geometric parameters necessitated adjustments to the relevant geometric model parameters, CAD model re-meshing, and interpolation of the temperature field during each iterative sub-step of the optimization process. Because the cross-sectional area remained basically unchanged, the variations in the vent hole had no significant impacts on the temperature field, and therefore, the temperature load of the finite element model remained unchanged during the optimization iterations. Given that the number of design variables was limited to only one, the NLPQL gradient algorithm [30] was employed in this study to expedite the optimization calculations. Figure 12 illustrates the comparison between the initial design scheme and the optimal design scheme achieved through geometric parameter-based optimization. Furthermore, Figure 13 depicts the distribution of the equivalent stress in the flow guide disk and the vent hole when employing the optimal design scheme based on the geometric parameters.

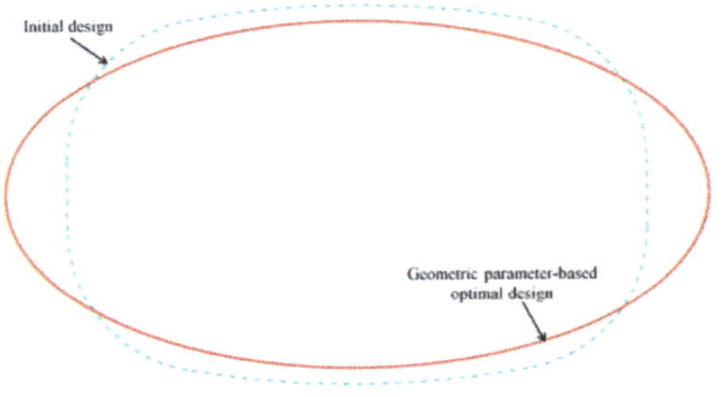

Figure 12. Comparison of the schemes between the initial design and the geometric parameter-based optimal design.

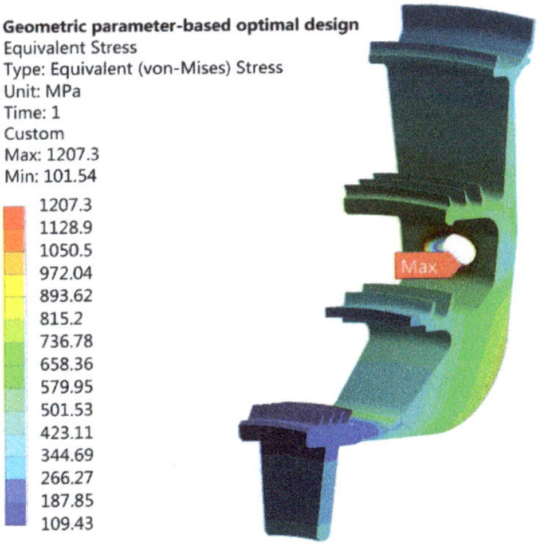

Figure 13. Equivalent stress distribution in the geometric parameter-based optimal design.

Table 2 presents a comparison of the optimization results between the initial design and the design achieved through geometric parameter-based optimization. The optimization results indicated that the semi-major axis (a) of the ellipse after optimization was 7.3 mm, corresponding to a semi-minor axis value of 3.67 mm. The area of the vent hole was 84.166 mm^2, which was 0.04% smaller than the initial value of 84.198 mm^2. This slight difference was acceptable within the engineering requirements for the design dimensions of the vent hole, which are typically specified to two decimal places, and it could be accommodated by the air system. The maximum equivalent stress at the hole edge in the optimal design based on the geometric parameters was 1207 MPa, reflecting a significant reduction of 20.2% compared to the initial design, which recorded a maximum equivalent stress of 1513 MPa. This improvement highlighted the effectiveness of the geometric parameter-based optimization approach in mitigating stress levels at the hole edge.

Table 2. Comparison of the results between the initial design and the geometric parameter-based optimal design.

Parameters	a/mm	S/mm^2	$\sigma_{eqv,\mathrm{max}}$/MPa
Initial design	-	84.198	1513
Geometric parameter-based optimal design	7.3	84.166	1207
Difference (%)	-	−0.04%	−20.2%

4.2. Non-Parametric-Based Optimization

For the non-parametric shape optimization of the vent hole, all nodes located along the hole's edge were chosen as design variables. To ensure sufficient mesh quality for the finite element analysis, adaptive mesh technology was employed in Tosca Structure. This technique enables the cells surrounding a design's variable nodes to move adaptively within a specified range, thereby maintaining a high-quality mesh around a vent hole. In this study, the optimization objective was to minimize the maximum equivalent stress at the vent hole's edge in the flow guide disk. The specific mathematical model for this optimization process can be expressed by Equation (2), as follows:

$$\begin{aligned} &\min\ \sigma_{eqv,\max} \\ &find\ \ Node_{i,x,y}\ \ (i = 1 \ldots j) \\ &s.t.\ \ S = S_0 \end{aligned} \tag{2}$$

In Equation (2), $\sigma_{eqv,\max}$ is the maximum equivalent stress at the hole edge, $Node_{i,x,y}$ is the location of node i of the hole edge, and S is the cross-sectional area of the vent hole, which is expected to be kept the same during the optimization iterations. To meet the air system's restriction requirements for the vent hole's cross-sectional area, the constraints were transformed to maintain the optimized structure volume equal to the initial design volume. Due to the manufacturability requirements of the vent hole, it needed to be aligned with the basic hole scheme in the axial direction. To meet the engineering manufacturing requirements, constraints were applied to the movement of these design variables. The nodes located at the same angular positions along the hole edges were selected as a group. Figure 14 illustrates the configuration of the four node groups (red nodes) established in this study, totaling 66 groups, ensuring the consistent direction and magnitude of movement for the nodes within the same groups. If this constraint was not imposed, the direction and magnitude of movement for each node within the same group would have been inconsistent, and the hole would not have been smooth along the axil direction. As a result, the optimization outcome could not have been manufactured.

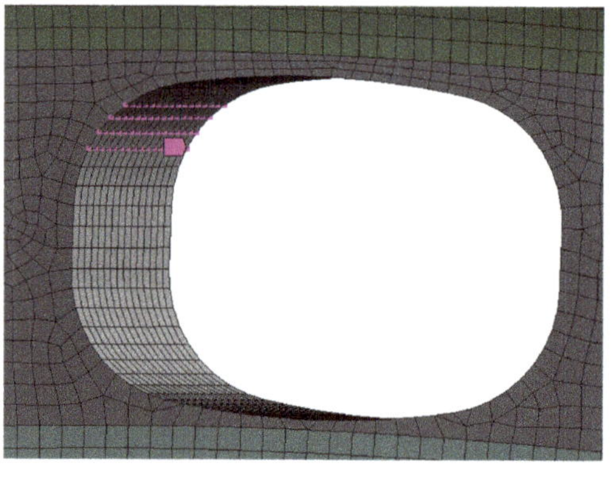

Figure 14. Equivalent stress distribution of the geometric parameter-based optimal design.

The variations in the equivalent (von-Mises) stress during the iteration process is shown in Figure 15. We can see that after 46 iterations in Tosca Structure, the equivalent (von-Mises) stress of vent hole converged to 1130 MPa, and we successfully obtained the optimal results for the vent hole in the flow guide disk.

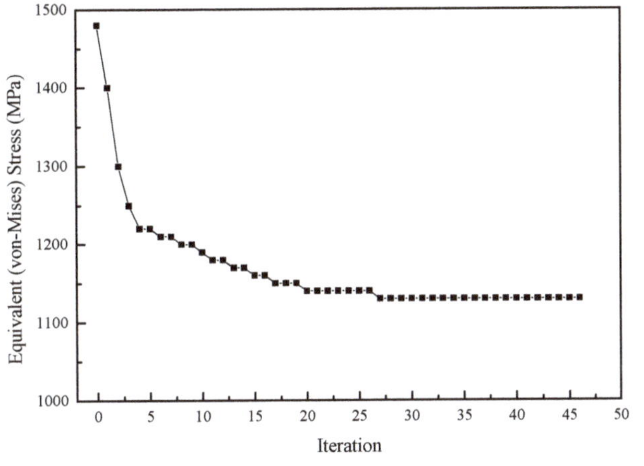

Figure 15. The variations in the equivalent (von-Mises) stress during the iterative process.

We generated an optimized 3D model by using a self-developed APDL script and spline curves. Figure 16 displays a comparison between the initial design scheme and the optimal design achieved through the non-parametric-based optimization approach. It was evident that the optimized scheme based on non-parametric optimization closely resembled the "ellipse-like" scheme derived from the geometric parameter-based optimization. Furthermore, Figure 17 presents the distribution of the equivalent stress in the flow guide disk and vent hole, highlighting the improvements achieved with the non-parametric-based optimal design.

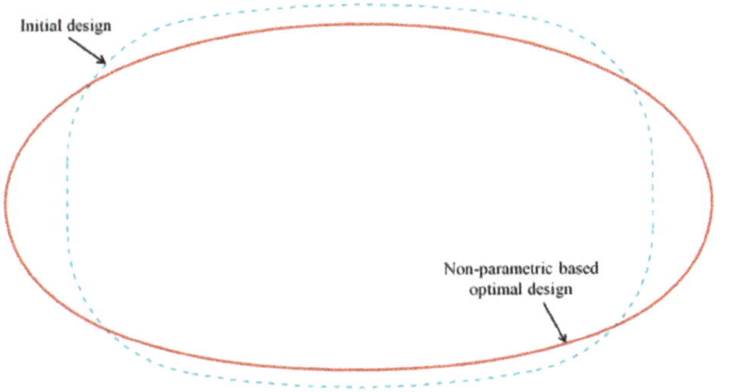

Figure 16. Comparison of the schemes between the initial design and the non-parametric-based optimal design.

Table 3 provides a comprehensive comparison of the optimization results between the initial design and the optimal design achieved through the non-parametric shape optimization approach. The maximum equivalent stress at the hole edge in the non-parameter-based optimal design was reduced by 24.5%, measuring 1142 MPa compared to the initial design's stress of 1513 MPa. The non-parametric-based optimal design showcased the successful reduction in stress concentrations, resulting in a more uniform stress distribution throughout the vent hole. These improvements highlighted the effectiveness of the non-parametric approach in achieving stress reductions and enhancing the overall performance of the design. After optimization, the cross-sectional area of the vent hole was approximately 84.304 mm^2, which was approximately 0.13% larger than its initial value of 84.198 mm^2. This discrepancy was attributed to the error introduced by using spline curves to describe the optimized vent hole mesh boundary. However, this could be accepted by the air system requirements.

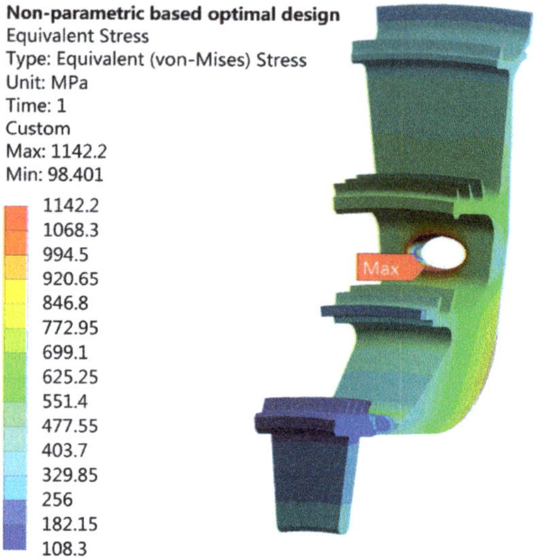

Figure 17. Equivalent stress distribution of the non-parametric-based optimal design.

Table 3. Non-parametric shape optimization-based method before and after the optimization results.

Parameters	S/mm^2	$\sigma_{eqv,\max}$/MPa
Initial design	84.198	1513
Non-parametric-based optimal design	84.304	1142
Difference (%)	+ 0.13%	−24.5%

5. Discussions

Two optimization methods, namely, geometric parameter-based optimization and non-parametric optimization, were employed to optimize the vent hole in a gas turbine flow guide disk. Figure 18 illustrates the design schemes of the initial design and the optimal designs obtained through the two different optimization methods. The results demonstrated significant changes in the optimal design schemes compared to the initial design. Notably, the optimal designs generated by both methods exhibited similarities, resembling elliptical shapes with slight variations in curvature at certain locations. This indicated that both optimization methods converged towards a similar optimal design solution, emphasizing the effectiveness of both approaches in achieving the desired design modifications.

Tables 2 and 3 present the results obtained using the two different optimization methods utilized in this study. The following observations could be made:

(1) Both optimization methods resulted in an optimal vent hole design that exhibited a more uniform stress distribution compared to the initial design.

(2) Both optimal designs effectively reduced stress concentrations. The geometric parameter-based optimal design achieved a 20.2% reduction in the maximum equivalent stress, while the non-parametric-based optimal design achieved a greater reduction of 24.5% compared to the initial design. It was worth noting that the non-parametric-based shape optimization method proved to be more efficient in optimizing the maximum equivalent stress at the hole edge.

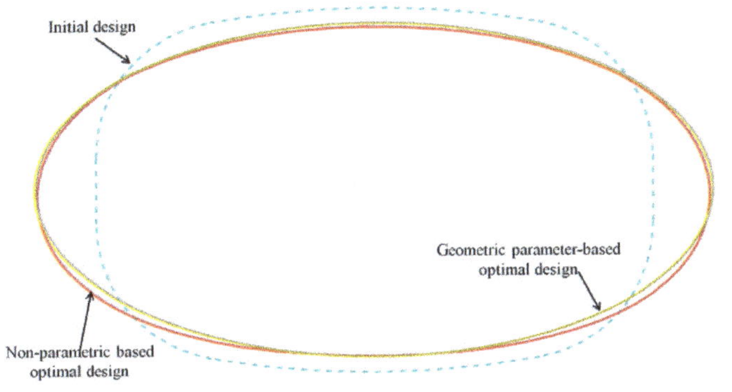

Figure 18. Comparison of the schemes between the initial design and two different optimization methods.

These findings demonstrated the effectiveness of both optimization methods in improving stress distribution and reducing stress concentration, with the non-parametric-based method showcasing a higher efficiency in stress optimization.

In order to gain a more detailed understanding of the differences between the two optimization methods, Figure 19 provides the equivalent stress distribution along the front edge of the vent hole for the initial design and the two optimal designs, and the following observations can be made:

(1) In comparison to the elliptical scheme, the non-parametric-based optimization resulted in a smaller curvature at the location where the maximum equivalent stress occurred. This led to a more uniform stress distribution and a lower maximum equivalent stress value.

(2) While the geometric parameter-based design achieved a 20.2% reduction in the maximum equivalent stress compared to the initial design, it is important to note that the establishment of the optimized hole shape boundary relied heavily on the engineering experience of the designer. Different designers may utilize different hole boundaries, which can result in varying design optimization outcomes.

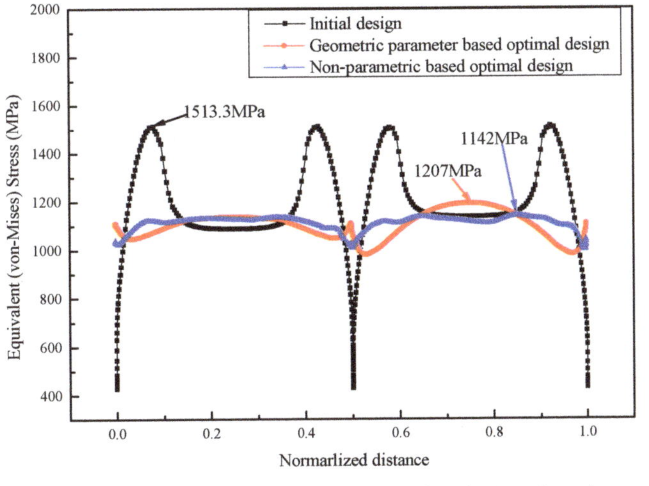

Figure 19. Comparison of the equivalent stress distribution along the vent hole edge between the different designs.

These findings highlighted the advantages of the non-parametric-based optimization method as it provided greater flexibility in achieving an optimal stress reduction and uniform stress distribution while mitigating the dependence on individual designer experience.

Based on the fundamental knowledge of fatigue theory, it is known that reducing the maximum equivalent stress and enabling a more balanced stress distribution along the vent hole edge through optimization is beneficial for improving the fatigue life of a vent hole. A comparison of fatigue life of the vent holes before and after optimization is presented in Table 4, and it is based on our institute's fatigue database for the FGH95 Ni-based superalloy. It can be observed that the geometric parameter-based optimal design resulted in an increase in fatigue life from 3701 cycles to 11,066 cycles, representing a 199% improvement, while the non-parametric-based optimal design yielded a fatigue life of 14,486 cycles, indicating a 291% improvement.

Table 4. Comparison of fatigue life between the different designs.

Parameters	Initial Design	Geometric Parameter-Based Optimal Design	Non-Parametric-Based Optimal Design
N_f/cycles	3701	11,066	14,486

6. Conclusions

This study established design optimization models and processes for vent holes using geometric parameter-based and non-parameter-based designs. Two shape optimization methods have been presented, and the differences between the two optimal designs have been studied. The key conclusions from this study are as follows:

(1) The stress concentration in the vent hole was reduced and the stress distribution became more uniform after optimization compared to the initial design. This optimization method can be used as a reference for the design optimization of similar engineering structures.

(2) The geometric parameter-based optimal design achieved a 20.2% reduction in maximum equivalent stress compared to the initial design, while the non-parameter-based optimal design achieved a 24.5% reduction.

(3) The non-parametric-based shape optimization method demonstrated better optimization results and effectively avoided the poor design optimization results caused by the lack of relevant structural and strength design experience among designers.

(4) Through the application of the two optimization methods, the geometric parameter-based approach resulted in an improvement in the fatigue life of the vent hole from 3701 cycles to 11,066 cycles. Additionally, the non-parametric-based shape optimization yielded a fatigue life of 14,486 cycles.

In conclusion, this study has successfully established effective design optimization models and processes for vent holes and has provided valuable insights into the differences between geometric parameter-based and non-parameter-based optimization methods. These findings will be useful for future engineering designs.

Author Contributions: Conceptualization, C.W., Y.Z. and D.M.; methodology, C.W., Y.Z. and H.Z.; investigation, Y.Z. and Z.Q.; writing—original draft preparation, C.W., Y.Z., D.M. and Z.Q.; writing—review and editing, C.W., Y.Z., D.M., Z.Q. and H.Z.; supervision, C.W. and D.M.; project administration, D.M.; funding acquisition, D.M. All authors have read and agreed to the published version of the manuscript.

Funding: This research was funded by the Independent Innovation Special Project of AECC (grant number: ZZCX-2018-017).

Data Availability Statement: Not applicable.

Conflicts of Interest: The authors declare no conflict of interest.

References

1. Lu, S.; Huang, Q. New method for damage tolerance analysis of turbine disk and its application. *J. Aerosp. Power* **2002**, *17*, 87–92.
2. Li, W.; Dong, L.W.; Cai, X.H.; Zhao, F.X.; Liu, D. Structure analysis and life evaluation of the pin holes in a turbine disk of a type of aero-engine. *J. Aerosp. Power* **2009**, *24*, 1699–1706.
3. Pedersen, N.L. Optimization of bolt thread stress concentrations. *Arch. Appl. Mech.* **2012**, *83*, 1–14. [CrossRef]
4. Yang, J.; Li, C.B.; Xie, S.S. Multiaxial low cycle fatigue life prediction and test verification for turbine disk. *J. Aerosp. Power* **2011**, *26*, 2220–2226.
5. Myneni, M.; Benjamin, C.C.; Rajagopal, K.R. Stress concentration factors around a circular hole in two fiber reinforced materials under large deformations. *Mech. Mater.* **2021**, *163*, 104089. [CrossRef]
6. Zheng, C.; Zhang, G.; Mi, C. On the reduction of stress concentrations in thick-walled hollow spheres under an arbitrary uniform axial symmetric outer traction. *Int. J. Press. Vessel. Pip.* **2022**, *194*, 104545. [CrossRef]
7. Chen, Q.; Guo, H.; Liu, X. Modeling and optimization of biaxially symmetric shaped hole structure of turbine disk. *J. Aerodyn.* **2013**, *28*, 1250–1256.
8. Chen, Q.; Guo, H.; Zhang, C.; Liu, X. Structural optimization of uniaxial symmetry non-circular bolt clearance hole on turbine disk. *Chin. J. Aeronaut.* **2014**, *27*, 1142–1148. [CrossRef]
9. Sun, S.P.; Hu, J.T.; Zhang, W.H. Optimization of slewing shell opening shape based on hyperelliptic equation and sequential response surface method. *J. Aeronaut.* **2015**, *36*, 3595–3607.
10. Yan, C.; Zhu, J.; Shen, X.; Fan, J.; Jia, Z.; Chen, T. Structural design and optimization for vent holes of an industrial turbine sealing disk. *Chin. J. Aeronaut.* **2021**, *34*, 265–277. [CrossRef]
11. Yan, C.; Yin, Z.; Shen, X.; Mi, D.; Guo, F.; Long, D. Surrogate-based optimization with improved support vector regression for non-circular vent hole on aero-engine turbine disk. *Aerosp. Sci. Technol.* **2020**, *96*, 105332. [CrossRef]
12. Nassiraei, H.; Rezadoost, P. Stress concentration factors in tubular X-connections retrofitted with FPR under compressive load. *Ocean. Eng.* **2021**, *229*, 108562. [CrossRef]
13. Nassiraei, H.; Rezadoost, P. SCFs in tubular X-connections retrofitted with FPR under in-plane bending load. *Compos. Struct.* **2021**, *274*, 114314. [CrossRef]
14. Nassiraei, H.; Rezadoost, P. SCFs in tubular X-joints retrofitted with FPR under out-of-plane bending moment. *Mar. Struct.* **2021**, *79*, 103010. [CrossRef]
15. Zhu, J.H.; Li, J.; Zhang, W.; Meng, L.; Wang, D. Application of modern shape optimization techniques in aero-engine component design. *Aerosp. Manuf. Technol.* **2012**, *23/24*, 30–35.
16. Movahedi, S.; Taghizadieh, N. Efficiency improvement of radial basis function meshless method in conjunction with bayesian theorem for electrical tomography of heterogeneous concrete. *Eng. Anal. Bound. Elem.* **2021**, *135*, 382–393. [CrossRef]
17. Luh, G.C.; Lin, C.Y. Optimal design of truss-structures using particle swarm optimization. *Comput. Struct.* **2011**, *89*, 2221–2232. [CrossRef]
18. Cao, Z.; Lin, G.; Shi, Q.; Cao, Q. Optimization analysis of NURBS curved variable stiffness laminates with a hole. *Mater. Today Commun.* **2022**, *31*, 103364. [CrossRef]
19. Gunasekaran, V.; Gulhane, S.; Gupta, S.; Pitchaimani, J.; Rajamohan, V.; Manickam, G. Structural–Acoustic Response Analysis of Variable Stiffness Laminates with Inherent Material Damping. *Int. J. Struct. Stab. Dyn.* **2022**, *22*, 2250127. [CrossRef]
20. Joseph, G.J.; Mohan, G.; Harikrishna, V.; Sandra, M.V.; Sajith, A.S. Size Optimization of Truss Structures Using Real-Coded Genetic Algorithm with a Novel Constraint Handling Method. In Proceedings of the International Conference on Structural Engineering and Construction Management, Angamaly, India, 1–3 June 2022; Springer: Cham, Switzerland, 2022; Volume 284, pp. 915–923.
21. Tian, D.; Shi, Z. MPSO: Modified particle swarm optimization and its applications. *Swarm Evol. Comput.* **2018**, *41*, 49–68. [CrossRef]
22. Hahn, Y.; Cofer, J.I. Study of Parametric and Non-Parametric Optimization of a Rotor-Bearing System. In *Turbo Expo: Power for Land, Sea, and Air*; American Society of Mechanical Engineers: New York, NY, USA, 2014. [CrossRef]
23. Chen, J.; Shapiro, V.; Suresh, K.; Tsukanov, I. Parametric and Topological Control in Shape Optimization. In Proceedings of the 32nd Design Automation Conference, Part A and B, Philadelphia, PA, USA, 10–13 September 2006; pp. 575–586.
24. Shimoda, M.; Nagano, T.; Shi, J.-X. Non-parametric shape optimization method for robust design of solid, shell, and frame structures considering loading uncertainty. *Struct. Multidiscip. Optim.* **2019**, *59*, 1543–1565. [CrossRef]
25. Shimoda, M.; Liu, Y.; Morimoto, T. Non-parametric free-form optimization method for frame structures. *Struct. Multidiscip. Optim.* **2014**, *50*, 129–146. [CrossRef]
26. Liu, Y.; Shimoda, M. Non-parametric shape optimization method for natural vibration design of stiffened shells. *Comput. Struct.* **2015**, *146*, 20–31. [CrossRef]
27. Lee, J.-R.; Harland, A.; Roberts, J. Non-Parametric Shape Optimization of a Football Boot Bottom Plate. *Proceedings* **2020**, *49*, 152.
28. Xu, G.; Wang, G.-Z.; Chen, X.-D. Free-Form Deformation with Rational DMS-Spline Volumes. *J. Comput. Sci. Technol.* **2008**, *23*, 862–873. [CrossRef]

29. Llorens, A.d.T.; Kiendl, J. An isogeometric finite element-boundary element approach for the vibration analysis of submerged thin-walled structures. *Comput. Struct.* **2021**, *256*, 106636.
30. Zhang, Y.; Xie, J.L. Truss structural optimization based on NLPQL and MIGA. *Appl. Mech. Mater.* **2013**, *365–366*, 150–154. [CrossRef]

metals

MDPI

Article

Mechanical Fracture of Aluminium Alloy (AA 2024-T4), Used in the Manufacture of a Bioproducts Plant

Luis Fabian Urrego [1], Olimpo García-Beltrán [2], Nelson Arzola [3] and Oscar Araque [1,*]

[1] Departamento de Desarrollo Tecnológico, Facultad de Ingeniería, Universidad de Ibagué, Ibagué 730002, Colombia; 2120151024@estudiantesunibague.edu.co
[2] Facultad de Ciencias Naturales y Matemáticas, Universidad de Ibagué, Carrera 22 Calle 67, Ibagué 730002, Colombia; jose.garcia@unibague.edu.co
[3] Research Group in Multidisciplinary Optimal Design, Departamento de Ingeniería Mecánica y Mecatrónica, Universidad Nacional de Colombia, Bogotá 111321, Colombia; narzola@unal.edu.co
* Correspondence: oscar.araque@unibague.edu.co

Abstract: Aluminium alloy (AA2024-T4) is a material commonly used in the aerospace industry, where it forms part of the fuselage of aircraft and spacecraft thanks to its good machinability and strength/weight ratio. These characteristics allowed it to be applied in the construction of the structure of a pilot plant to produce biological extracts and nano-encapsulated bioproducts for the phytosanitary control of diseases associated with microorganisms in crops of *Theobroma cacao* L. (Cacao). The mechanical design of the bolted support joints for this structure implies knowing the performance under fatigue conditions of the AA2024-T4 material since the use of bolts entails the placement of circular stress concentrators in the AA2024-T4 sheet. The geometric correction constant (Y) is a dimensionless numerical scalar used to correct the stress intensity factor (SIF) at the crack tip during propagation. This factor allows the stress concentration to be modified as a function of the specimen dimensions. In this work, four compact tension specimens were modeled in AA2024-T4, and each one was modified by introducing a second circular stress concentrator varying its size between 15 mm, 20 mm, 25 mm, and 30 mm, respectively. Applying a cyclic load of 1000N, a load ratio R=-1 and a computational model with tetrahedral elements, it was determined that the highest SIF corresponds to the specimen with a 30 mm concentrator with a value close to 460 MPa.mm$^{0.5}$. Where the crack propagation had a maximum length of 53 mm. Using these simulation data, it was possible to process each one and obtain a mathematical model that calculates the geometric correction constant (Y). The calculated data using the new model was shown to have a direct relationship with the behavior obtained from the simulation.

Keywords: crack; fatigue; geometric factor; support vector regression; pilot plant

Citation: Urrego, L.F.; García-Beltrán, O.; Arzola, N.; Araque, O. Mechanical Fracture of Aluminium Alloy (AA 2024-T4), Used in the Manufacture of a Bioproducts Plant. *Metals* **2023**, *13*, 1134. https://doi.org/10.3390/met13061134

Academic Editors: Haitao Cui and Qinan Han

Received: 9 May 2023
Revised: 10 June 2023
Accepted: 13 June 2023
Published: 16 June 2023

1. Introduction

Theobroma cacao (Cocoa) is considered one of the most important raw materials in international trade; it is a source of foreign exchange in 58 producing countries, highlighting that 89% of this production is found in Ivory Coast, Ghana, Indonesia, Nigeria, Brazil, Ecuador, Malaysia and Cameroon [1]. Cocoa production is sometimes affected by environmental, physical, and chemical factors and inadequate pest and disease control [2]. This crop is mainly infected by disease-causing microorganisms, among which *Moniliophthora roreri* and *Phytophthora* spp. stand out, which are the two main risk factors that directly affect annual cocoa production [3,4]. This is why management alternatives for these diseases, such as the use of plant extracts and essential oils (EO), should be sought, being favorable for environmental sustainability and human health [5]. This implies the development of an infrastructure capable of producing the raw material used in the production of bio-based products.

The study of the phenomenon of fracture has its origin in work proposed by Griffith (1921, 1924); the researcher Irwin (1957) made an important advance by proposing the analysis of fracture toughness as a function of stress, the toughness of a material is obtained from the applied stress and the crack length, but given the different test configurations, these values are known as a function of the failure modes called Stress Intensity Factor (SIF) K_I (Opening), K_{II} (in-plane shear) and K_{III} (out-of-plane shear). Estimation of the SIF in materials with linear elastic behavior (LEFM) is possible by quantifying the nominal stress and crack size. The use of the finite element method (FEM) allows testing for various configurations due to the versatility of the method; using the SMART component integrated into the latest versions of ANSYS finite element software, it is possible to perform simulations to verify the behavior of the geometric factor accompanying the determination of the SIF in the different failure modes.

Researchers have worked on this method, Nairn [6] has proposed the analysis of the geometric factor as a correction factor of the general SIF equation and that it can be expressed as a function of crack length (a) and width (W), on the other hand, for the author Mecholsky, the geometric factor (Y) for semi-elliptical cracks in materials of high hardness and brittleness is used to explain the position and shape of the crack because it is a function of an angle (θ) between the surface of the crack front and any peripheral point above it [7], otherwise, authors Taylor, Cornetti and Pugno, cataloged this geometric factor not only in terms of geometry but also in terms of the crack notch [8], on the other hand, when analyzing fatigue crack propagation, the geometrical factor is included in an expression known as the initial value of propagation for short cracks (a_0), which according to the authors B. Atzori, P. Lazzarin and G. Meneghetti, occurs at the point of intersection between the change in realized stress ($\Delta\sigma$) and the different values of toughness (ΔK), where the geometrical factor is calculated using a simulation in ANSYS 5.6 software [9].

The authors Smith and Scattergood have analyzed ceramic materials defining toughness as a sum of two different toughnesses: (K_{bend}), which is defined as a toughness that is a function of the stress intensity factor and the residual toughness ($K_{residual}$) that results from the residual stress field due to strain, an equation is obtained to determine the value of the toughness (K_{bend}) by an exponential equation, which involves a shape factor and the crack depth [10]. However, when analyzing materials with a higher degree of ductility, the empirical approaches of authors J.C. Newman and I.S. Raju obtained an equally accepted behavior for the determination of toughness [11].

On the other hand, when testing chromium steels, authors Nix and Lindley determined that the behavior of the shape factor (C_s) was also exponential in nature, where the basis again was the ratio (a/c), where (a) is the crack depth, and (c) is the crack length, where the values of this ratio were previously calculated by subjecting chromium steel specimens On the other hand, when testing chromium steels, the authors Nix and Lindley established that the behavior of the form factor, identifying that for the tests developed, the fracture toughness equation must involve a factor called (M_f), which is a crack front correction factor [12].

These correction factors have been the product of a rigorous algebraic analysis, where the basis of each of the equations that describe these factors, part of the simulations in finite element programs, which provide the input values necessary to apply the respective analysis algorithms, as proposed by the authors Clarke, Griebsch and Simpson, who explain how it is possible to glimpse different situations of mechanical nature by means of the Support Vector Regression (SVR), algorithm, this algorithm allows from some input values in the Cartesian plane, to obtain an equation that adjusts to the distribution of given values, considering the dispersion (ξ) of the same one, and the margin (ε) between the support vectors [13]. This algorithm uses the principles of Lagrangian optimization, simultaneously involving Kernel functions, which can have a polynomial, Gaussian or Sigmoidal nature, and as explained by researchers Schölkopf and Smola [14], authors Heydari and Choupani, indicate a correction factor for fracture toughness of a logarithmic nature based on the rate of energy release [15], authors El-Desouky and El-Wazery, indicated fracture toughness for materials with a high degree of brittleness, using a fifth-degree polynomial (F_1), whose variable is the relation (a/W), where (a) is the crack length and (W) is the length of the cross-sectional area [16].

The models for the geometric factor for compact specimens that are currently planned in the literature are based on specimens that have a single stress concentrator.

Aluminum alloys are widely used in industry to reduce the weight of structural components in machine design. Their light weight, easy machinability and excellent fatigue strength make these alloys ideal materials for manufacturing in the modern world. One of the most recent applications is in the high-speed rail industry, where 5083P-O Aluminium alloy is involved in the design of train bodies, leading to increased running speed and improved assembly performance [17].

The goal of the present work is to evaluate the behavior of this parameter when there are two stress concentrators arranged in the specimen geometry since it has been visually observed that the crack rupture direction is affected by the size and location of the concentrator with respect to the propagation plane. In the same way, the study of the geometrical factor implies a study of the stress intensity factor at the crack tip, which means that this last parameter must also be analyzed for the geometrical conditions proposed. The main differentiating feature lies in determining the stress intensity factor by applying the FEM method and then processing this data by means of the SVR algorithm. Applied to non-standardized specimens with variable diameter stress concentrators. Allowing the data obtained by the simulation to be used within the Nadaraya–Watson estimator to find a mathematical model that explains the behavior under variable tensile loading and crack propagation.

The importance of the model to be developed consists in its use as an element for joining bolted joints in structures with diameter variation with the purpose of establishing life projection parameters useful for design processes.

2. Materials and Methods

2.1. Material and Specimens

In this article, a modified compact tension specimen (MCTE) was used based on the ASTM-E399 standard, where the value of W was set at 100 mm and its thickness at 10 mm. In the same way, a stop hole of variable diameter was set in; in the same way, a stop hole of variable diameter was fixed in 4 different specimens; therefore, 4 simulations were performed for each modified compact specimen. The dimensions are shown in Figure 1. For the configuration of the SMART method used in ANSYS, it was necessary to characterize the Aluminium alloy (AA 2024-T4). This is hardened by a thermal aging process and presents the mechanical properties contained in [18] and presented in Table 1; similarly, the author Zyad Nawaf Haji [19] characterized the fatigue deformation parameters for AA2024-T4. These parameters are determined from the Ramberg–Osgood and Coffin–Mason equations that describe well the cyclic behavior of the material, but they are not physical laws [20]. These parameters are presented in Table 2.

Table 1. Mechanical properties AA2024-T4 [18].

Property	Value
Density (Kg/m^3)	2770
Coefficient of thermal expansion (1/C)	0.000023
Young's Modulus (MPa)	71,000
Poisson's Ratio	0.33
Shear Modulus (MPa)	26,692
Bulk Modulus (MPa)	69,608

The mechanical and fatigue properties of the AA2024-T4 used for the ANSYS simulation are shown below.

The authors B.M. Faisal, A.T. Abass and A.F. Hammadi [21] subjected to fatigue tests several specimens of AA2024-T4, determining its characteristic S-N curve, which is presented in Figure 2. Table 3 shows the values for the C and m constants for AA2024-T4 determined by authors Yang Guang, Gao Zengliang, Xu Feng and Wang Xiaogui [22] by subjecting eleven specimens to fatigue tests.

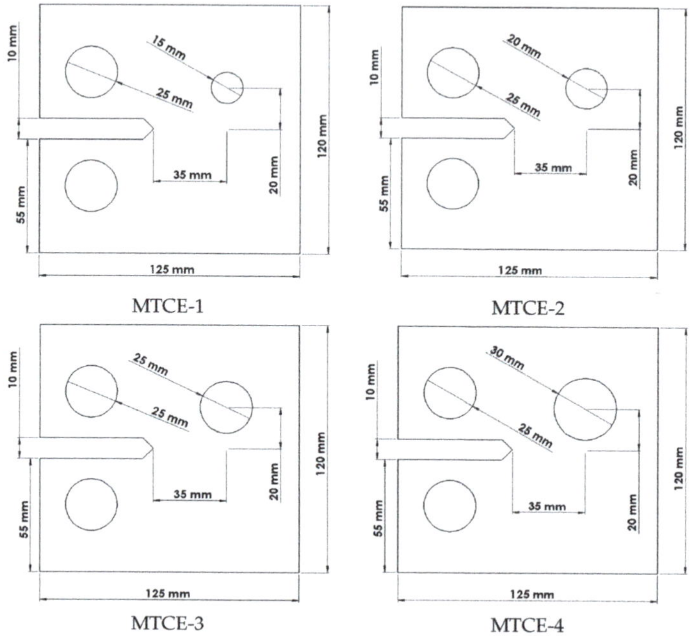

Figure 1. Dimensions of modified compact specimens used.

Table 2. Fatigue Parameters AA2024-T4 [20].

Parameter	Value
Strength Coefficient (MPa)	714
Strength Exponent	−0.078
Ductility Coefficient	0.166
Ductility Exponent	−0.538
Cyclic Strength Coefficient (MPa)	502
Cyclic Strain Hardening Coefficient	0.15

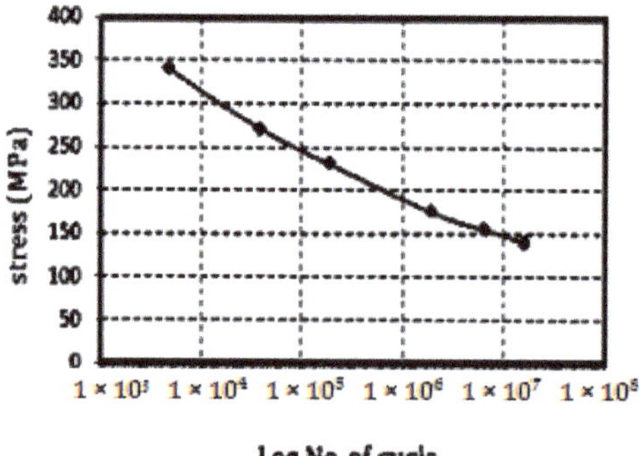

Figure 2. Characteristic fatigue behavior of the material AA 2024-T4 [21].

Table 3. Paris' Law Constants [22].

Constant	Value
C	5.75×10^{-8}
m	3.09

All modified compact tension specimens were subjected to failure mode 1 (opening), which causes a unidirectional cyclic stress perpendicular to the crack plane; using a force of 1000 N, using a Stress Ratio (R) = -1, since according to author C.M. Hudson fatigue cracks in Aluminium alloy AA 2024 propagate at a faster rate with $R = -1$ than with $R = 0$ when the same load was applied in both tests. Apparently, the compression portion of the loading cycle accelerates crack growth in this material [23]. Therefore, we proceeded to perform the simulation using ANSYS Mechanical ADPL 21R1, from which we calculated the K_I values and the crack extension values (a) in each sub-step of the solution. It was taken from Equation (1), formulated by R.P. Wei [24] for high-strength aluminum subjected to cyclic axial loading.

$$\Delta K_{I\ max} = Y\Delta\sigma\sqrt{\pi a} \quad (1)$$

Expressing ΔK_I as $(K_{I\ max} - K_{I\ min})$ and $\Delta\sigma$ as $(\sigma_{max} - \sigma_{min})$, it is possible to derive Equation (2) algebraically.

$$K_{I\ max} = Y\sigma_{max}(1-R)\sqrt{\pi a} \quad (2)$$

Subtracting for Y gives:

$$Y = \frac{K_{I\ max}}{\sigma_{max}(1-R)\sqrt{\pi a}} \quad (3)$$

Which is a dimensional expression that, for the purposes of the investigation, will be compared with the relative crack length, which is expressed by a ratio a/W and is also dimensionless.

2.2. Computational Model Using SMART Method

To achieve a better analysis of the SIF at the crack tip, the meshing was defined using a Patch Conforming Method defined by Tetrahedral elements, these elements, according to the ANSYS usage guide [25], are unique to the SMART method. Tetrahedral elements are 3-dimensional in nature, have a quadratic displacement behavior and are well suited for modeling irregular meshes. The element is defined by 10 nodes that have 3 degrees of freedom in each of the $x, y,$ and z nodal directions see Figure 3.

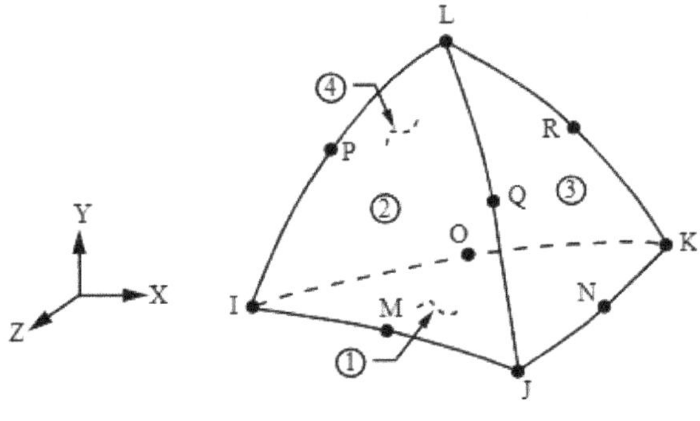

Figure 3. Tetrahedral elements.

The element has plasticity, hyperelasticity, creep, tensile stiffness, good deflection, and tension capabilities. It also has mixed formulation capabilities to simulate deformations of nearly incompressible elastoplastic materials and fully incompressible hyperelastic materials [26]. Similarly, the Refinement method was used to perform sectorized refinements in the crack front and its adjacent areas, as well as in the areas surrounding the circular stress concentrators (stop holes). The grids, the number of elements and nodes of each computational model used are shown below.

The Smart Crack Growth component used in the current research does not require the placement of a pre-crack geometry; however, it does require the involvement of 3 parameters defined within the Fracture-Premesh Crack block that are defined based on the geometry of the notch placed in the specimen. These parameters must be nodal surfaces adjacent to the crack front. The first is defined as a function of the nodes located on the edge of the crack front, which in the present work, we call the front, and interprets the place at which the crack propagation would start; the second, which we call the top, interprets the nodes located on the upper adjacent surface to the crack front; the third, which we call the bottom, interprets the nodes located on the lower adjacent part of the crack front.

Similarly, Figure 4a illustrates the nodes on the crack front (front), Figure 4b illustrates the nodes on the top surface adjacent to the crack front (top), Figure 4c illustrates the nodes on the bottom surface adjacent to the crack front (bottom) and Figure 4d illustrates the location of the pre-crack conditions with the coordinate axes, inferring that the propagation will be along the x-axis.

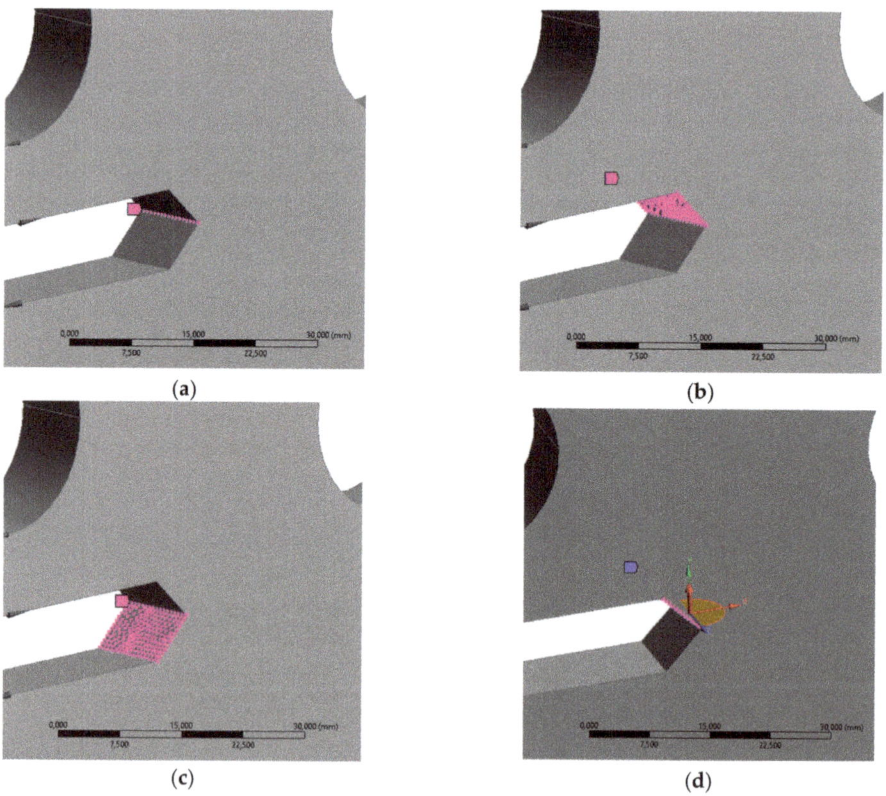

Figure 4. (**a**) illustrates the nodes on the crack front (front), (**b**) illustrates the nodes on the top surface adjacent to the crack front (top), (**c**) illustrates the nodes on the bottom surface adjacent to the crack front (bottom) and (**d**) illustrates the location of the pre-crack conditions with the coordinate axes, inferring that the propagation will be along the x-axis.

In Figures 5–8, the working specimen is presented, showing the number of elements and nodes for each of the variations in the size of the diameter of the stress concentrator elements.

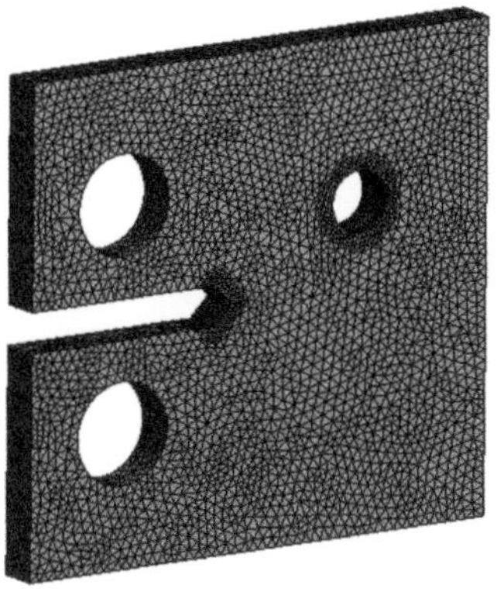

Figure 5. MTCE-1 mesh with 101742 elements and 151424 nodes.

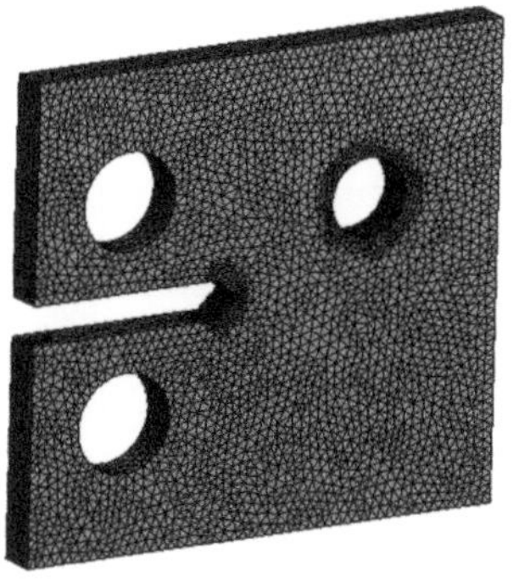

Figure 6. MTCE-2 Mesh with 104899 elements and 156293 nodes.

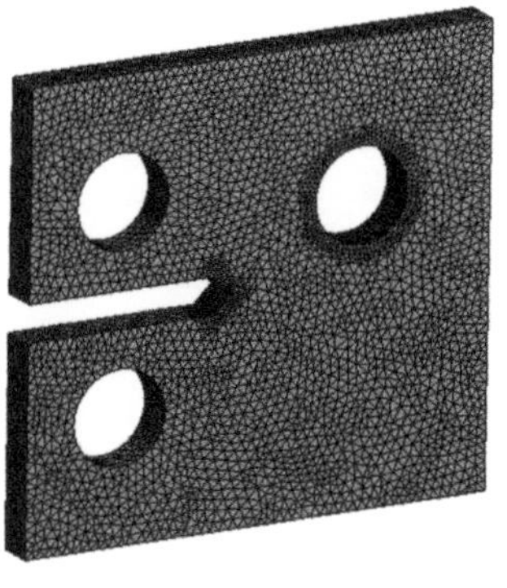

Figure 7. MTCE-3 Mesh with 107059 elements and 159635 nodes.

The numerical model was defined using the patch-forming method with tetrahedral elements. A general refinement was applied to the entire geometry of each specimen by setting the size of each element to 2.5 mm, and then a surface refinement was applied to each of the concentrators. Within the analysis setup, a time-defined solution was established with 10 s as the time limit and 0.5 s as the time in each solution step.

Figure 8. MTCE-4 Mesh with 99726 elements and 149780 nodes.

2.3. Data Processing Using Support Vector Regression (SVR) and Nadaraya-Watson Estimator (NWE)

The SVR algorithm is an algorithm for linear or non-linear regression of points in the Cartesian plane, whose purpose is to find the equation of the hyper-plane that interpolates all the points, based on the use of Kernel functions $K(x)$ (see Table 4). Furthermore, considering the margin between vectors (ε), the dispersion that exists between the margins and the points furthest from it (ζ) and the arithmetic mean of the data (μ). These are variables that can be imposed during regressions in this way. The quality of the regressions performed can be measured in terms of the coefficient of determination (R^2), since the closer this value is to 1, the higher the quality of the regression. The rationale of the SVR algorithm lies in the quadratic optimization of a Gram matrix, which is maximized subject to a real domain condition to find the Lagrange multipliers [26].

Table 4. Kernel Equations.

Kernel Equations	
Linear	$K(\vec{X}_n, \vec{X}_n') = \vec{X}_n^T \cdot \vec{X}_n'$
Polynomial	$K(\vec{X}_n, \vec{X}_n') = \left(\vec{X}_n \cdot \vec{X}_n'\right)^d$
Gaussian	$K(\vec{X}_n, \vec{X}_n') = e^{-\frac{\lVert \vec{X}_n - \vec{X}_n' \rVert^2}{2\sigma^2}}$
Sigmoidal	$K(\vec{X}_n, \vec{X}_n') = \tanh\left(\left(\vec{X}_n \cdot \vec{X}_n'\right) + \varphi\right)$
Epanechnikov	$K(\vec{X}_n, \vec{X}_n') = \frac{3}{4}\left(1 - (\vec{X}_n \cdot \vec{X}_n')^2\right)$

The calculation of the weight values w_n that are calculated by means of the Lagrange multipliers were obtained by means of a code in Python language and from which the approximation was carried out by means of the Nadaraya-Watson estimator. To determine which of the Kernel equations shows similar behavior.

According to author Larroca, F., the Nadaraya–Watson estimator is a type of non-parametric estimator that uses an equation, $\hat{m}(x)$, and a fit parameter, ε_o, to interpolate a series of data in the Cartesian plane, this equation, $\hat{m}(x)$, is given as a function of a weighted average of a Kernel density equation (see Table 4), which is assigned to each of the points taken [27]. However, the author Cai. Z. [28] proposes that the function $\hat{m}(x)$ is altered with weight values, w_n, to improve its fit, therefore, its general form is given by:

$$Y_{pred} = \hat{m}(x) \pm \varepsilon_o \tag{4}$$

According to the authors Demir, S. and Toktamis, O. [29], the use of the Kernel within ENW implies changing the dot product between vectors by the expression $\frac{x_n - x}{h}$, then, the function $\hat{m}(x)$ is posed as:

$$\hat{m}(x) = \frac{\sum_{i=1}^{n} w_n \cdot K\left(\frac{x_n - x}{h}\right) y_n}{\sum_{i=1}^{n} w_n \cdot K\left(\frac{x_n - x}{h}\right)} \tag{5}$$

According to the author Fan, J., K symbolizes a Kernel density equation, x_n is the x-axis coordinate of each point in the Cartesian plane, y_n is the y-axis coordinate of each point in the Cartesian plane, n is the number of points in the plane, and h is a parameter known as Bandwidth that controls the accuracy of the interpolating equation with respect to the Cartesian plane data; this last parameter is considered optimal when the highest accuracy is obtained, that is, the smallest mean square error (MSE). This error is calculated as [30].

$$MSE = \frac{1}{n}\left(Y_{pred} - y_n\right)^2 \tag{6}$$

Then, the equation interpolating the values is expressed as follows:

$$Y_{pred} = \frac{\sum_{i=1}^{n} w_n \cdot K\left(\frac{x_n - x}{h}\right) y_n}{\sum_{i=1}^{n} w_n \cdot K\left(\frac{x_n - x}{h}\right)} \pm \varepsilon_o \tag{7}$$

The analysis procedure used to process the data obtained by using the finite element method is illustrated in Figure 9, where it is shown that the desired Kernel equation and the data obtained from the simulation are used to complete the Gram matrix. This same data will then be used in the NWE to multiply the vector weights obtained from the Lagrangian optimization problem. From this multiplication, a mathematical model is finally obtained. It should be noted that the definition of the dependent and independent variables is maintained from the beginning of the processing and does not change at the end of the processing.

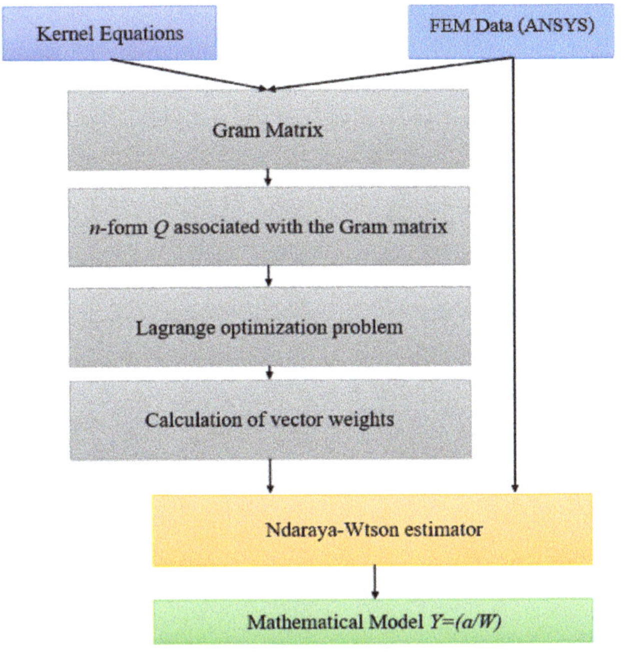

Figure 9. Methodology of analysis applied to FEM data.

3. Results

Figure 10 illustrates the behavior of the SIF in failure mode I with respect to crack length (a):

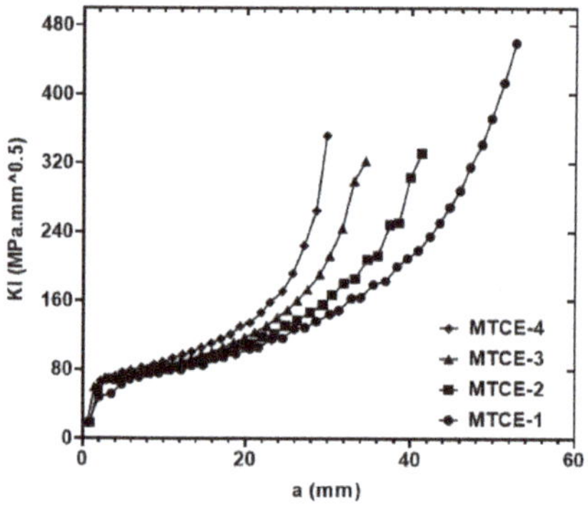

Figure 10. Mode I SIF with respect to crack length (a) for each specimen tested.

Based on the results illustrated in Figure 11, it is observed that MTCE-1 does not present a failure in the concentrator as observed in the other specimens. Therefore, the propagation length was higher in this specimen; this can be explained by the function of a greater cross-sectional area existing in MTCE-1 since the diameter of its circular concentrator is 15 mm, which is the smallest among the other specimens; Figure 11 shows the crack propagation obtained in each specimen.

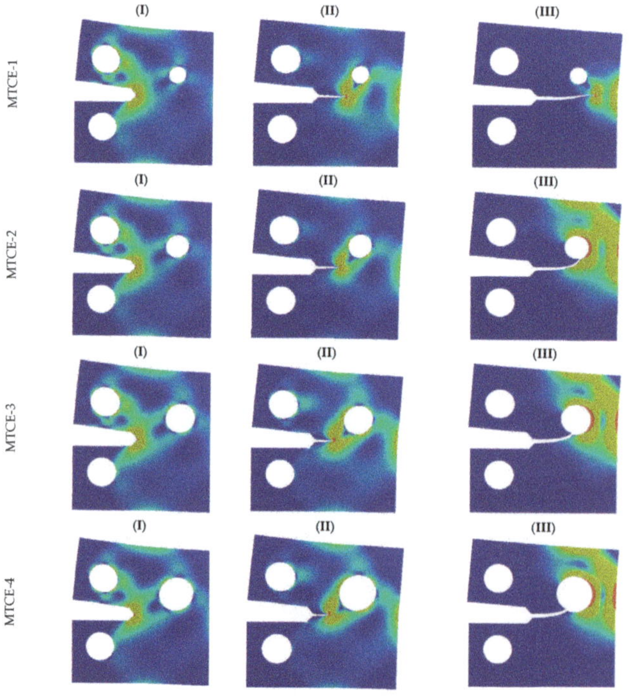

Figure 11. Crack propagation obtained in each specimen, **(I)** illustrates the solution in the first sub-step of the solution, **(II)** illustrates the propagation in the middle sub-step of the solution and **(III)** illustrates the propagation in the last sub-step of the solution.

Figure 12 shows the behavior of the geometric correction factor ($\boldsymbol{Y}$) as a function of the relative crack length ($\boldsymbol{a}/\boldsymbol{W}$).

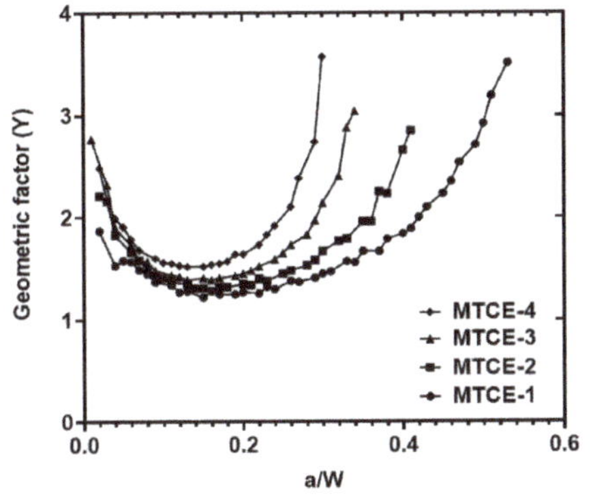

Figure 12. Geometric factor compared to the relative crack length.

From these calculated data, a regression was developed by applying the concepts of the SVR and the Nadaraya-Watson estimator (NWE), applying the Epanechnikov Kernel since, according to the authors Chu, C. Y., Henderson, D. J., and Parmeter, C. F, this is the equation that shows the highest efficiency when interpolating data placed in a Cartesian plane [30].

This model was found using a value for Bandwidth (h) equal to 0.47 and a value for ε_0 of 0.4, whose associated curve is illustrated in Figure 13.

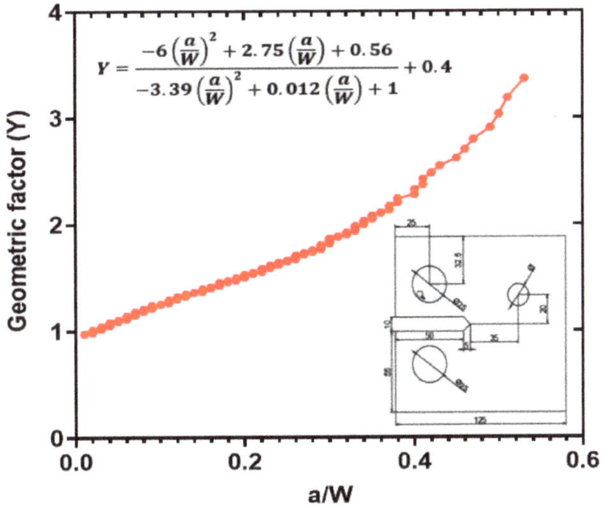

Figure 13. Mathematical model relating the geometric correction factor as a function of relative crack length.

This model has an MSE of 11.2%, which may be due to the high dispersion presented when $0.2 \leq a/W \leq 0.4$ as from this value, in Figure 12. A high dispersion of the data is observed; this dispersion is caused by the variation in the diameter of the circular stress concentrator used in each MTCE analyzed, so it is prudent to infer that when the specimens analyzed present unique geometric characteristics, the mathematical models obtained must be similarly unique for each MTCE, since for the specimens that are standardized, there are already validated models for their geometries.

4. Discussion

With the model found in Figure 13, the design of a bolted joint can be proposed from Equation (2), where the hole through which the bolt will pass is assumed to be a circular stress concentrator whose diameter is among those used in each of the MTCE specimens. Then. Starting from Equation (2), we have that:

$$K_{I\ max} = Y\sigma_{max}(1-R)\sqrt{\pi a} \tag{8}$$

By subtracting for σ_{max}, the expression is left as:

$$\sigma_{max} = \frac{K_{I\ max}}{Y(1-R)\sqrt{\pi a}} \tag{9}$$

Figure 14a illustrates one of the 10 mm diameter holes used in the assembly of the pilot plant structure, and Figure 14b illustrates the force direction (black) in the hole due to the estimated weight of the equipment. The load value used is 1000 N, corresponding to the force exerted on the joint under cyclic loading with R = −1.

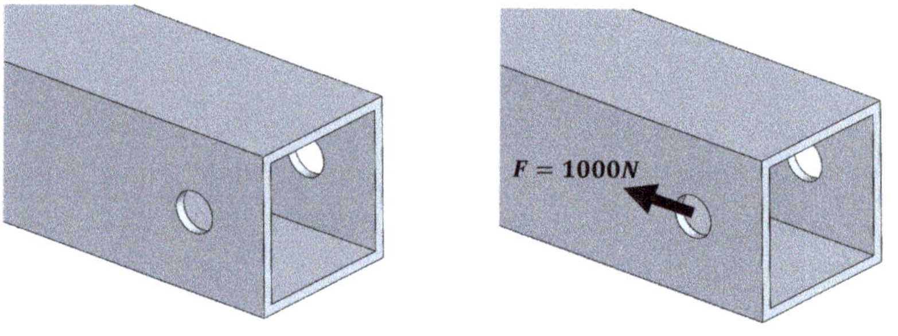

(**a**) (**b**)

Figure 14. (**a**) support bolt hole, (**b**) forces assumed for joint design.

Assuming that the material of the structure will be AA-2024-T4, then the joint dimensions can be calculated by the following expression:

$$A = \frac{(\tau_{AI})\left(\frac{-6\left(\frac{a}{w}\right)^2+2.75\left(\frac{a}{W}\right)+0.56}{-3.39\left(\frac{a}{W}\right)^2+0.012\left(\frac{a}{W}\right)+1}+0.4\right)(1-R)\sqrt{\pi a}}{K_{Imax}} \tag{10}$$

By performing the calculations using Equation (10), we obtain that the fatigue resistance area of the joint should not be less than 76.2 MPa. Figure 15 illustrates a rendering of the structure model designed.

It is observed that both the behavior of the SIF and the crack length (a) found in this research have a similarity with the data found by the authors Alshoaibi, Abdulnaser M. [31], in this study, the behavior of crack propagation is analyzed in aluminum alloy 7075-T6 specimens, which were subjected to a force of 20 KN and a stress ratio of 0.1, finding crack lengths of almost 30 mm, and with values for the SIF of about 7000 KPa.mm$^{0.5}$.

The reason for the difference between this study and the current one lies firstly in the composition of each of the alloys, the 7075-T6 alloy being stronger, and secondly, the high magnitude of the applied force with respect to the force proposed in the present investigation, which was 1KN, however, the behavior of the SIF is like that found in this article. Figure 16 illustrates a comparison of the graphs of SIF vs. crack length (a) between the works of the authors Alshoaibi, Abdulnaser M. (a) and the present one (b).

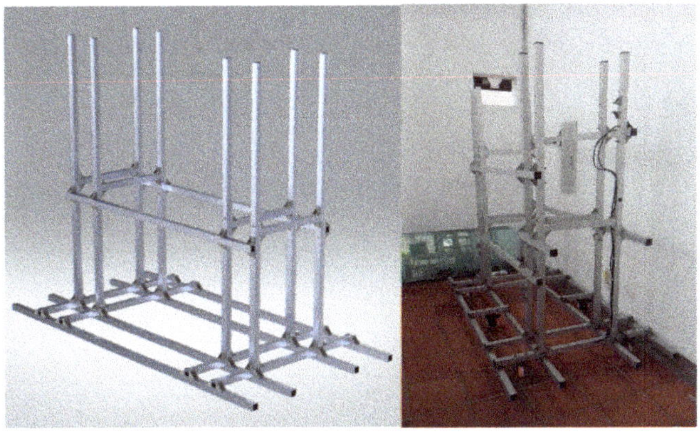

Figure 15. Model of the designed structure.

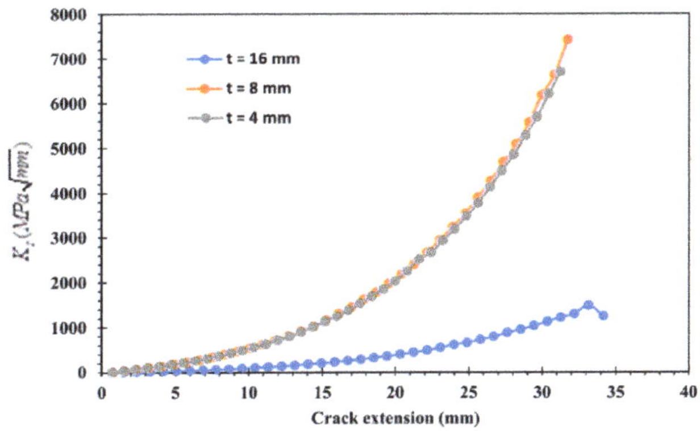

(a)

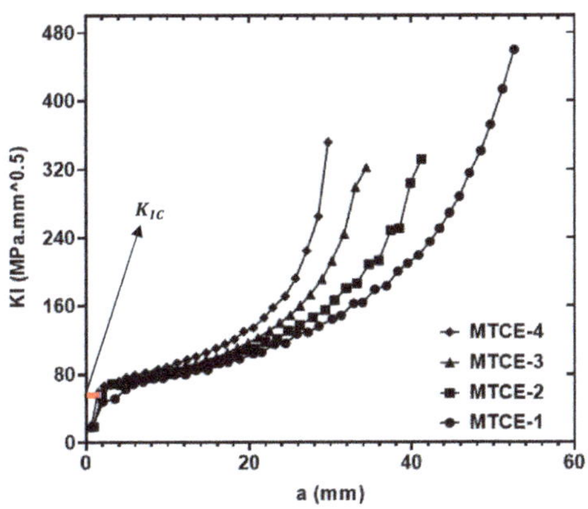

(b)

Figure 16. SIF data comparison. (**a**) Illustrates a comparison of the graphs of SIF vs. crack length between the works of the authors Alshoaibi, Abdulnaser M. and (**b**) the present one.

Looking at the scale of the values obtained in Figure 16. It is possible to compare them with the results found by the authors J. M. D. Rahmatabadi, M. Pahlavani, A. Bayati and R. Hashemi [32], who, in their work, propose the use of a standardized compact stress specimen with a size between 23.75 mm × 22.8 mm using a dual phase melt at 770 °C of Mg LZ71 and Mg LZ91 alloys subjected to quasi-static loading which varied between 0 N and 900N. In the results, the authors show that for a load of around 900 N a $K_{IC} = 18.9$ MPa·m$^{1/2}$ is generated. If we look at Figure 16, we have that for all the studied MTCEs, the approximate K_{IC} value is 70 MPa·mm$^{1/2}$, which dimensionally equals 2.21 MPa·m$^{1/2}$. This clear difference between the K_{IC} values could be explained by the thickness of the plates used, being the thinner plate used by the authors J. M. D. Rahmatabadi, M. Pahlavani, A. Bayati and R. Hashemi, which is 1.7 mm thicker than the one with the smallest stress cross-sectional area, while the one used in this work is 10 mm thick, which implies that much less stress will be concentrated on the 10 mm plate.

5. Conclusions

Computational models were generated for each proposed MTCE using a high density of elements in the meshes, with an average of 103356 elements. Likewise, the use of sectored refinements in the notch and in the circular stress concentrator facilitated the analysis of each of the models. Observing the behavior of the SIF in each MTCE, it can be deduced analytically that the presence of a circular stress concentrator alters the area where the stress applied to the specimen affects and therefore alters the SIF. This coincides with the behavior observed in Figure 9 since the presence of a smaller diameter concentrator implies a larger area of incidence of the cyclic stress; therefore, if there is a larger area of incidence, the stress along the cross-sectional area is smaller, and therefore, the SIF takes longer to reach high values, which results in a longer propagation length. When determining the geometric factor Y, it was observed that it increases as the diameter of the circular stress concentrator also increases; this is closely related to the behavior of the SIF since this factor was determined based on the data obtained from the SIF in failure mode I. The mathematical model found presented an MSE of 11.2%, fitting the geometric factor (Y) with a function that depends on the relative crack length (a/W).

From the propagation instants in Figure 10, it can be observed that the direction in which the crack propagates changes as the concentrator hole becomes larger. Based on this idea, it can be deduced that as the radius of the hole is farther away from the crack front plane, the propagation angle tends to maintain values close to 0°, while if the distance between the hole radius and the midline of the specimen becomes shorter, the propagation angles tend to be much higher.

It would be important to analyze the behavior of the propagation angle direction under the same geometrical conditions of the specimens shown in the present work, as well as to establish a relationship between the size of the second concentrator and the crack propagation direction.

The behavior of the crack propagation direction is determined by the crack length. If it is observed in Figure 10 (III) for each of the specimens, the propagation angle with respect to a horizontal axis becomes tighter as the crack length tends to be smaller. This behavior could be explained as a function of the reduction of the cross-sectional area because of the diameter change in the second circular concentrator.

The joint designed using Equation (10) for the bioplant structure proved to be useful at the time of construction since the structure will be subjected to vibration, which produces alternating stresses that fall on the bolted joint. The nomenclature used in the manuscript is shown in Table 5.

Table 5. Nomenclature.

	Nomenclature	Greek Symbols	
K_I	Stress Intensity Factor Mode I	π	Pi number
a	Crack length	σ	Axial stress
W	Length measured from the point of grip to the end	ε_o	ENW Setting Parameter
a/W	Relative crack length	ε	Amplitude between support vectors
Y	Geometric correction factor	ζ	External margin between support vectors
h	Bandwidth	μ	Arithmetic mean of data
R	Stress ratio	**Subscripts**	
$K(x)$	Kernel Equation	NWE	Nadaraya–Watson Estimator
C	Paris law material constant	MTCE	Modified Tension Compact specimen
m	Paris law material constant	SIF	Stress Intensity Factor
$\hat{m}(x)$	Nadaraya—Watson equation	SVR	Support Vector Regression
w_n	Weight value of each vector	MSE	Mean Square Error

Author Contributions: Conceptualization, L.F.U., O.G.-B., N.A. and O.A.; Formal analysis, L.F.U., O.G.-B., N.A. and O.A.; Investigation, O.G.-B., N.A. and O.A.; Methodology, L.F.U., O.G.-B., N.A., and O.A.; Resources, O.G.-B. and O.A.; Writing—original draft, L.F.U.; Writing—review&editing, O.G.-B., N.A. and O.A. All authors have read and agreed to the published version of the manuscript.

Funding: The authors would like to thank the funding provided for this research from the Department of Research of the University of Ibagué, the Ministry of Science, Technology and Innovation, the Ministry of Education, the Ministry of Industry, Commerce and Tourism, and ICETEX, Program Ecosistema Científico-Colombia Científica, from the Francisco José de Caldas Fund, Grand RCFP44842-212-20184.

Institutional Review Board Statement: Not applicable.

Informed Consent Statement: Not applicable.

Data Availability Statement: Not applicable.

Acknowledgments: The authors would like to thank the University of Ibague research department, the funding provided for this research from the Ministry of Science, Technology and Innovation, the Ministry of Education, the Ministry of Industry, Commerce and Tourism, and ICETEX, Program Ecosistema Científico-Colombia Científica, from the Francisco José de Caldas Fund, Grand RCFP44842-212-20184.

Conflicts of Interest: The authors declare that they have no known competing financial interests or personal relationships that could have appeared to influence the work reported in this paper.

References

1. Sanchez-Capa, M.; Viteri-Sanchez, S.; Burbano-Cachiguango, A. New Characteristics in the Fermentation Process of Cocoa (*Theobroma cacao* L.) 'Super Á rbol' in La Joya de los. *Sustainability* **2022**, *14*, 7564. [CrossRef]
2. Gómez, E.H.; Campo, I.; Rosario, E.; Tapachula, K.C. Factores socieconómicos y parasitológicos que limitan la producción del cacao en Chiapas, México Socioeconomic and parasitological factors that limits cocoa production in Chiapas, Mexico, 2015. *Rev. Mex. Fitopatol.* **2015**, *33*, 232–246.
3. Chitiva-Chitiva, L.C.; Ladino-Vargas, C.; Cuca-Suárez, L.E.; Prieto-Rodríguez, J.A.; Patiño-Ladino, O.J. Antifungal Activity of Chemical Constituents from Phytopathogen Fungi of Cocoa. *Molecules* **2021**, *26*, 3256. [CrossRef] [PubMed]
4. Guerrini, A.; Sacchetti, G.; Rossi, D.; Paganetto, G.; Muzzoli, M.; Andreotti, E.; Tognolini, M.; Maldonado, M.E.; Bruni, R. Bioactivities of *Piper aduncum* L. and *Piper obliquum* Ruiz & Pavon (Piperaceae) essential oils from Eastern Ecuador. *Environ. Toxicol. Pharmacol.* **2009**, *27*, 39–48. [CrossRef]
5. Guerrero, R.; Risco, G.; Cevallos, O.; Villamar, R.; Peñaherrera, S. Extractos vegetales: Una alternativa para el control de enfermedades en el cultivo de cacao (*Theobroma cacao*). *Ing. Innovación* **2020**, *8*, 2326. [CrossRef]

6. Nairn, J.A. Direct comparison of anisotropic damage mechanics to fracture mechanics of explicit cracks. *Eng. Fract. Mech.* **2018**, *203*, 197–207. [CrossRef]
7. Mecholsky, J.J. Fracture mechanics principles. *Dent. Mater.* **1995**, *11*, 111–112. [CrossRef]
8. Taylor, D.; Cornetti, P.; Pugno, N. The fracture mechanics of finite crack extension. *Eng. Fract. Mech.* **2005**, *72*, 1021–1038. [CrossRef]
9. Atzori, B.; Lazzarin, P.; Meneghetti, G. Fracture mechanics and notch sensitivity. *Fatigue Fract. Eng. Mater. Struct.* **2003**, *26*, 257–267. [CrossRef]
10. Smith, S.M.; Scattergood, R.O. Crack-Shape Effects for Indentation Fracture Toughness Measurements. *J. Am. Ceram. Soc.* **1992**, *75*, 305–315. [CrossRef]
11. Newman, J.C., Jr.; Raju, I. An empirical stress-intensity factor equation for the surface crack. *Eng. Fract. Mech.* **1981**, *15*, 185–192. [CrossRef]
12. Nix, K.J.; Lindley, T.C. The Application of Fracture Mechanics to Fretting Fatigue. *Fatigue Fract. Eng. Mater. Struct.* **1985**, *8*, 143–160. [CrossRef]
13. Clarke, S.M.; Griebsch, J.H.; Simpson, T.W. Analysis of Support Vector Regression for Approximation of Complex Engineering Analyses. *J. Mech. Des.* **2004**, *127*, 1077–1087. [CrossRef]
14. Smola, A.J.; Scholkopf, B. A tutorial on support vector regression. *Stat. Comput.* **2004**, *14*, 199–222. Available online: http://citeseerx.ist.psu.edu/viewdoc/download;jsessionid=1CAD92EF8CCE726A305D8A41F873EEFC?doi=10.1.1.114.4288&rep=rep1&type=pdf%0Ahttp://download.springer.com/static/pdf/493/art%3A10.1023%2FB%3ASTCO.0000035301.49549.88.pdf?auth66=1408162706_8a28764ed0fae9 (accessed on 10 April 2023). [CrossRef]
15. Heydari, M.H.; Choupani, N. A New Comparative Method to Evaluate the Fracture Properties of Laminated Composite. *Int. J. Eng.* **2014**, *27*, 991–1004. [CrossRef]
16. El-Desouky, A.R. Mixed Mode Crack Propagation of Zirconia/Nickel Functionally Graded Materials. *Int. J. Eng.* **2013**, *26*, 885–894. [CrossRef]
17. Guo, K.; Gou, G.; Lv, H.; Shan, M. Jointing of CFRP/5083 Aluminum Alloy by Induction Brazing: Processing, Connecting Mechanism, and Fatigue Performance. *Coatings* **2022**, *12*, 1559. [CrossRef]
18. USA Department of Defense. *MIL-HDBK-2097, Military Handbook: Acquisition of Support Equipment and Associated Integrated Logistics Support*; USA Department of Defense: Washington, DC, USA, 1997.
19. Haji, Z. Low cycle fatigue behavior of aluminum alloys AA2024-T6 and AA7020-T6. *Diyala J. Eng. Sci.* **2010**, 127–137.
20. Meggiolaro, M. Statistical evaluation of strain-life fatigue crack initiation predictions. *Int. J. Fatigue* **2004**, *26*, 463–476. [CrossRef]
21. Faisal, B.M.; Abass, A.T.; Hammadi, A.F. Fatigue Life Estimation of Aluminum Alloy 2024-T4 and Fiber Glass-Polyester Composite Material. *Int. Res. J. Eng. Technol.* **2016**, *2016*, 1760–1764. Available online: www.irjet.net (accessed on 10 April 2023).
22. Yang, G.; Gao, Z.L.; Xu, F.; Wang, X.G. An Experiment of Fatigue Crack Growth under Different R-Ratio for 2024-T4 Aluminum Alloy. *Appl. Mech. Mater.* **2011**, *66–68*, 1477–1482. [CrossRef]
23. Hudson, M.; Scardina, J. Effect of stress ratio on fatigue crack growth in 7075-T6 Al alloy sheet. *Natl. Symp. Fract. Mech.* **1967**.
24. Wei, R.P. Fatigue-crack propagation in a high-strength. *Int. J. Fract. Mech.* **1968**, *4*, 159–168. [CrossRef]
25. ANSYS. *Meshing Guide*; Finite Elem. Simulations Using ANSYS; Ansys: Canonsburg, PA, USA, 2015; Volume 15317, pp. 407–424. [CrossRef]
26. Araque, O.; Arzola, N. Weld Magnification Factor Approach in Cruciform Joints Considering Post Welding Cooling Medium and Weld Size. *Materials* **2018**, *11*, 81. [CrossRef] [PubMed]
27. Shawe-Taylor, J.; Cristianini, N. *Kernel Methods for Pattern Analysis*; Cambridge University Press: Cambridge, UK, 2004.
28. Demir, S.; Toktamiş, Ö. On the adaptive Nadaraya-Watson kernel regression estimators. *Hacet. J. Math. Stat.* **2010**, *39*, 429–437.
29. Fan, J.; Gijbels, I. *Local Polynomial Modelling and Its Applications*; Applied Pr. New York; CRC Press: Boca Raton, FL, USA, 1996.
30. Chu, C.-Y.; Henderson, D.J.; Parmeter, C.F. On discrete Epanechnikov kernel functions. *Comput. Stat. Data Anal.* **2017**, *116*, 79–105. [CrossRef]
31. Alshoaibi, A.M. Computational Simulation of 3D Fatigue Crack Growth under Mixed-Mode Loading. *Appl. Sci.* **2021**, *11*, 5953. [CrossRef]
32. Rahmatabadi, D.; Pahlavani, M.; Bayati, A.; Hashemi, R.; Marzbanrad, J. Evaluation of fracture toughness and rupture energy absorption capacity of as-rolled LZ71 and LZ91 Mg alloy sheet. *Mater. Res. Express* **2018**, *6*, 036517. [CrossRef]

metals

Article

Fatigue Reliability Prediction Method of Large Aviation Planetary System Based on Hierarchical Finite Element

Ming Li [1], Yuan Luo [1] and Liyang Xie [2,*]

1 School of Mechatronics Engineering, Shenyang Aerospace University, Shenyang 110136, China
2 School of Mechanical Engineering and Automation, Northeastern University, Shenyang 110819, China
* Correspondence: liyoulu4166@163.com

Abstract: The reliability of planetary equipment determines the economic affordability and service safety, to a large extent, for a helicopter transmission system. However, with the continuous improvement of the progressiveness and large-scale degree of new aviation planetary equipment, the contradiction between reliability design indexes and R&D economy is also gradually highlighted. This paper takes the large aviation planetary system as a research object, aims to accurately evaluate the system reliability level formed in design processes, and deeply excavates the inherent characteristics of the planetary system in functional realization and builds a system fatigue reliability evaluation model accordingly. An advanced hierarchical finite element technology is used to calculate dangerous tooth load histories under the influence of system global elastic behavior, and the tooth probability fatigue strength is obtained through the gear low-cycle fatigue test and life distribution transformation method, so as to provide economic load and strength input variables, respectively, for the reliability model. This prediction method can provide targeted structural optimization guidance in the development and design of the large aviation planetary system and significantly reduce the cost of reliability index realization for this kind of large-scale, high-end equipment in design iteration processes.

Keywords: planetary transmission; hierarchical analysis; finite element method; reliability modeling; fatigue test

Citation: Li, M.; Luo, Y.; Xie, L. Fatigue Reliability Prediction Method of Large Aviation Planetary System Based on Hierarchical Finite Element. *Metals* **2022**, *12*, 1785. https://doi.org/10.3390/met12111785

Academic Editor: Alberto Campagnolo

Received: 27 August 2022
Accepted: 19 October 2022
Published: 23 October 2022

1. Introduction

The heavy-lift helicopter is military and civilian general-purpose strategic equipment related to core national interests, as well as an important symbol of the aviation science and technology levels and even the comprehensive national strength. The high-power transmission system technology is a core technology field for improving the performance of the heavy-lift helicopter by reducing its noise and vibration levels and controlling its life-cycle cost. Both the United States and Russia have listed the reliability and economic affordability of the high-power transmission system as key technical indicators in their respective R&D plans for the advanced heavy-lift helicopter, and they have put forward specific low maintenance design requirements for the transmission system [1]. Among the largest number of heavy-lift helicopters currently in service, the large aviation planetary mechanism, as the basis and core of their transmission systems, determines the scientific and technological levels of the transmission systems to a great extent and is one of the bottlenecks restricting the development of transmission system technologies in the heavy-lift helicopter [2].

As a deceleration terminal directly connected with the main rotor, the large aviation planetary mechanism is a power transmission link with the worst load environment and highest strength requirements in the heavy-lift helicopter transmission system [3]. With the continuous improvement of performance standards, such as the power density level and dry running capacity of the advanced heavy-lift helicopter main reducer in developed countries, the speed and bearing limits of the large aviation planetary mechanism continue to break through, and the structure and power sharing forms become more complex. These increase the complexity and uncertainty in structural strength design and analysis for the planetary system and lead to a higher risk of structural failure when the new generation of heavy aviation planetary equipment is pursuing lightweight [4]. The transmission system is one of the three key dynamic systems (engine system, transmission system, and rotor system) in a helicopter. Considering weight or space, a redundancy design for the transmission system cannot be realized. Therefore, the reliability of its core link will directly determine the service safety and life-cycle cost of the helicopter [5].

High reliability and long life have become the core technology development directions for large aviation planetary equipment in the future [4,6]. Compared with small and medium-sized helicopters, the planetary mechanism in the heavy-lift helicopter has larger parts and components, higher requirements for structural mechanics indexes, and more stringent design standards for important geometric elements. If only relying on the design level and experience parameters of existing related products to develop a larger-sized isomorphic mechanism, it may directly lead to the failure to guarantee the reliability and durability for large aviation planetary equipment. In order to meet performance matching requirements of the new generation heavy-lift helicopter main reducer, a system reliability index prediction method that adapts to the structure and operation characteristics of the large-scale aviation planetary equipment is urgently needed. This method will enable developers to accurately evaluate the reliability level of products in the early stage of design, and then provide targeted reference data for the structural optimization and reliability improvement of products. This will significantly reduce the cost of expensive large-scale aviation planetary mechanism products for the realization of reliability indicators in development iteration processes.

In a gear transmission system, the failure of any gear or tooth will affect the transmission capacity of the entire system, so it is generally assumed that the gear transmission system is a series system, with the gear or tooth as basic functional units in the reliability analysis [7,8]. The loads on different individual gears in the system, or even different teeth on the same gear, may be different, but they all have a certain mathematical relationship with the system input power. This load correlation and general load randomness make the failure behavior of each unit no longer independent, so it is unreasonable to simply think that the reliability of the series system is equal to the product of each unit's reliability. Especially for the large aviation planetary gear train with more complex structure, this "unit independent failure assumption" will even lead to serious errors in the task of system reliability analysis [9]. In addition, most relevant research works simplify the gear transmission system into a general series system or series parallel hybrid system, which completely fails to reflect the special attribute that the functional form of this kind of system changes with time [10]. Different from the "chain type" series system, in which each potential failure unit is loaded at the same time, the energy transfer in the gear system is completed by the alternating loading of each tooth. Therefore, the gear system is not only a series system in the sense of spatial structure, but also a series system in the time domain. The structure and motion forms of the large aviation planetary gear train are more complex, and the teeth meshing timing characteristic will be more prominent. Therefore, the important influence of this behavior mechanism needs to be fully considered in a strict and effective reliability evaluation.

Tooth root bending fatigue strength is one of the most important strength check indexes in high-speed and heavy-duty gear systems. For the planetary mechanism in heavy-lift helicopters, its extreme load conditions and harsh reliability requirements will put forward a higher standard for tooth root bending fatigue resistance. Gear researchers are always trying to find a trade-off solution between obtaining accurate results from gear stress analyses and high-efficiency computing. For both factors, accuracy on the one hand, and high-efficiency computing on the other hand, the accuracy and efficiency of tooth root bending stress calculation determines the validity of fatigue reliability assessment of large aviation planetary wheel systems. At present, the main calculation method of tooth root bending stress is to use general finite element tools [11–13], which are flexible in technical processing and are not limited by input conditions, such as geometric characteristics and material properties, and the analysis results are relatively comprehensive (depending on software post-processing ability). Several studies have investigated how to better analyze gear stresses using finite element models. A strategy for the mesh refinement of finite element models of gear drives, based on the application of multi-point constraints, has been applied. It allows for a considerable reduction of the computational cost in the analysis of the whole cycle of meshing. In the model, the refined area mesh and the non-refined area mesh were connected by multi-point constraint (MPC) at the same time, in order to save time on the FEM solution, on the premise of ensuring the accuracy of model analysis [14–16]. Chen investigated the effects of a gear tooth spalling in a helical gear by finite elements. The results showed that spalling also caused the rapid increase of tooth root stress in the spalling meshing area [17]. Franco Concli analyzed early crack propagation in single tooth bending fatigue by combining finite element analysis with critical plane fatigue criterion. The results showed that the crack propagation direction at the ρ_{fp} did not follow the plane of maximum alternating shear stress, but the plane of maximum damage parameter [18]. However, the general finite element method has high computational cost in model setting and solution operations and is usually only suitable for isolated solutions of gear parts or several teeth, but it is difficult to perform global operations at the system-level. Moreover, it is difficult to predict tooth bending conditions or add corresponding boundary conditions to the model, especially for thin-walled rim gears or gears directly installed on bearings (such as the planetary gears with rotation and revolution characteristics). In addition, the mesh density on the tooth surface cannot be refined to the micron level to reflect the micro-geometric state of the tooth surface, so it is difficult to analyze the influence of tooth surface modification on tooth root stress by the general finite element method.

This paper takes the large aviation planetary system as a research object, aims to accurately evaluate the system reliability level formed in design processes, and deeply excavates the inherent characteristics of the planetary system in functional realization and builds a system fatigue reliability evaluation model accordingly. Advanced hierarchical finite element technology is used to calculate tooth dangerous load histories under the influence of system global elastic behavior, and the tooth probability fatigue strength is obtained through the gear low-cycle fatigue test and life distribution transformation method, so as to provide economic load and strength input variables, respectively, for the reliability model. This prediction method can provide targeted structural optimization guidance in the development and design of the large aviation planetary system and significantly reduce the cost of reliability index realization for this kind of large-scale, high-end equipment in the design iteration processes. At the same time, it can provide important reference data for the first renovation period of relevant finalized products and then provide the technical support for their economic guarantee in the whole life-cycle.

2. Tooth Root Stress Calculation Based on Hierarchical Finite Element Technology

2.1. Principle Analysis of Hierarchical Finite Element Technology

Most tooth root stress calculation standards are analyzed and processed according to the geometric characteristics of tooth root, referring to empirical data, and considering working conditions. The biggest disadvantage of these methods is that they are limited to the specified geometries, materials, and working conditions. For the unconventional tooth geometries commonly existing in the aviation field, the calculation results obtained by these standard methods may not be accurate. In particular, the urgent need for lightweight components for the new generation of large aviation planetary equipment makes the components in the system widely adopt lighter and thinner structural design forms. The introduction of such a large number of flexible features means that the transmission shafts, support frames, gears, and other critical transmission parts will undergo significant elastic deformation under heavy loads. The resulting stiffness problem makes it necessary to fully consider the nonlinear mechanical behaviors for the whole system, such as elastic deformation and meshing misalignment, in order to accurately calculate the tooth root stress.

In order to solve the contradiction between the calculation accuracy and calculation cost of the general finite element method, when facing the mechanical simulation analysis task for the large complex gear system, an advanced hierarchical finite element method for root stress analysis of the large aviation planetary mechanism is proposed in this paper. When calculating the tooth root stress, it is directly based on a detailed tooth 3D finite element sub-model, so it does not need to rely on empirical data and industry experts. In the system-level model, the dynamic load line on the tooth surface was obtained by quasi-static static analysis and tooth surface micro-geometric analysis, and then it was loaded on the tooth surface of the finite element sub-model as a load boundary condition. At the same time, system elastic deformation results, including the deformation behavior of the gear rim and tooth, were extracted and loaded into the finite element sub-model as a displacement boundary condition. In this way, the calculation results of tooth root stress could include the influence of the elastic deformation (including deformations, such as rim distortion and tooth bending and so on), meshing misalignment, and tooth surface micro-geometry in the whole system.

The analysis and solution for the system-level model took the analytical algorithms in the international standard as the default calculation basis, and they have a strong auxiliary calculation ability in the tooth surface micro-geometric analysis. At the same time, the general finite element method was used as a backup advanced analysis method, and the introduction of finite element components into the system-level model made the system elastic deformation results more reliable. In the system-level model, there is no need to perform detailed root stress calculation, which greatly saves the calculation cost for system-level analysis. In addition, tooth surface micro-geometric analysis was run in the system-level model, so it is no longer necessary to include micro-geometric parameters in the secondary finite element sub-model. In the quasi-static analysis results of the system-level model, the physical effects, such as rotation, centrifugation, and thermal expansion of transmission components, were considered, so the boundary conditions imposed on the secondary sub-model were relatively simple. In the face of advanced simulation and analysis tasks for a large aviation planetary system, only considering the convenience of modeling and boundary condition setting, the computational efficiency of the hierarchical finite element method will be much higher than that of the general finite element method. In addition, different from the commercial finite element software using a nonlinear equation solver, the hierarchical finite element method uses an improved simplex solver to ensure convergence within a set number of iterations. Although the total number of freedom degrees in the two-level finite element model may be very large, this hierarchical analysis approach keeps the amount of CPU time and memory required for analytical calculation within the capabilities of a personal computer.

It can be seen that the hierarchical finite element method is a more accurate, fast, and easy method for calculating tooth root stress, and it is especially suitable for the strength checking and analysis tasks of large aviation gear systems. In this paper, a high-fidelity mechanical simulation model was constructed by the hierarchical finite element method, based on the structure and material performance parameters of a certain type of large aviation planetary mechanism products. Based on this, the dangerous tooth root stress under system quasi-static elastic mechanical behavior was calculated, and the effective load input variables were provided for the system reliability evaluation model.

2.2. *System-Level Elastic Mechanical Behavior Simulation Analysis*

2.2.1. Overall Configuration of System Model

A semi-analytical finite element technique was used to construct the system-level elastic mechanical simulation model of the large aviation planetary mechanism and accurately evaluate the elastic deformation of the large thin-walled parts, the meshing misalignment between teeth, and the micro-modification on tooth surface, and then provide detailed load and displacement boundary conditions for the secondary sub-model of the tooth root stress analysis. The semi-analytical finite element method was used in the system-level analysis process. Although fine mesh densities were used for some large structural members, the effective combination of analytical and finite element calculations resulted in significant improvements in the computational efficiency at the system-level. The system model of the planetary mechanism and the structural parameters of the gear train are shown in Figure 1 and Table 1, respectively. The power flow path inside the system starts from the input shaft, passes through the sun gear and planet gear to a planet carrier, and finally, the planet carrier transmits the motion and power to a main rotor shaft, after a deceleration of 3.334 times. In addition, the rated working condition parameters of the planetary mechanism mainly include an input power of 5000 kW, input speed of 500 rpm, and working temperature of 70 °C.

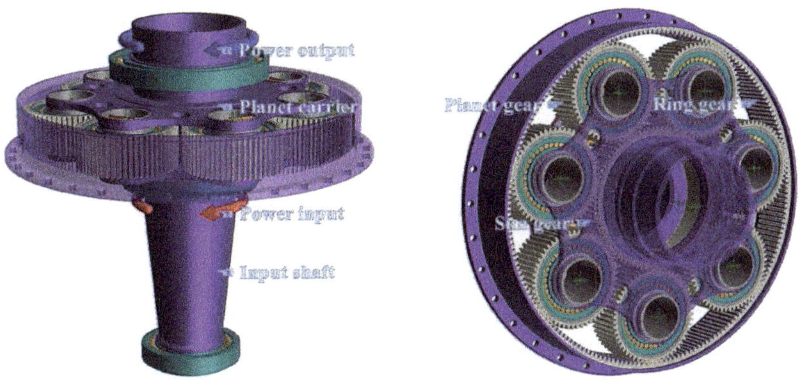

Figure 1. System model of planetary mechanism.

Table 1. Geometric parameter of planetary gear train.

Parameters	Sun Gear	Planet Gear	Ring Gear
Module (mm)	5.012	5.012	5.012
Number of teeth	84	56	196
Number of gears	1	7	1
Pressure angle (°)	20	20	20
Helix angle (°)	0	0	0
Effective face width (mm)	120	120	120
Base circle diameter (mm)	394.671	263.114	920.899
Base circle pitch (mm)	14.761	14.761	14.761
Root fillet radius (mm)	2.757	2.870	2.657
Tooth surface hardness	60 HRC	60 HRC	60 HRC
Tooth core hardness	35 HRC	35 HRC	35 HRC
Elastic modulus (MPa)	2.07×10^5	2.07×10^5	2.07×10^5
Poisson ratio	0.3	0.3	0.3

2.2.2. Deformable Planet Carrier

An FE model and the geometry of a planet carrier are illustrated in Figure 2, which shows an example of an FE grid for a seven-planet system, with its supporting conditions simulated by lumped stiffness elements. Since there are no relative displacements between the axis of rotation of the planet pins and the planet nodes (planet centers), a fixed interface component mode synthesis method can be used to reduce the size of the carrier model. The stiffness of the springs is considered to be significantly higher than that of the bearings and pins, thus creating a rigid link in the radial directions between the contour nodes and planet centers. The classic reduction process of Craig and Bampton [19] was employed, in which the displacements were expressed, in terms of static and dynamic modes, as

$$\begin{bmatrix} \mathbf{X}_{\mathrm{PCint}} \\ \mathbf{X}_{\mathrm{PCbou}} \end{bmatrix} = \begin{bmatrix} [\boldsymbol{\phi}_{\ \mathrm{D}}] & [\boldsymbol{\phi}_{\ \mathrm{S}}] \\ [\mathbf{0}] & [\mathbf{I}] \end{bmatrix} \cdot \begin{bmatrix} \mathbf{q}_{\mathrm{PC}} \\ \mathbf{X}_{\mathrm{PCbou}} \end{bmatrix} \quad (1)$$

with $\mathbf{X}_{\mathrm{PCint}}$ as the vector of the internal degrees of freedom, $\mathbf{X}_{\mathrm{PCbou}}$ as the vector of the degrees of freedom at the contour nodes, $[\boldsymbol{\phi}_{\ \mathrm{D}}]$ as the truncated modal matrix of the fixed-interface planet-carrier structure, $[\boldsymbol{\phi}_{\ \mathrm{S}}]$ as the static mode matrix, $[\mathbf{I}]$ as the identity matrix, $[\mathbf{0}]$ as the nil matrix, and $\mathbf{q}_{\mathrm{PC}}$ as the vector of the planet-carrier modal unknowns.

Figure 2. Example of FE model and geometry of planet carrier.

2.2.3. Planet Bearing Element

This connecting part was composed of seven lumped spring elements across the planet bore, in order to connect node O_j of planet j to the corresponding three contour nodes of the planet-carrier substructure (denoted as N_1, N_2, and N_3 in Figure 3). Planets were modeled as rigid disks with 6 degrees of freedom (DOFs), which are the infinitesimal generalized displacements superimposed on rigid-body motions and represented by screws of coordinates.

$$\{\tau_j\}\begin{Bmatrix} \mathbf{u}_\mathrm{j}(O_j) = v_j\mathbf{S}_\mathrm{j} + w_j\mathbf{T}_\mathrm{j} + u_j\mathbf{Z} \\ \boldsymbol{\omega}_\mathrm{j} = \varphi_j\mathbf{S}_\mathrm{j} + \psi_j\mathbf{T}_\mathrm{j} + \theta_j\mathbf{Z} \end{Bmatrix} \quad (2)$$

where $\mathbf{S}_\mathrm{j}$, $\mathbf{T}_\mathrm{j}$, and $\mathbf{Z}$ are the unit vectors of the frame fixed to the sun gear/planet j mesh (Figure 4).

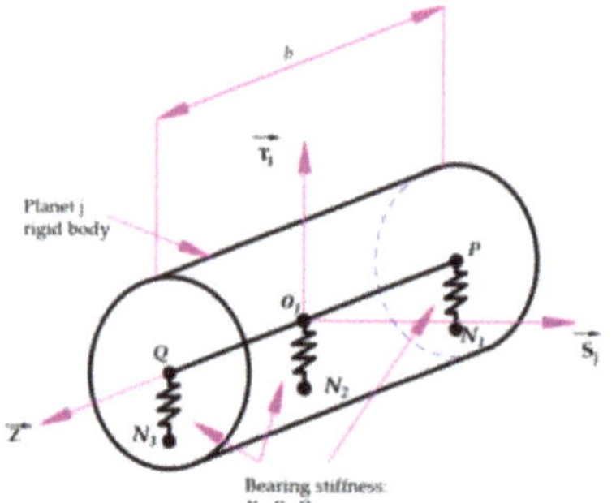

Figure 3. Planet bearing element.

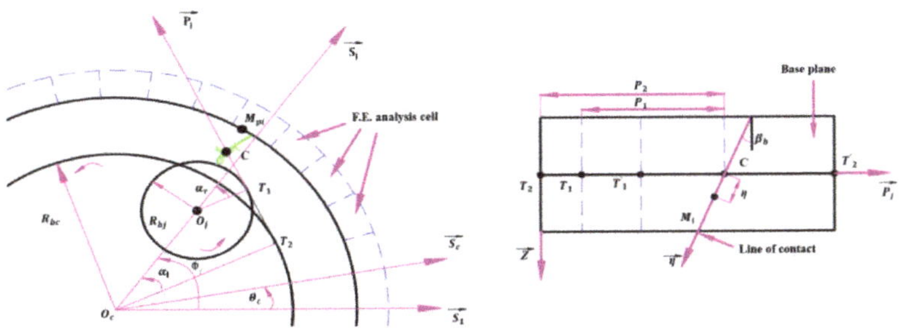

Figure 4. Geometrical parameter for internal gear modeling.

Considering two points, P and Q, belonging to planet j, their displacements were expressed, using the shifting property of the moment of screw $\{\tau_j\}$, as

$$\begin{aligned} \mathbf{u_j}(P) &= \mathbf{u_j}(O_j) + \boldsymbol{PO}_j \times \boldsymbol{\omega}_\mathbf{j} \\ \mathbf{u_j}(Q) &= \mathbf{u_j}(O_j) + \boldsymbol{QO}_j \times \boldsymbol{\omega}_\mathbf{j} \end{aligned} \tag{3}$$

Denoting K_v, K_w, and K_u the stiffness in the $\mathbf{S_j}$, $\mathbf{T_j}$, and $\mathbf{Z}$ directions between P and N_1, Oj, and N_2, and Q and N_3, and the strain energy stored in the spring element 1 connecting node N_1 and point P reads

$$U_1 = \frac{1}{2}\mathbf{X}^T\left[K_u\mathbf{V_u}\cdot\mathbf{V_u^{\mathit{T}}} + K_v\mathbf{V_V}\cdot\mathbf{V_V^{\mathit{T}}} + K_W\mathbf{V_W}\cdot\mathbf{V_W^{\mathit{T}}}\right]\mathbf{X} = \frac{1}{2}\mathbf{X}^T[\mathbf{K}_1]\mathbf{X} \tag{4}$$

where $\mathbf{X}^T = (u_j, v_j, w_j, \varphi_j, \psi_j, \theta_j, u_{N1}, v_{N1}, w_{N1})$ is the vector containing the degrees of freedom associated with nodes O_j (6 DOFs) and N_1 (3 DOFs). The structural vectors $\mathbf{V_u}$, $\mathbf{V_v}$, and $\mathbf{V_w}$ are expressed, in terms of the planet width and its angular position Φ_j (Figure 4), as

$$\begin{aligned} \mathbf{V}_u^T &= \langle 1,0,0,0,0,0,-1,0,0\rangle \\ \mathbf{V}_v^T &= \left\langle 0, \cos\Phi_j, -\sin\Phi_j, -\tfrac{b}{2}\sin\Phi_j, -\tfrac{b}{2}\cos\Phi_j, 0,0,-1,0\right\rangle \\ \mathbf{V}_w^T &= \left\langle 0, \sin\Phi_j, \cos\Phi_j, \tfrac{b}{2}\cos\Phi_j, -\tfrac{b}{2}\sin\Phi_j, 0,0,0,-1\right\rangle \end{aligned} \tag{5}$$

A similar procedure was used for the other spring elements attached to the same planet, thus leading to the 15 × 15 stiffness matrix of the planet j bearing element, which connected the 6 degrees of freedom at the planet center $(u_j, v_j, w_j, \varphi_j, \psi_j, \theta_j)$ and those at nodes $N_1(u_{N1}, v_{N1}, w_{N1})$, $N_2(u_{N2}, v_{N2}, w_{N2})$, and $N_3(u_{N3}, v_{N3}, w_{N3})$.

2.2.4. Deformable Ring Gears

Ring gears and their supports were modeled by using FE analysis, combined with lumped parameter elements, to account for the ring-gear bearing stiffness and for the contribution of the teeth modeled as lumped masses. The ring-gear FE model was reduced using a component mode synthesis method based on the mode shapes of the undamped isolated structure. After separating the internal displacements, $\mathbf{X}_{\mathrm{RGint}}$, and the degrees of freedom at the boundary nodes, $\mathbf{X}_{\mathrm{RGbou}}$, i.e., on the root cylinder and in correspondence with the planet ring-gear contacts, the following approximation was used:

$$\begin{bmatrix} \mathbf{X}_{\mathrm{RGint}} \\ \mathbf{X}_{\mathrm{GDbou}} \end{bmatrix} = [\Phi_{\mathrm{N}}]\mathbf{q}_{\mathrm{RG}} \tag{6}$$

with $\mathbf{X}_{\mathrm{RGint}}$ as the vector of the internal degrees of freedom, $\mathbf{X}_{\mathrm{RGbou}}$ as the vector of the degrees of freedom at the contour nodes, $[\Phi_{\mathrm{N}}]$ as the truncated modal matrix of the undamped ring-gear structure, and $\mathbf{q}_{\mathrm{RG}}$ as the vector of the modal unknowns.

2.2.5. System Support Condition

In this analysis task, it was assumed that the reducer box had an absolutely rigid body; the input shaft was supported by two tapered roller bearings (TRB) with O-type layout, and their outer rings were rigidly connected to the box. All planet gears in the system were evenly distributed on the planet carrier along a circumferential direction. The double-row tapered roller bearing (DRTRB), with an X-type layout, was assembled inside each planet gear, and the bearing inner ring was rigidly connected with the planet shaft. A radial ball bearing (RBB) fixed the planet carrier and made it have a certain floating amount (its size and direction were controlled by the bearing clearance) to counteract the unequal load sharing among planet gears, and its outer ring was rigidly connected with the box. The structural forms and parameters of various main bearings in the system are shown in Figure 5 and Table 2, respectively.

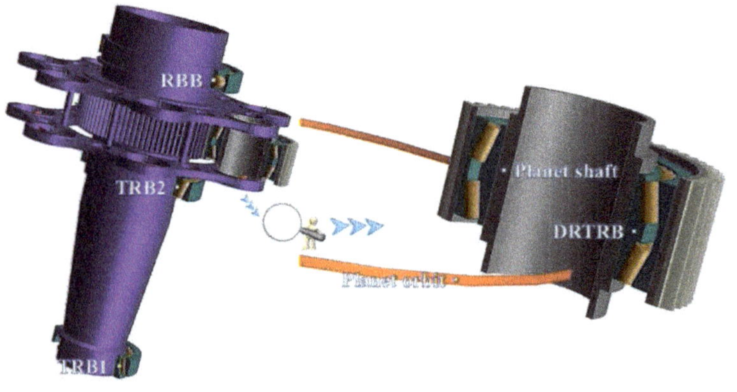

Figure 5. Bearing structure in planetary system.

Table 2. Bearing parameters in planetary system.

Parameters	TRB1	TRB2	DRTRB	RBB
External diameter (mm)	310	420	240	480
Internal diameter (mm)	200	300	160	360
Width (mm)	70	76	102	56
Number of rollers	31	40	72	24
Roller diameter (mm)	23	26	16	40
Roller length (mm)	50	55	38	-
Contact angle (°)	15.945	14.931	17.049	0

2.2.6. Finite Element Component Modeling

The solid model creation of standard parts, such as gears and bearings, can be realized with the help of professional software, such as RomaxDesigner or GearTrax (GearTrax2020, Camnetics Inc., Oregon, WI, USA). The powerful parametric modeling function for the standard parts in the software can enable designers to complete this work easily. In addition, four key finite element components were also included in the planetary mechanism system model, which are the planet carrier, the ring gear, the planet gear, and the input shaft, integrated with the sun gear. Based on the advantages of 3D modeling software, such as NX (UG2020, Siemens PLM Software, Plano, TX, USA) and SolidWorks (SW2021, Dassault Systemes, Concord, MA, USA), and general finite element software, such as HyperMesh (HyperMesh2020, Altair Troy, MI, USA) and ANSYS (ANSYS 19.0, ANSYS Corporation, Pittsburgh, PA, USA), a high-quality parametric modeling function for finite element components can be realized, and the core technology implementation can refer to the literature [20,21]. The final modeling results and mesh quality parameters are shown in Table 3. It should be emphasized that creating complete teeth features in the gear finite element models can make the system-level flexibility analysis results cover the accurate mechanical behavior characteristics, such as the rim distortion and tooth bending deformations, which can provide more effective displacement boundary conditions for the secondary sub-model of the tooth root stress calculation. Additionally, it is now possible to keep CPU requirements down to about 8 s per time step and memory needs down to 128 MB. This has been made possible by accepting programming complexity, in exchange for an increase in speed.

Table 3. Finite element model and mesh quality parameter.

Finite Element Models	Effective Size	Minimum Angle
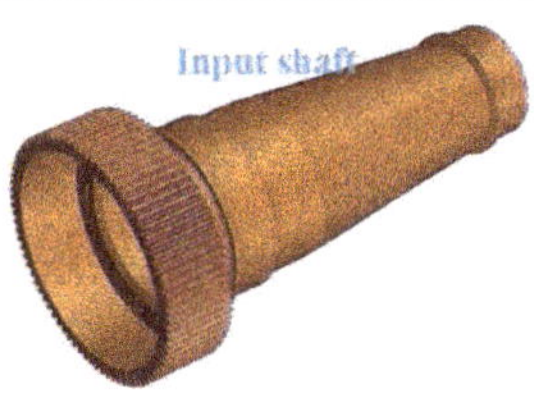	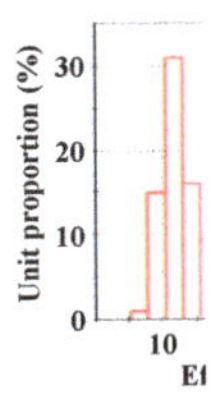	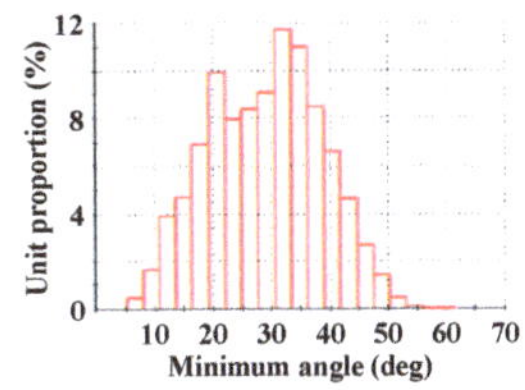
Number of surface mesh node: 31,248	Number of surface mesh elements: 15,589	
	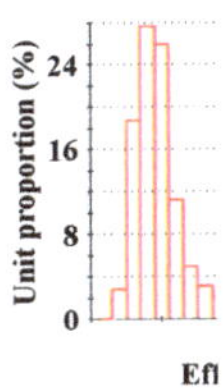	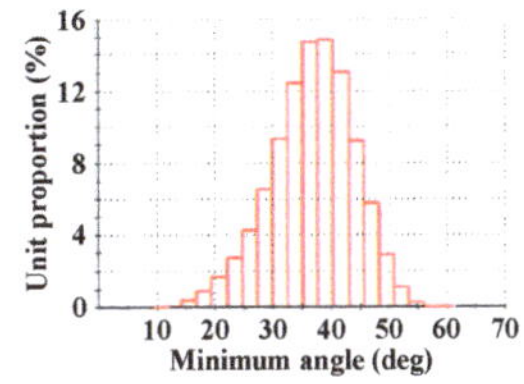
Number of surface mesh node: 9273	Number of surface mesh elements: 4606	
	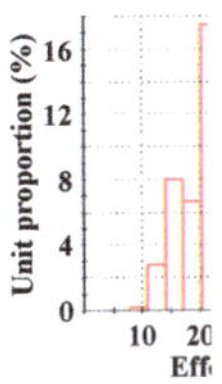	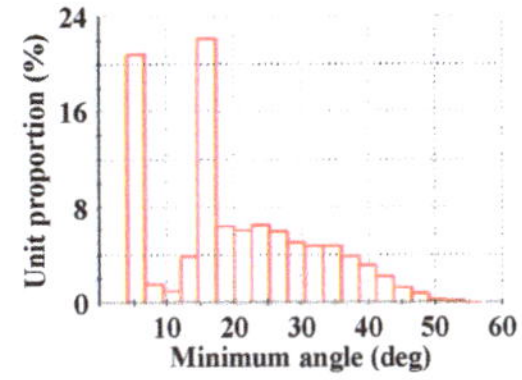
Number of surface mesh node: 16,111	Number of surface mesh elements: 8021	
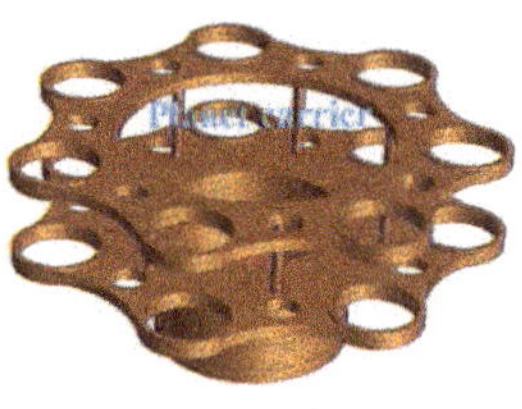	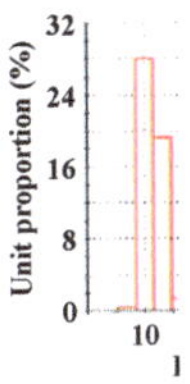	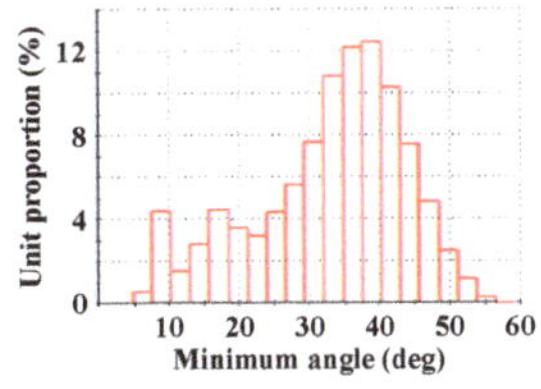
Number of surface mesh node: 23,956	Number of surface mesh elements: 11,942	

2.2.7. Setting Node Connection

The above-mentioned solid models, such as the gears and bearings, as well as finite element models, such as the input shaft and planet carrier, were imported into the RotationMaster (RM) software platform, respectively. It can seamlessly integrate a variety of

CAD and CAE software, and it is an advanced simulation platform for the comprehensive analysis and calculation of large and complex transmission systems. It has been widely used in related engineering technology fields. The imported solid models and finite element models were positioned and assembled in a unified global coordinate system, and the node connection relationships between the two types of models were established. We comprehensively adjusted the control parameters, such as the search criteria and selection modes, and used RM's tolerance search technology to screen the connection node groups between the finite element models and the solid models. The analytical schematic diagram of two search criteria in the technology is shown in Figure 6.

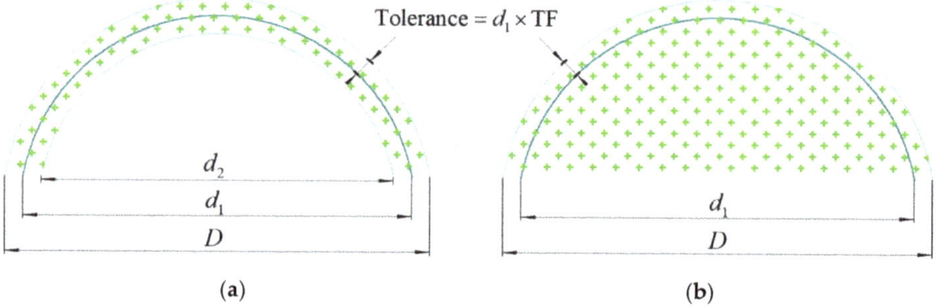

Figure 6. Node search criteria: (**a**) shell search; (**b**) solid search. Where d_1 is pitch circle diameter (PCD), and D and d_2 indicate the search range. $D = d_1 \times (1 + 2 \times \text{TF})$, $d_2 = d_1 \times (1 - 2 \times \text{TF})$.

The two semicircles in the figure show the geometric dimensions of the cross-sections of the hollow and solid cylinders, respectively, which will be used to define the search spaces corresponding to the two search criteria. Within the range of the specified cylinder height, "shell search" will search for all the nodes between the inner and outer walls of the hollow cylinder and on the walls. This search criterion is suitable for the node connection setting between the solid model of a bearing inner ring and the finite element model of a shaft. "Solid search" will search out all the nodes inside the solid cylinder, including its outer surface. This search criterion is suitable for establishing the node connection relationship between finite element teeth and their meshing lines. RM can automatically generate the pitch circle diameter (PCD), according to the geometric characteristics at the node connection location. The tolerance factor (TF) was used to adjust the diameter parameters of the inner and outer walls of the cylinders to optimize the connection attributes, such as the number of connection nodes and the connection stiffness. In addition, RBE2 element type (detailed in the NASTRAN help file) can bundle the selected connection nodes into a rigid node group, which means that all the nodes will have the same displacement parameters, so it can simplify the model and improve calculation efficiency. RBE3 element type distributes loads evenly across the selected connection nodes. It reduces the stiffness of the connection area by increasing the flexibility around the nodes, so that the elastomer deformation behavior obtained from the analysis will be more accurate, but the calculation cost will also increase significantly. Table 4 shows the node connection results and corresponding node parameters between the sun gear teeth on the input shaft and the meshing lines (pitch circle position), between the planet gear and the outer ring of a tapered roller bearing, between the mounting holes on the ring gear and the reducer box, and between the pin holes on the planet carrier and the planet shafts (single side).

2.2.8. System Elastic Behavior Analysis

The finite element models in the system are polycondensed to extract the corresponding mass and stiffness matrices, and the calculation process can be performed on general finite element software, such as ANSYS or ABAQUS(ABAQUS 2021, Dassault SIMULIA, Providence, RI, USA) or the RM software platform. At the same time, the deformation smoothness was used as an evaluation index, and the performance of each stiffness matrix was tested by means of the load transfer behavior among polycondensation nodes. Finally, the loading boundary conditions were applied to the system model, and the global calculation of quasi-static elastic behavior was performed to obtain the node displacement response of each elastic member under rated working conditions. In the system-level simulation analysis, the planet gear transferred the power received from the sun gear to the double-row tapered roller bearing (DRTRB) and the planet shaft, in turn, and finally, acted on the load on a pin hole of the planet carrier. Figure 7 shows the calculation results of the planet carrier elastic deformation and the load distribution on each DRTRB. Under the rated working conditions of the system, the maximum node resultant displacement (NRD) at the pin hole of the planet carrier exceeded 1000 μm, which will lead to significant movement position error for the planet shaft in working state, and seriously affect the tooth root stress response in the planetary gear train.

Table 4. Node connection between finite element model and solid model.

Connection Nodes	Connection Objects	Node Parameters
Input shaft 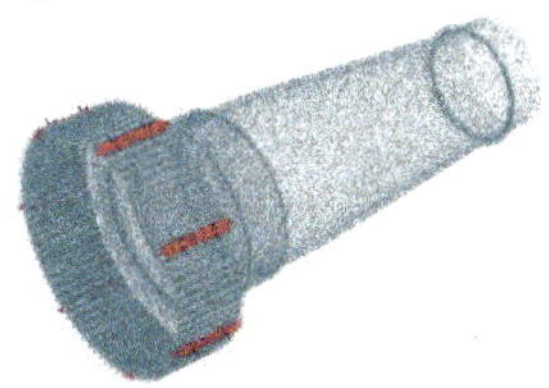	Meshing lines 	Number of connection nodes: 1830 Number of node groups: 7 Node search criterion: solid search Node selection method: space Element type: RBE2 PCD = 10 mm, TF = 0.01
Planet gear 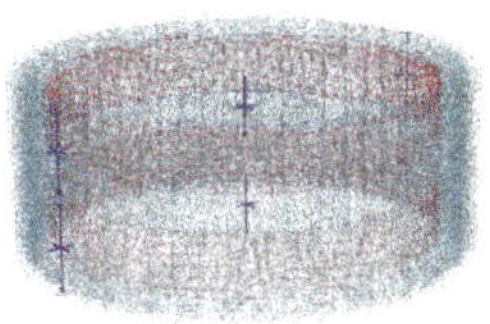	DRTRB 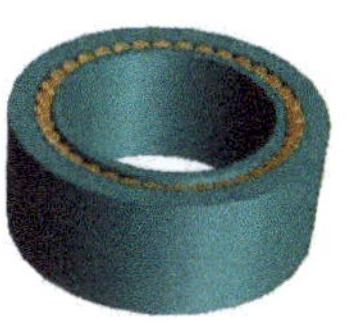	Number of connection nodes: 520 Number of node groups: 2 Node search criterion: solid search Node selection method: surface Element type: RBE3 PCD = 240 mm, TF = 0.01
Ring gear 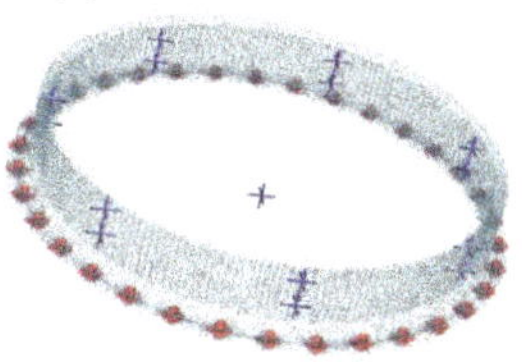	Main reducer box 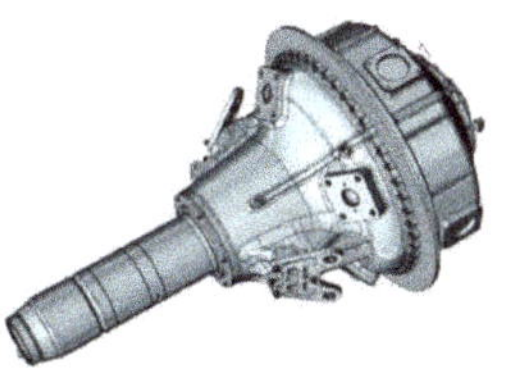	Number of connection nodes: 930 Number of node groups: 34 Node search criterion: shell search Node selection method: surface Element type: RBE2 PCD = 20 mm, TF = 0.01
Planet carrier 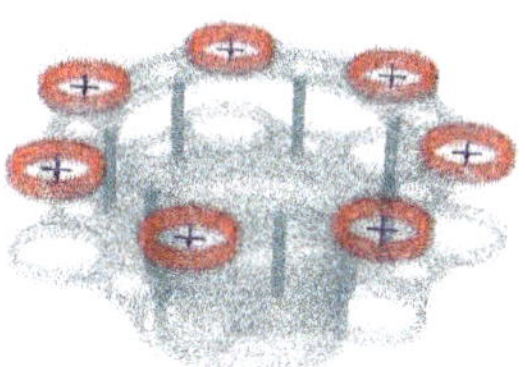	Planet shafts 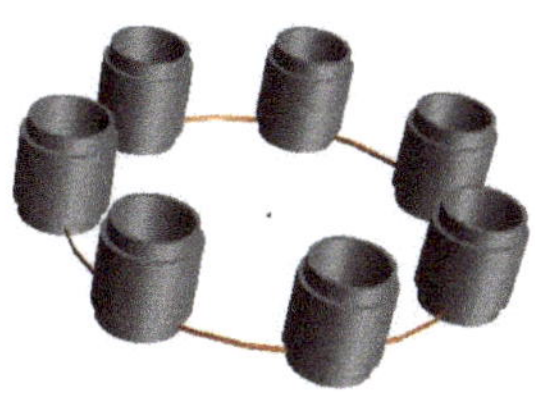	Number of connection nodes: 1410 Number of node groups: 7 Node search criterion: solid search Node selection method: surface Element type: RBE3 PCD = 140 mm, TF = 0.01

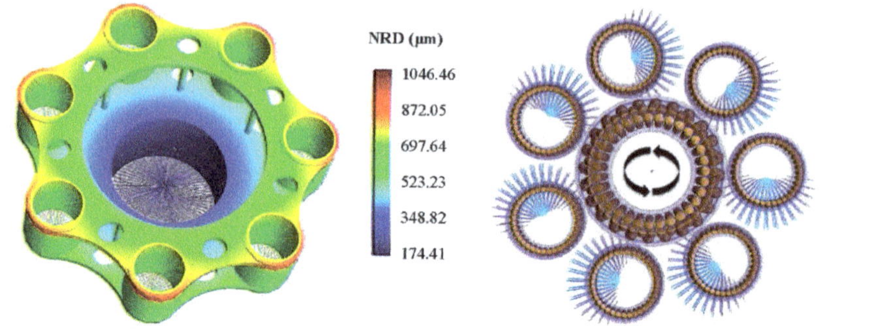

(a) (b)

Figure 7. Mechanical analysis result: (**a**) displacement nephogram of planet carrier; (**b**) roller force diagram of DRTRB.

In the system-level simulation analysis, it is necessary to obtain the node displacement response of the planet gear rim and teeth, so as to provide necessary displacement boundary conditions for the secondary sub-model of tooth root stress calculation. Figure 5 shows the calculation results of the elastic deformation and stress distribution of a planet gear under rated working conditions. The structural element integrity of the system model ensures that the elastic deformation of the planet carrier and planet shaft, as well as the bearing clearance and other factors, can be taken into account in the results. Comparing the original contour of the planet gear with its loaded deformation contour (the deformation proportion has been enlarged), as shown in Figure 8a, it can be seen that the elastic deformation of the planet gear rim presents an elliptical shape. Figure 8b shows the stress distribution nephogram of the planet gear after being loaded, which identifies the two teeth meshing with sun gear and ring gear, respectively. The high stress zone was located at the tooth root of the loaded side of the two teeth, respectively, and the maximum von Mises stress (vMS) reached 298 MPa. It can also be found from the figure that the tooth root stress value of the teeth that were not engaged in meshing was not zero, which is mainly due to the significant elliptical deformation of the rim structure. The imprecise tooth root stress behavior was included in the system-level simulation analysis results, and more accurate tooth root stress calculation results need to be further obtained from the secondary sub-model.

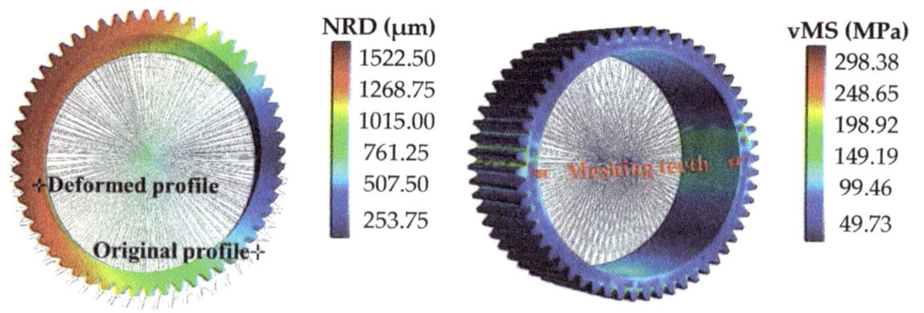

(a) (b)

Figure 8. Mechanical analysis result of planet gear: (**a**) displacement nephogram; (**b**) stress nephogram.

In the system-level simulation analysis, the node displacement response of the sun gear rim and teeth are reflected in the displacement nephogram of the input shaft, and the calculation result is shown in Figure 9a. It can be seen from Figure 9 that the sun gear and the ring gear have the same deformation trend, and there are obvious elastic deformation zones near the teeth meshing with the planet gears. The flexible design of the sun gear rim structure makes each high deformation zone evenly cover multiple teeth in a circumferential direction, and the thin-walled rim and cantilever support of the ring gear make each high deformation zone offset by the same amount in axial direction. These elastic behavior characteristics will have a significant impact on the tooth root stress distribution and stress level for the corresponding gears.

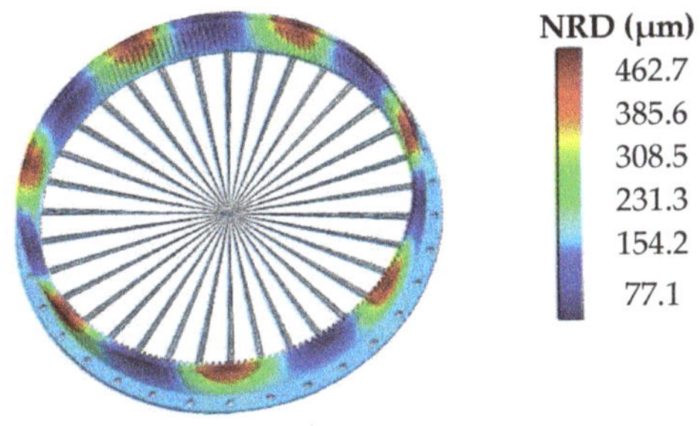

Figure 9. Elastic deformation result: (**a**) displacement nephogram of input shaft; (**b**) displacement nephogram of ring carrier.

2.2.9. Calculation of Tooth Surface Load Line

The tooth surface micro-geometric analysis was carried out at the system-level, which used the boundary conditions of the whole system model. Based on the calculation results of system elastic deformation and meshing misalignment, and considering tooth surface micro-geometric information, the dynamic load line on tooth surface was calculated, which provided the load boundary conditions for the secondary sub-model of tooth root stress calculation. The micro-geometric analysis modules in the RotationMaster and RomaxDesigner software (RomaxDesigner20, Romax Technologies Ltd., Nottingham, UK) can carry out very accurate mathematical definition for the tooth surface, including micro-geometric parameters, such as involute shape and tooth surface modification, and the nonlinear contact model of the tooth surface can be automatically created by accurately evaluating the geometry and meshing direction on the tooth surface. The tooth surface modification geometric information in the planetary mechanism is digitally expressed, and the visualized modeling results are shown in Figure 10, which will be used as one of the key input conditions for the system-level tooth surface micro-geometry analysis. The contour maps of the modification amount in the figure, respectively, describes the modification state on the working tooth surface for each gear, which, respectively, cover the micro-modification data, such as inclination, drum amount, and top trimming, in both directions of tooth direction and tooth profile.

The tooth surface dynamic load line results of each gear in the system can be obtained through tooth surface micro-geometric analysis. The results consider the influence of the elastic deformation and meshing misalignment of the system and record the information, such as the meshing line movement trajectory on tooth surface and the tooth surface load distribution under each rotation step. The calculation results of the tooth surface load line of the sun gear, planet gear, and ring gear are shown in Figure 11, which show the tooth surface load distribution state, when the load line moved to the center of tooth height. It can be seen from the figure that the load on each tooth surface was basically evenly distributed along the direction of tooth width, and a peak load was located in the center area of tooth width. The load lines (contact spots) in Figure 11c,d were wider, which was the result of internal meshing forming between the ring gear and planet gear.

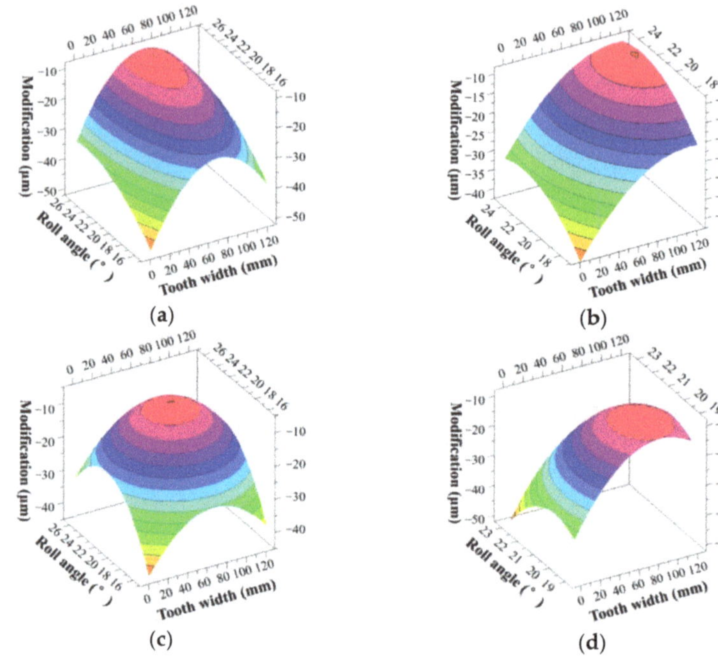

Figure 10. Contour map of tooth surface modification: (**a**) planet gear tooth surface meshed with sun gear; (**b**) working tooth surface of sun gear; (**c**) planet gear tooth surface meshing with ring gear; (**d**) working tooth surface of ring gear.

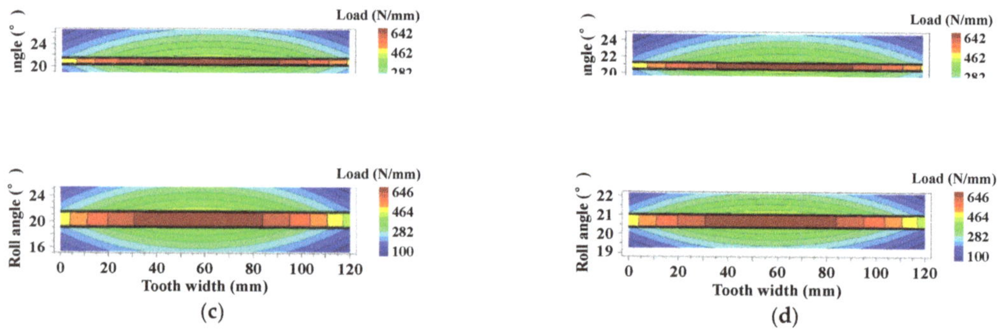

Figure 11. Load line on tooth surface: (**a**) planet gear tooth surface meshed with sun gear; (**b**) working tooth surface of sun gear; (**c**) planet gear tooth surface meshing with ring gear; (**d**) working tooth surface of ring gear.

2.3. Secondary Sub-Model Construction and Tooth Root Stress Analysis

2.3.1. Construction of Secondary Sub-Model

In the system-level model, the dynamic load lines on tooth surface were obtained through quasi-static static analysis and tooth surface micro-geometric analysis (see Figure 11), which could be loaded on the tooth surface of the secondary sub-model as load boundary conditions. At the same time, the system elastic deformation results, including the deformation behavior of the gear rim and teeth (see Figures 7–9), could be extracted and loaded into the sub-model as displacement boundary conditions. In this way, the tooth root stress analysis results of the sub-model naturally included the elastic deformation of the entire system (including rim distortion deformation and tooth bending deformation, etc.), meshing misalignment, and tooth surface micro-geometry, as well as other factors. In addition, the dynamic behavior factors, such as the gear rotation and meshing, were also considered in the system model. When these important factors are fully considered in the system model, the overall structure of the sub-model can be significantly simplified. It could only contain a few teeth and the corresponding rim feature; the configuration is shown in Figure 12, and its modeling and computational costs focused more on the tooth root geometric details. In terms of model quality, the tooth root geometric elements of the sub-model were more comprehensive than those of the system model, and its quality indicators, such as accuracy level and mesh quality were also higher.

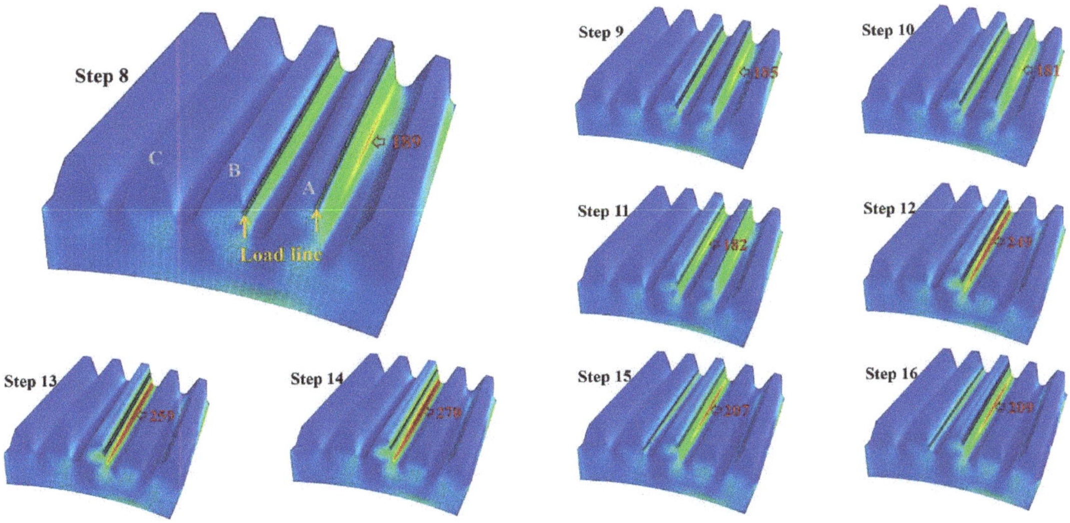

Figure 12. Evolution process of tooth root stress distribution on sun gear.

The secondary sub-model of the tooth root stress analysis with high-fidelity was established by a general finite element method. The parametrically defined detailed gear solid models in the system-level model were auxiliary modeling elements for the 3D finite element sub-models, which solved the problem of difficult modeling for the general finite element method, to a certain extent. In addition, in order to further alleviate the contradiction between calculation accuracy and operation speed, a first-order hexahedron element was adopted in the finite element sub-model. When the meshes are relatively coarse, the tooth top may appear "jagged" in appearance, due to shear self-locking or hourglass effect from the first-order element, but this does not affect the analysis results of tooth root stress. RM software platform can recommend economical and adaptive mesh density here, which can accurately capture the steep stress gradient at tooth root and is enough to make the calculation of the sub-model converge within predetermined number of iterations. In system-level analysis, the meshing force on teeth is applied through condensation nodes and finite element nodes, but in a secondary sub-model, the meshing force is applied by the tooth surface load line obtained from the system-level analysis.

2.3.2. Result Analysis of Tooth Root Stress

If the planetary system is considered an absolutely rigid system, and the misalignment of the gear mesh will be caused mainly by the clearance between the moving parts. This will generate phase-wise meshing interference between the planetary gears and, thus, will significantly increase the stress level in the tooth roots. Introducing the elastic mechanism into the system model can alleviate the meshing interference behavior between gears, to some extent (the more significant the flexible characteristics of the system within a certain range, the more pronounced this mitigation effect), reducing tooth root stress.

Firstly, the tooth root stress analysis program of the sun gear sub-model was run, and the maximum principal stress value at the tooth root was calculated according to the first strength theory. The calculation results of the system deformation and tooth surface load line are applied to the sub-model as input conditions, and then the model is solved according to the set rotation steps. According to the coincidence degree of gear pairs in the system, the meshing in and meshing out processes of a tooth are divided into 16 rotation steps, and the sub-model performed a finite element solution for each rotation step. Although this made the processes of stiffness decomposition and load vector inverse substitution relatively complex (involving multiple recursive traversals at the substructure layer), it was worthwhile to significantly reduce the calculation cost by appropriately increasing the program complexity.

Figure 13a shows the calculation results of the tooth root stress of the sun gear under 16 rotation steps. The stress results, without considering system elastic deformation, were 11.9%–17.3% higher than those considering this factor, which indicates that ignoring the flexible behavior characteristics of the large aviation planetary mechanism may directly lead to an over-conservative design scheme for corresponding structural strength. Figure 12 shows the evolution process of tooth root stress distribution of sun gear teeth during meshing (corresponding to steps 8~16 in Figure 13a). Load lines successively passed through three teeth, A, B, and C, in the sub-model, and the figure indicates the specific value and position of the maximum tooth root stress (the maximum mentioned here refers to the maximum in position) under each rotation step. As can be seen from the action of load lines in the first three steps that tooth A gradually disengaged meshing, and its root stress kept decreasing, while the adjacent tooth B gradually entered meshing, and its root stress kept increasing. When step 11 was reached, the maximum tooth root stress position jumped from tooth A to tooth B. When the 12th step was reached, the sun gear changed from double-tooth meshing to single-tooth meshing, and the root stress at tooth B increased significantly, then reached a stress peak state at the 14th rotation step. It was the maximum value state that tooth root stress could reach during the meshing process, and it was also one of the most dangerous load states that caused the sun gear to suffer from bending fatigue tooth breakage; then, the gear entered double-tooth meshing again and continuously circulated the above process. In addition, the stress peak of tooth root bending fatigue obtained by the calculation model in ISO6336 is also shown in Figure 13 a, and the calculation model is as follows:

$$\sigma_F = \frac{F_t}{b \cdot m} \cdot K_A \cdot K_V \cdot K_{F\beta} \cdot K_{F\alpha} \cdot Y_F \cdot Y_S \cdot Y_\beta \cdot Y_B \cdot Y_{DT} \quad (7)$$

The simulation result, without considering system elastic deformation, is in good agreement with this calculation result, which proves the validity of the load boundary conditions and other model parameters in the simulation analysis process.

Figure 13b shows the calculation results of the tooth root stress of the planet gear meshing with the sun gear and the ring gear, respectively, in 16 rotation steps, and it also marks the demarcation line of the tooth root stress grade under single-tooth meshing and double-tooth meshing. One side tooth surface of the planet gear tooth was meshed with the sun gear, and the stress peak at the tooth root reached 272 MPa; the other side's tooth surface meshed with the ring gear, and the stress peak at the tooth root was 276 MPa. In addition, the calculation result of the tooth root stress peak of the ring gear was 280 MPa. These stress results were obtained under system-rated working conditions, and the effects of the elastic deformation, meshing misalignment, and tooth surface micro-geometry of the whole system were fully considered.

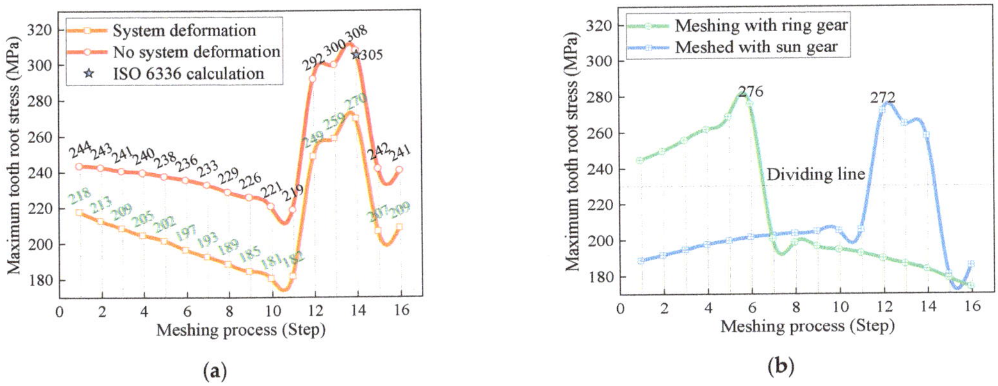

Figure 13. Maximum tooth root stress history: (**a**) sun gear; (**b**) planet gear.

2.3.3. Load Input Variable Conversion

Based on the hierarchical finite element method, the system power peak distribution was transformed into the gear stress peak distribution, which provided direct load input variables for the system fatigue reliability evaluation model. In a typical service mission, with a long flight time, the input power history of this kind of planetary system was collected, as shown in Figure 14a. The power peaks in the load history were extracted as load statistical characteristic and fitted with normal distribution, and the power peak distribution result is shown in Figure 14b. Based on the hierarchical finite element method, a mapping relationship of the same probability quantile from power distribution to stress distribution was established, and the power peak distribution was transformed into the tooth root stress peak distribution of each gear in the system. Final calculation results are shown in Figure 14d, where the PS represents the tooth root stress on the planet gear tooth (the tooth side meshed with the sun gear), and the other codes in the figure have similar meanings.

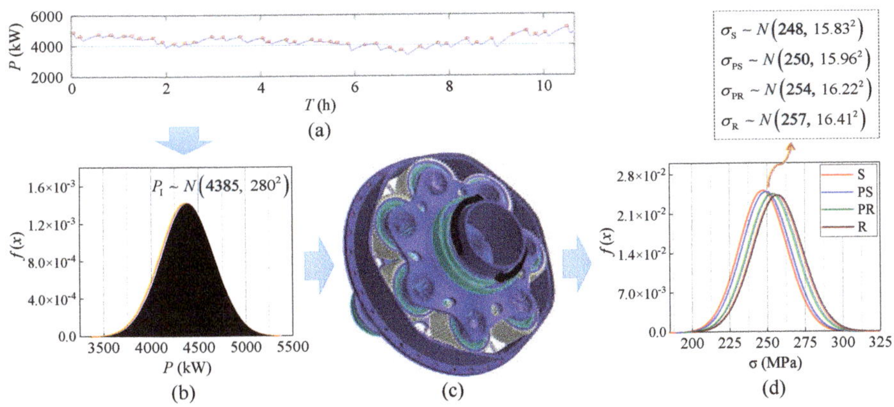

Figure 14. Load input variable conversion: (**a**) input power history; (**b**) power peak distribution; (**c**) hierarchical finite element simulation; (**d**) stress peak distribution.

3. Tooth Probability Strength Fitting Based on Gear Fatigue Test

3.1. Gear Bending Fatigue Test

A gear bending fatigue accelerated life test was carried out by a power flow closed gear rotation testing machine, which provided strength input variables for the fatigue reliability prediction model of the large aviation planetary system. The overall layout of the test platform is shown in Figure 15a, and its working principle and verification standard are shown in the previous relevant research work [2]. For power flow input inside the testing machine, a conical friction surface-type mechanical loading device was adopted, as shown in Figure 15d, which ensured a reliable seal of the power flow under an ultra-high cycle. A vibration monitoring system (see Figure 15b,e) enabled the testing machine to have the function of real-time automatic shutdown, in case of sudden fatigue tooth breakage, which better ensured the consistency of the failure states for all the gear samples. The final shapes and sizes of each tooth root crack were basically the same, as shown in Figure 15h. At the same time, tooth root fatigue fracture surface morphology is shown in Figure 15i, and the macro morphological features, such as fatigue expansion zones and fatigue instantaneous fracture zones, can be clearly seen. A top-view of test gearbox inside is shown in Figure 15g. The gear pair adopted an assembly form of full tooth width contact, and the lubrication and cooling for the gear pair were realized by oil injection during the test. The structural details and specific parameters of the gear samples are shown in Figure 15f and Table 5, respectively, and the preparation process strictly followed the principle of "teeth fatigue strength equivalent", that is, to make the sample parameters as equal, or similar, to the gear parameters in the planetary mechanism as possible, in terms of material properties, tooth geometry, machining, and heat treatment, etc. A reliable guarantee for this similarity level will help to improve fatigue reliability prediction accuracy for the planetary system. In addition, all the gear samples came from the same batch production process to minimize the dispersion of test data.

Figure 15. Gear bending fatigue test: (**a**) test platform; (**b**) vibration monitoring; (**c**) revolution counter; (**d**) loading device; (**e**) acceleration sensor; (**f**) gear sample; (**g**) gear installation; (**h**) root crack; (**i**) fracture morphology.

Table 5. Parameter list of test gear.

Items	**Parameters**	**Items**	**Parameters**
Module (mm)	5	ISO quality grade	5
Number of teeth	25	Material brand	1Cr18Ni9Ti
Pressure angle (°)	20	Carburized depth (mm)	0.8 ± 0.13
Helix angle (°)	0	Tooth surface hardness	59–63 HRC
Face width (mm)	32	Tooth core hardness	35–48 HRC
Root fillet radius (mm)	2.7	Precision machining	Grinding

3.2. Tooth Probability Strength Fitting

The tooth root stress peak was taken as an evaluation index for the stress grade, and the bending fatigue performance of the tooth root was tested by the group method under four stress grades. The selected stress levels and the number of test points under various stress levels were 649 MPa (17 points), 618 MPa (22 points), 586 MPa (29 points), and 555 MPa (38 points), respectively. During the test, if any tooth on the gear sample failed first, the testing machine would automatically stop, and the direct data obtained from this was the gear life, rather than the tooth life. It represents the ability of the individual gear to maintain excellent transmission function under current stress level, so the gear life was also the "first broken tooth" life. From the perspective of probability, the more teeth on a gear, the more potential failure links. Therefore, under the same stress and revolution conditions, the failure risk of the gear will increase with the increase of the number of teeth. In the process of this test, a complete rotation of the sample was recorded as one gear life. In this conventional counting mode, the statistical characteristics of the number of teeth led to the difference in the probability life between the gear and tooth. In order to obtain direct strength input variables for the reliability prediction model, a probabilistic statistical transformation method was proposed to fit the tooth P-S-N curves, based on the gear life data.

The probability life relationship between the gear and tooth was established based on the concept of minimum order statistics. The fracture of any tooth on a gear will cause the gear to lose excellent transmission capacity. For this reason, it can be considered that the life of a gear depends on the minimum life of its teeth. Suppose $X_1, X_2, \ldots, X_n$ is a set of samples from a parent X, then $X_{min} = \min(X_1, X_2, \ldots, X_n)$ is the minimum order statistics of the parent. This probability model will be applied to the life transformation calculation under the failure mode of tooth root bending fatigue, and then the gear probability life is equal to the minimum order statistics of tooth probability life.

Assuming that the cumulative distribution function of random variable X is $F(x)$ and its probability density function is $f(x)$, then the probability density function of the minimum order statistics of X can be expressed as

$$g_{\min}(x) = z \cdot [1 - F(x)]^{z-1} \cdot f(x) \tag{8}$$

If the two-parameter Weibull distribution is adopted to express tooth probability life, then the cumulative distribution function can be expressed as

$$F(x) = 1 - \exp\left[-(x/\theta)^{\beta}\right] \tag{9}$$

and the probability density function is

$$f(x) = \left(\beta \cdot x^{\beta-1}/\theta^{\beta}\right) \exp\left[-(x/\theta)^{\beta}\right] \tag{10}$$

where β and θ are, respectively, the shape parameter and scale parameter of the tooth life distribution.

Equations (9) and (10) are brought into Equation (8) to obtain the following equation

$$g_{\min}(x) = \left[\beta \cdot x^{\beta-1} / \left(\theta/z^{1/\beta}\right)^{\beta}\right] \exp\left[-\left(x \cdot z^{1/\beta}/\theta\right)^{\beta}\right] \tag{11}$$

If the number of teeth on a gear is z, then $g_{\min}(x)$ directly represents the probability density function of the gear life distribution. From the expression of this function, it can be seen that the gear probability life also follows two-parameter Weibull distribution, and the shape parameter and scale parameter are as follows

$$\begin{cases} \beta_{\mathrm{Gear}} = \beta \\ \theta_{\mathrm{Gear}} = \theta/z^{1/\beta} \end{cases} \tag{12}$$

In the statistical processing for the test data, the two-parameter Weibull distribution function was adopted to fit the probability distribution of gear life points under each stress level. The probability life transformation between the gear and tooth was then performed by model (12). Finally, a least square method was used to linearly fit the same probability quantiles of the tooth life distribution under each stress level in a single logarithmic coordinate system, and the results of tooth bending fatigue P-S-N curves obtained are shown in Figure 16. Under deterministic loading, the dispersion of fatigue life generally increases as stress level decreases; therefore, in a linear coordinate system, the P-S-N curve family will appear as an "umbrella" shape with a small upper opening and a large lower opening. However, due to the single logarithmic coordinate system, the curve family presents a corresponding inverted shape.

(a)

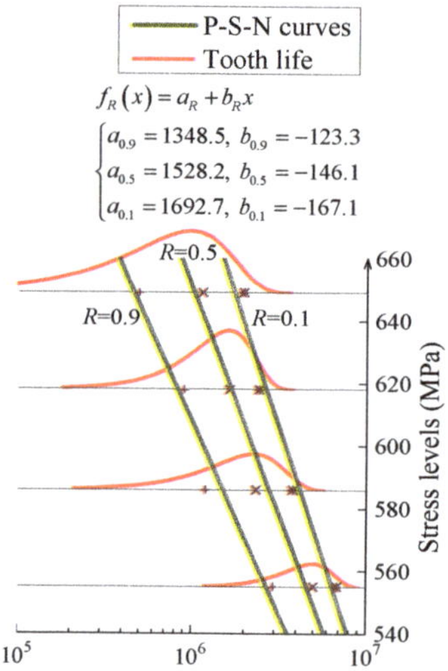

(b)

Figure 16. Statistical transformation of test data: (**a**) probability life transformation for tooth; (**b**) P-S-N curves fitting for tooth.

4. System Reliability Modeling Considering Planetary Transmission Characteristic

4.1. Conditional Probability Expectation Algorithm for Part Fatigue Reliability Calculation

The traditional "load and strength interference" analysis method was extended to establish a conditional probability expectation algorithm for calculating the fatigue reliability of the parts, based on the probability distribution of the stress level and the life distribution under the specified stress level. For the static strength failure of parts under one load, the reliability can be regarded as a function of stress, and the conditional reliability model under specified stress is established. That is, under the condition of stress σ, the probability calculation formula of static strength S greater than the stress is

$$\zeta(\sigma) = \int_{\sigma}^{\infty} f(S)\mathrm{d}S \tag{13}$$

where $f(S)$ is the probability density function of static strength S.

Furthermore, the static strength reliability model of parts that can reflect the effect of load uncertainty can be expressed as

$$R = \int_{0}^{\infty} h(\sigma) \cdot \zeta(\sigma)\mathrm{d}\sigma \tag{14}$$

where $h(\sigma)$ is the probability density function of stress σ.

If Equation (14) is the traditional "load and strength interference model", then the basic meaning is extended from probability perspective. Based on the total probability calculation principle of continuous random variable, Equation (14) can be interpreted as the mathematical expectation of random function $\zeta(\sigma)$ in the definition domain of random variable σ.

For fatigue reliability, because it is difficult to obtain the fatigue strength distribution under specified life directly, a mathematical expression for calculating the fatigue reliability of parts directly, based on life distribution, can be constructed, according to above extended thinking. Assuming that $\zeta(n,\sigma)$ is the conditional fatigue reliability function under the specified stress level, that is, under the condition of stress level σ, the probability calculation formula of the life N greater than the number of load cycles n is

$$\zeta(n,\sigma) = \int_n^{\infty} f(N|\sigma)\mathrm{d}N \tag{15}$$

where $f(N|\sigma)$ is the life probability density function at stress level σ.

Correspondingly, under the action of random stress level σ, the fatigue reliability model of parts can be expressed as

$$R(n) = \int_0^{\infty} h(\sigma)\cdot\zeta(n,\sigma)\mathrm{d}\sigma \tag{16}$$

The fatigue reliability index of parts under random constant amplitude cyclic load can be directly calculated by Equation (16).

4.2. Fatigue Reliability Evaluation Model of Series System Considering Failure Dependence

The fracture of any tooth in a planetary system will affect the transmission capacity of the whole system. Therefore, if each tooth in the system is regarded as a potential failure unit, then the system is a typical series system. In the service process, the load borne by each tooth has significant dependence on the system input power, and the load dependence and general load randomness make the failure of each element not independent of each other, so the reliability of the series system cannot be simply considered as the reliability product of each unit.

The fatigue reliability evaluation model of a series system is established based on conditional probability expectation algorithm and considering the failure dependence among unit parts. First of all, only under the action of deterministic load, the failure of each part in the system is independent of each other, so the conditional reliability of the series system is equal to the conditional reliability product of each part. Then, considering the load uncertainty effect at the system-level, the probability that r parts in the system will not fail, that is, the fatigue reliability evaluation model of the series system is

$$R_{\mathrm{SYS}}(n) = \int_0^{\infty} h_0(\sigma)\prod_{i=1}^{r}\left[\int_n^{\infty} f_i(N|(s_i\sigma+\mu_i))\mathrm{d}N\right]\mathrm{d}\sigma \tag{17}$$

where $h_0(\sigma)$ is the probability density function of standard normal distribution, and $f_i(N|(s_i\sigma+\mu_i))$ is the life probability density function of the ith part under stress level $s_i\sigma+\mu_i$.

The model assumes that the load follows normal distribution and realizes load normalization for each part through the mathematical relationship transformation between normal distribution and standard normal distribution, so as to consider the load uncertainty effect at the system-level.

4.3. Structural Optimization of Reliability Model Considering Sequence Characteristic

The kinematic equation of planetary transmission was deduced according to a periodic operation law in the planetary gear train; at the same time, the single tooth meshing times reliability level of the planetary gear system. When the ring gear rim is thin, the increase in the thickness of the planet carrier base plate will accelerate the change in the reliability of the system caused by the thickening of the rim, while, when the ring gear rim is thick, the thickness of the planet carrier base plate will have the opposite effect. This "thin vs. thick" dimensional range will be determined by the other structural elements in the planetary mechanism, as well as the level of external loads. Based on the base plate thickness–rim thickness–reliability curve cluster, the best matching results for two key structural dimensions in a specified type of large aviation planetary system to meet the reliability and lightweight requirements were determined. The results of this study will provide important reference data for the structural optimization design of a large aviation planetary system.

Author Contributions: Conceptualization, M.L.; methodology, L.X.; software, M.L. and Y.L.; validation, M.L.; formal analysis, M.L.; investigation, M.L. and L.X.; resources, M.L.; data curation, M.L. and Y.L.; writing—original draft preparation, M.L.; writing—review and editing, M.L., Y.L. and L.X.; visualization, L.X.; supervision, M.L.; project administration, M.L.; funding acquisition, M.L. All authors have read and agreed to the published version of the manuscript.

Funding: This research was funded by the National Natural Science Foundation of China (NSFC) (grant No.52005350), Scientific Research Foundation of Education Department of Liaoning Province (grant No. LJKZ0196), and National Defense Key Laboratory Open Foundation of Shenyang Aerospace University (grant No. SHSYS202103).

Institutional Review Board Statement: Not applicable.

Informed Consent Statement: Not applicable.

Data Availability Statement: Not applicable.

Acknowledgments: The authors would like to thank Liyang Xie for the reliability calculation model.

Conflicts of Interest: The authors declare no conflict of interest.

Nomenclature

ρ_{fp}	Tooth root radius
CPU	Central processing unit
FE	Finite element
TRB	Tapered roller bearings
DRTRB	Double-row tapered roller bearing
RBB	Radial ball bearing
d_1	Pitch circle diameter
PCD	Pitch circle diameter
TF	Tolerance factor
NRD	Node resultant displacement
PS	The tooth root stress on the planet gear tooth (the tooth side meshed with the sun gear)
P (kw)	The input power history of this kind of planetary system has been collected
X_i	A set of samples from a parent X
$F(x)$	The cumulative distribution function of random variable X
$f(x)$	Probability density function
$g_{\min}(x)$	The probability density function of the minimum order statistics of X
β	The shape parameter of the tooth life distribution
θ	The Scale parameter of the tooth life distribution
z	Number of teeth on a gear
σ	Condition of stress
$\zeta(\sigma)$	Under the condition of stress σ, the probability calculation formula of static strength S greater than the stress
$f(S)$	The probability density function of static strength S
$h(\sigma)$	The probability density function of stress σ
$R(n)$	Under the action of random stress level σ, the fatigue reliability model of parts

$h_0(\sigma)$	The probability density function of standard normal distribution
ω	Represents the angular velocity of the member
$i_{\mathrm{ab}}^{\mathrm{c}}$	Represents the ratio of the relative rotational speed of member a and member b respectively relative to member c
p	The kinematic characteristic parameter of the planetary gear train
k_{P}	The number of planet gears in the system
$n_{\mathrm{PS}}(t)$	The meshing times between the target single tooth of a planet gear and the sun gear within time interval t
ζ_i	The calculation factors of tooth element conditional fatigue reliability of various gears
FD	Failure dependence
MS	Meshing sequence

References

1. Yin, Z.Y.; Fu, B.B.; Xue, T.B.; Wang, Y.H. Development of helicopter power transmission system technology. *Appl. Mech. Mater.* **2011**, *86*, 1–17. [CrossRef]
2. Li, M.; Xie, L.; Ding, L. Load sharing analysis and reliability prediction for planetary gear train of helicopter. *Mech. Mach. Theory* **2017**, *115*, 97–113. [CrossRef]
3. Xue, H.L.; Liu, G.; Yang, X.H.; Han, B. Key technologies research of helicopter transmissions. *Appl. Mech. Mater.* **2015**, *743*, 55–60. [CrossRef]
4. McFarland, J.M.; Riha, D.S. Uncertainty quantification methods for helicopter fatigue reliability analysis. In Proceedings of the American Helicopter Society 65th Annual Forum, Grapevine, TX, USA, 27–29 May 2009; American Helicopter Society International: Grapevine, TX, USA, 2009.
5. Li, T.; Zhao, Z.; Sun, C.; Yan, R.; Chen, X. Adaptive channel weighted CNN with multisensor fusion for condition monitoring of helicopter transmission system. *IEEE Sens. J.* **2020**, *20*, 8364–8373. [CrossRef]
6. Brown, M.A.; Chang, J.H. Analytical techniques for helicopter component reliability. In Proceedings of the American Helicopter Society 64th Annual Forum, Montreal, QC, Canada, 29 April–1 May 2008; American Helicopter Society International, 2008.
7. Li, M.; Xie, L.Y.; Li, H.Y.; Ren, J.G. Life distribution transformation model of planetary gear system. *Chin. J. Mech. Eng.* **2018**, *31*, 1–8. [CrossRef]
8. Chen, J.; Li, W.; Sheng, L.; Jiang, S.; Li, M. Study on reliability of shearer permanent magnet semi-direct drive gear transmission system. *Int. J. Fatigue* **2020**, *132*, 105387. [CrossRef]
9. Yan, Y. Load characteristic analysis and fatigue reliability prediction of wind turbine gear transmission system. *Int. J. Fatigue* **2020**, *130*, 105259. [CrossRef]
10. Xie, L.; Wu, N.; Qian, W. Time domain series system definition and gear set reliability modeling. *Reliab. Eng. Syst. Saf.* **2016**, *155*, 97–104. [CrossRef]
11. Wang, X.; Yang, Y.; Wang, W.; Chi, W. Simulating coupling behavior of spur gear meshing and fatigue crack propagation in tooth root. *Int. J. Fatigue* **2020**, *134*, 105381. [CrossRef]
12. Zorko, D. Investigation on the high-cycle tooth bending fatigue and thermo-mechanical behavior of polymer gears with a progressive curved path of contact. *Int. J. Fatigue* **2021**, *151*, 106394. [CrossRef]
13. Vučković, K.; Čular, I.; Mašović, R.; Galić, I.; Žeželj, D. Numerical model for bending fatigue life estimation of carburized spur gears with consideration of the adjacent tooth effect. *Int. J. Fatigue* **2021**, *153*, 106515. [CrossRef]
14. Alshoaibi, A.M.; Fageehi, Y.A. Simulation of Quasi-Static Crack Propagation by Adaptive Finite Element Method. *Metals* **2021**, *11*, 98. [CrossRef]
15. Mohsenzadeh, R.; Soudmand, B.H.; Shelesh-Nezhad, K. Failure analysis of POM ternary nanocomposites for gear applications: Experimental and finite element study. *Eng. Fail. Anal.* **2022**, *140*, 106606. [CrossRef]
16. Li, Q.; Xie, L. Analysis and optimization of tooth surface contact stress of gears with tooth profile deviations, meshing errors and lead crowning modifications based on finite element method and Taguchi method. *Metals* **2020**, *10*, 1370. [CrossRef]
17. Chen, K.; Ma, H.; Che, L.; Li, Z.; Wen, B. Comparison of meshing characteristics of helical gears with spalling fault using analytical and finite-element methods. *Mech. Syst. Signal Process.* **2019**, *121*, 279–298. [CrossRef]
18. Concli, F.; Maccioni, L.; Fraccaroli, L.; Bonaiti, L. Early crack propagation in single tooth bending fatigue: Combination of finite element analysis and critical-planes fatigue criteria. *Metals* **2021**, *11*, 1871. [CrossRef]
19. Craig, R.R.; Bampton, M.C.C. Coupling of Substructures for Dynamic Analysis. *AIAA J.* **1968**, *6*, 1313–1319. [CrossRef]
20. Ericson, T.M.; Parker, R.G. Experimental measurement and finite element simulation of elastic-body vibration in planetary gears. *Mech. Mach. Theory* **2021**, *160*, 104264. [CrossRef]
21. Wang, C.; Dong, B.; Parker, R.G. Impact of planet mesh phasing on the vibration of three-dimensional planetary/epicyclic gears. *Mech. Mach. Theory* **2021**, *164*, 104422. [CrossRef]
22. Henry, Z.S. *Bell Helicopter Advanced Rotorcraft Transmission (ART) Program*; Textron Bell Helicopter: Fort Worth, TX, USA, 2013.

exposed to the actual marine atmosphere for 15, 30, 90 and 180 days, respectively. After the outdoor exposure tests, the obtained atmospherically pre-corroded specimens were withdrawn from the Wanning test site and kept in a desiccator (DZF-6020, Hefei Kejing Materials Technology Co., Ltd., Hefei, China). Afterwards, fatigue tests were carried out on the atmospherically pre-corroded specimens that had been exposed for various duration under constant amplitude axial loading on a QBG-50 fatigue testing machine (Changchun Qianbang, Changchun, China) in laboratory air at room temperature to determine their fatigue life. Sine wave load and a maximum cyclic stress (S_{max}) of 240 MPa were applied in all the fatigue tests. Moreover, the stress ratio (R) was set as 0.1 and the loading frequency was about 80 Hz. The fatigue tests were repeated at least five times for each type of marine atmospherically pre-corroded specimen.

In addition, selected as-received (uncorroded) and atmospherically pre-corroded aluminum alloy specimens were cross-sectioned, ground, polished and then etched for microstructural observation using an optical microscope (OM, DMI5000M, Leica, Wetzlar, Germany) and a scanning electron microscope (SEM, Evo18, Carl Zeiss, Germany) equipped with energy disperse spectroscopy (EDS). Selected samples were also cut from the as-received and pre-corroded specimens in order to analyze their surface morphologies and fatigue fracture characteristics using SEM (secondary electron and backscattered electron) and EDS. The thin foils were mechanically thinned to a thickness of 100 μm before being twin-jet electro-polished in a solution consisting of 5% perchloric acid and 95% ethanol for microstructural analysis of the as-received specimens using a transmission electron microscope (TEM, JEM-2100, JEOL Ltd., Tokyo, Japan).

3. Results and Discussion

3.1. Microstructures of As-Received 7075 Aluminum Alloy

Figure 2 displays the OM microstructure of as-received 7075 aluminum alloy. As indicated, the microstructure contains many elongated grains that are parallel to the extrusion direction of the aluminum alloy. In addition, several second-phase constituent particles can be observed within the grain or near the grain boundary.

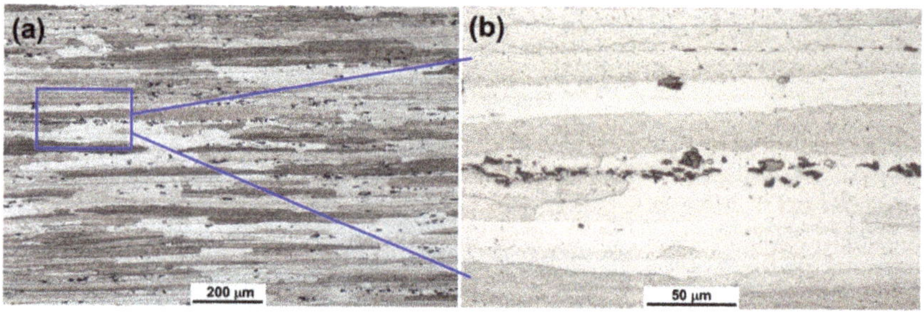

Figure 2. OM microstructures of as-received 7075 aluminum alloy. Image (**b**) shows the magnified microstructure in the frame area of image (**a**).

Figure 3 shows the surface morphology of the as-received 7075 aluminum alloy. It can be seen that the as-received alloy surface is free from corrosion and relatively smooth. Moreover, some second-phase constituent particles can also be observed on the alloy surface. According to previous studies [25–27], the constituent particles in 7075 aluminum alloy might be Fe-rich phase ($Al_{23}CuFe_4$ or Al_7Cu_2Fe). The EDS results in Figure 3b further indicate that the chemical composition of these constituent particles was $Al_{23}CuFe_4$.

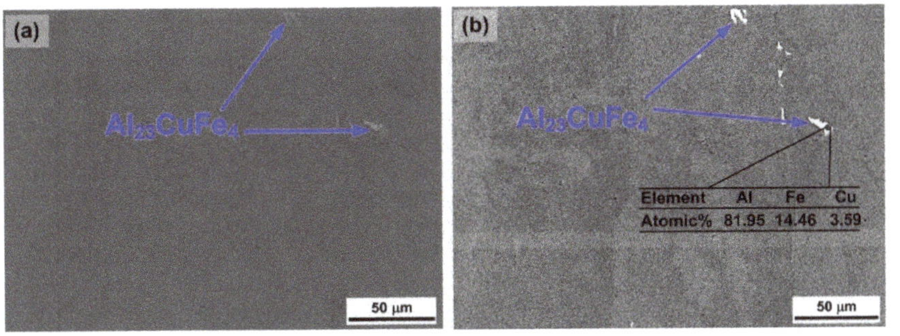

Figure 3. Surface morphology of as-received 7075 aluminum alloy: (**a**) secondary electron image and (**b**) backscattered electron image.

Figure 4 shows the TEM microstructure of the 7075-T73 aluminum alloy. It is clear that a lot of fine second-phase precipitates are distributed homogeneously within the aluminum alloy matrix. In addition, a continuous chain of precipitates can also be observed at the grain boundaries. The electron diffraction pattern (Figure 4b) reveals that the grain boundary precipitates are $MgZn_2$ (η) phase.

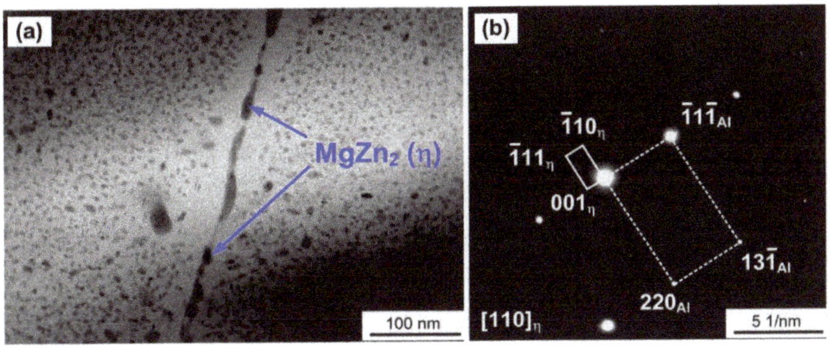

Figure 4. TEM microstructure of the 7075-T73 aluminum alloy: (**a**) bright-field image and (**b**) electron diffraction pattern of grain boundary precipitates.

3.2. Corrosion Characteristics

Figure 5 shows the surface morphologies of 7075 aluminum alloy specimens exposed to marine atmosphere for different numbers of days. After 15 days of outdoor exposure, some pits with various sizes and irregular shape can be observed on the surface of the aluminum alloy specimen, suggesting that localized pitting corrosion has occurred. The atmospheric corrosion products in the pitting corrosion regions can be seen as having a black appearance in the SEM backscattered electron images (Figure 5b,d). The EDS results show that these atmospheric corrosion products mainly contain the elements O and Al. According to the ratio of Al to O, it can be deduced that the corrosion products of the 7075 aluminum alloy specimens suffering from marine atmospheric pre-corrosion mainly consist of $Al(OH)_3$ [23]. Furthermore, as shown in Figure 5c,d, some atmospheric corrosion products might be washed away by rainwater, and the mud-crack characteristic can be observed on the surface of the remaining corrosion products. After 30 days of outdoor exposure, the number and size of the pits increase, and the mud-crack characteristic of corrosion product surface becomes more obvious, as shown in Figure 5e–h. When the outdoor exposure time is increased to 90 days, the severity of the localized atmospheric corrosion on the surface of the aluminum alloy specimens increases. The adjacent pits tend to interact with each other, and the size of the pits further increases, as shown in Figure 5i–l.

After 180 days of outdoor exposure, most of the aluminum alloy specimen surface suffers from atmospheric corrosion. As indicated in Figure 5m–p, the size of the pits increases significantly, which can be attributed to adjacent pits converging into larger pits. Moreover, there are some newly formed small pits on the specimen surface. It is worth noting that, as shown in Figure 5, several corrosion pits are distributed close to the Fe-rich constituent particles ($Al_{23}CuFe_4$). Accordingly, it can be inferred that the occurrence of localized pitting corrosion on the surface of 7075-T73 aluminum alloy specimens exposed to the marine atmosphere is closely associated with the Fe-rich constituent particles, which might have higher electrochemical potential than the surrounding aluminum alloy matrix [25].

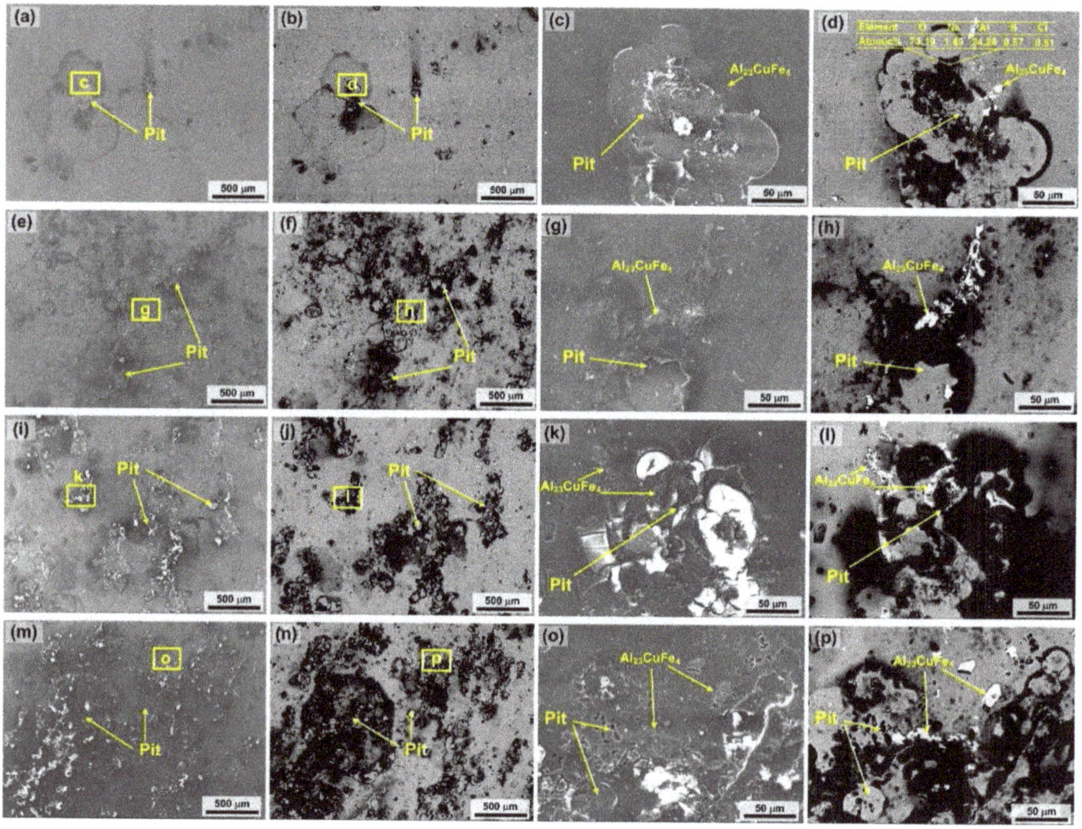

Figure 5. SEM secondary electron and backscattered electron images of the surface morphologies of 7075 aluminum alloy specimens exposed outdoor for different numbers of days: (**a–d**) 15 days; (**e–h**) 30 days; (**i–l**) 90 days; (**m–p**) 180 days.

Figure 6 shows the cross-sectional microstructures of 7075 aluminum alloy specimens subjected to outdoor exposure for different numbers of days. As indicated, there is no obvious corrosion in the as-received specimen (see Figure 6a). After 15 days of outdoor exposure, as shown in Figure 6b,f, localized intergranular corrosion can be seen to have occurred in the aluminum alloy matrix, and lamellar cracks extend along the elongated grain boundary parallel to the alloy surface. The EDS results presented in Figure 6f show that the intergranular corrosion products are also mainly $Al(OH)_3$. After 30 days of outdoor exposure, both the length and depth of the intergranular corrosion have increased (see Figure 6c). With further increases in the duration of outdoor exposure, the degree of severity of the intergranular corrosion progressively increases. When the duration of outdoor exposure is increased to 180 days, the corrosion layers near the surface of the aluminum alloy specimens come apart and are even exfoliated (see Figure 6e), indicating that exfoliation corrosion is occurring.

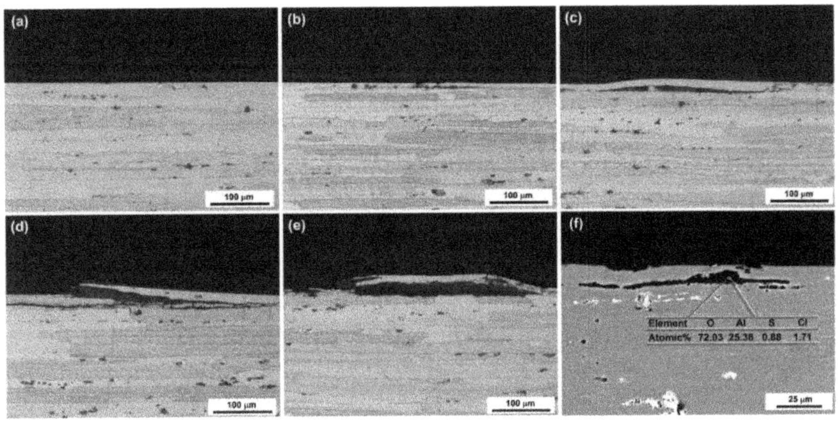

Figure 6. (**a**–**e**) OM and (**f**) backscattered electron images for the cross-sectional microstructures of 7075 aluminum alloy specimens subjected to outdoor exposure for different numbers of days: (**a**) 0 days; (**b**,**f**) 15 days; (**c**) 30 days; (**d**) 90 days; (**e**) 180 days.

As mentioned above, pitting, intergranular, and exfoliation corrosion may occur in the 7075-T73 aluminum alloy specimens exposed to the actual marine atmosphere at the Wanning test site, which has a high humidity and high chlorine ion (Cl^-) concentration. Under the conditions of the marine atmospheric environment, the oxide film on the surface of the 7075 aluminum alloy specimen, especially close to the Fe-rich constituent particles, might be damaged by electrolyte water films containing Cl^-, and then the aluminum alloy matrix is exposed directly to marine atmosphere. Because the electrochemical potential of the Fe-rich constituent particles is different from that of the aluminum alloy matrix [25], electrochemical reactions between them may occur. Compared with the Fe-rich constituent particles, the surrounding aluminum alloy matrix with lower electrochemical potential acts as an anode, and would be dissolved preferentially, leading to the occurrence of pitting corrosion.

In addition, intergranular and exfoliation corrosion of the exposed 7075-T73 aluminum alloy specimens might be related to the precipitates at the grain boundaries. As mentioned above, in the 7075-T73 aluminum alloy, the precipitates at the grain boundaries were identified as $MgZn_2$ phase (see Figure 4). Under the action of corrosive media in the actual marine atmosphere, an electrochemical corrosion system might be formed by the $MgZn_2$ grain boundary precipitates and the aluminum alloy matrix. It has been reported that $MgZn_2$ phase has a lower electrochemical potential than the aluminum alloy matrix [28]. Therefore, in the above electrochemical corrosion system, the $MgZn_2$ precipitates at the grain boundaries act as the anode and are dissolved preferentially, resulting in the occurrence of intergranular corrosion. As the intergranular corrosion proceeds along the elongated grain boundaries parallel to the specimen surface, the corrosion products will accumulate at the grain boundaries. The volume of corrosion products ($Al(OH)_3$) will be larger than that of the consumed aluminum alloy matrix [29,30]. As a result, wedging force might be generated at the elongated grain boundaries, which could cause the rupture and exfoliation of the alloy surface. Accordingly, exfoliation corrosion occurs in the 7075-T73 aluminum alloy specimen subjected to outdoor exposure for 180 days, as shown in Figure 6e.

3.3. Fatigue Performance

Figure 7 shows the average fatigue lives of the 7075-T73 aluminum alloy specimens subjected to outdoor exposure in marine atmosphere for different numbers of days. The maximum cyclic stress (S_{max}) was 240 MPa. As indicated, the average fatigue life (N_f) of the uncorroded aluminum alloy specimens was about 125.16 $\times$ 10^4 cycles. After 15 days of outdoor exposure, the average N_f of the pre-corroded aluminum alloy specimens had significantly decreased to about 16.58 $\times$ 10^4 cycles, indicating that low-level marine atmospheric corrosion damage was able to dramatically reduce the fatigue life of the 7075-T73 aluminum alloy specimens. After the initial sharp degradation, with increasing outdoor exposure duration, the average fatigue life of the aluminum alloy specimens still progressively decreased, but the rate of decrease became much smaller. Thus, the average N_f of pre-corroded aluminum alloy specimens decreased from about 16.58 $\times$ 10^4 cycles to about 6.21 $\times$ 10^4 cycles when number of days of exposure increased from 15 to 180.

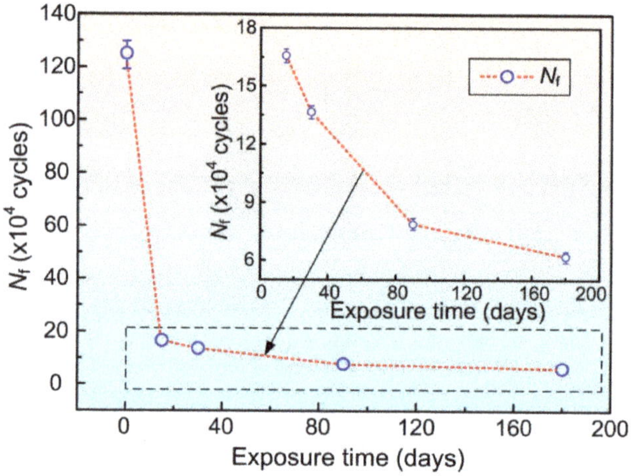

Figure 7. Average fatigue life (N_f) of the 7075 aluminum alloy specimens subjected to outdoor exposure for different numbers of days.

Figure 8 shows the fatigue fracture morphologies of 7075 aluminum alloy specimens subjected to outdoor exposure for different numbers of days. It can be seen that, in the absence of pre-corrosion, the fatigue crack initiated at a Fe-rich constituent particle ($Al_{23}CuFe_4$) near the surface of the aluminum alloy specimen, as shown in Figure 8a–c. The Fe-rich constituent particles are harder than the aluminum alloy matrix and can give rise to stress concentration. Thus, they could act as an initiation site for fatigue cracking. After 15 days of outdoor exposure, the fatigue crack originated from a shallow localized corrosion region near the corner of the pre-corroded aluminum alloy specimen, as shown in Figure 8d–f. When the outdoor exposure duration increased to 180 days, the initiation site of the fatigue crack revealed an irregular deep localized corrosion feature that was also situated near the corner of the alloy specimen, as shown in Figure 8g–i.

As can be seen from the experimental results, marine atmospheric corrosion can penetrate the surface of aluminum alloy specimens and change the initiation site of fatigue cracking. The stress concentration caused by the generation of localized atmospheric corrosion regions during outdoor exposure in the marine atmosphere could significantly accelerate the initiation of fatigue cracks. Therefore, the fatigue life of atmospherically pre-corroded 7075-T73 aluminum alloy specimens with 15 days of exposure was dramatically reduced compared with that of the uncorroded specimens (see Figure 7). With the prolongation of the duration of atmospheric pre-corrosion, the depth of the localized pre-corroded region increased progressively, as shown in Figure 8d–i, which might further increase the severity of stress concentration and thus reduce the initiation time of fatigue cracking. However, the contribution of the increase in the depth of localized pre-corrosion region to the decrease in the fatigue life might be weaker than that of the generation of the localized pre-corrosion region. As a consequence, the fatigue life of the atmospherically pre-corroded 7075-T73 aluminum alloy specimens decreased slowly when the duration of outdoor exposure was increased from 15 days to 180 days.

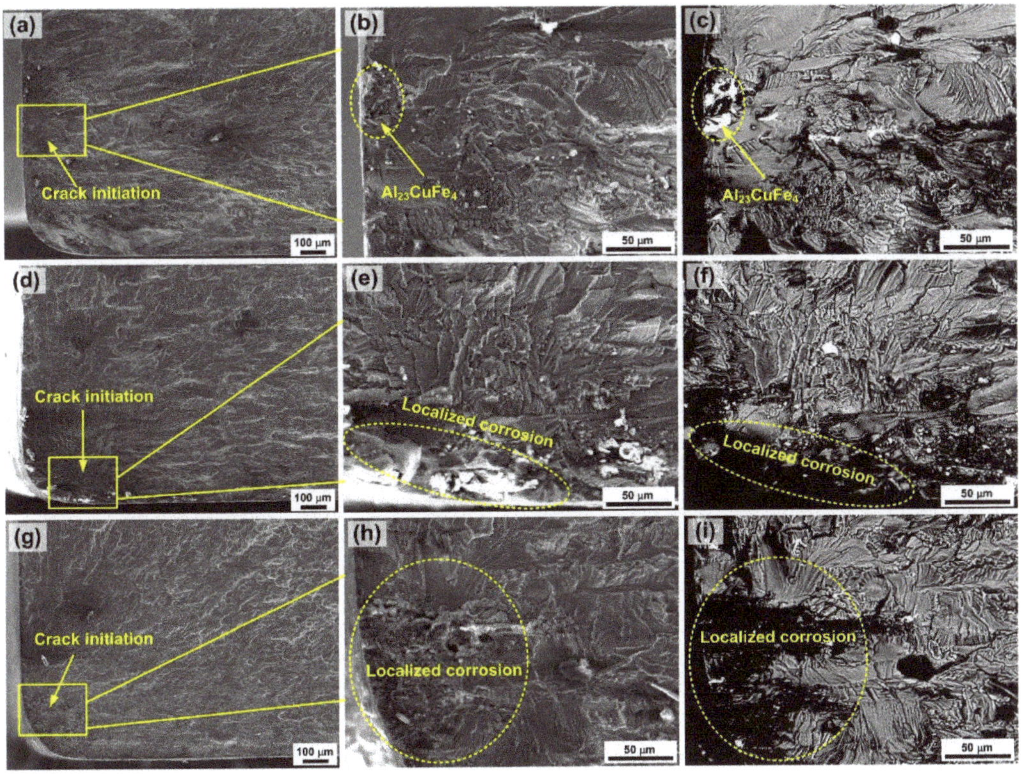

Figure 8. SEM secondary electron and backscattered electron images of fatigue fracture morphologies of 7075-T73 aluminum alloy specimens exposed outdoor for different days: (**a–c**) 0 days; (**d–f**) 15 days; (**g–i**) 180 days.

4. Conclusions

(1) When the 7075-T73 aluminum alloy specimen was subject to outdoor exposure in actual marine atmosphere for a short duration (no more than 15 days), localized pitting and intergranular corrosion occurred near the alloy surface. The degree of severity of marine atmospheric corrosion increased with increasing exposure duration. After 180 days of outdoor exposure, exfoliation corrosion also took place.

(2) The formation of corrosion pits was mainly attributed to the anodic dissolution of the matrix around the Fe-rich constituent particles ($Al_{23}CuFe_4$). The intergranular corrosion and exfoliation corrosion were mainly related to the preferential anodic dissolution of $MgZn_2$ (η) phase precipitates at the elongated grain boundaries of the aluminum alloy.

(3) The fatigue cracking of the as-received (uncorroded) specimens was mainly initiated near the $Al_{23}CuFe_4$ particles, causing stress concentration. With increasing exposure time, the fatigue life (N_f) of the alloy specimens first decreased dramatically and then continued to decrease more slowly. The initial sharp decrease in N_f can be attributed to the generation of a localized pre-corrosion region, inducing more serious stress concentration

and thus significantly accelerating the initiation of fatigue cracking. The following slow decrease of N_f might be related to the increasing depth of the localized pre-corrosion region with the prolongation of exposure time.

Author Contributions: L.S. and L.X. conceived, designed, and performed the experiments; L.S., L.X., J.T. and Q.C. analyzed the data; J.L. and Y.Z. contributed reagents/materials/analysis tools; L.S. wrote the paper. All authors have read and agreed to the published version of the manuscript.

Funding: The work was supported by the Southwest Institute of Technology and Engineering Cooperation Fund (Grant No. HDHDW5902020105) and the Basic and Advanced Research Project of CQ CSTC (Grant No. cstc2018jcyjAX0035).

Institutional Review Board Statement: Not applicable.

Informed Consent Statement: Not applicable.

Data Availability Statement: The data presented in this study are available on request from the corresponding author. The data are not publicly available due to privacy.

Conflicts of Interest: The authors declare no conflict of interest.

References

1. Georgantzia, E.; Gkantou, M.; Kamaris, G.S. Aluminium alloys as structural material: A review of research. *Eng. Struct.* **2021**, *227*, 111372. [CrossRef]
2. Zhou, B.; Liu, B.; Zhang, S.G. The advancement of 7xxx series aluminum alloys for aircraft structures: A review. *Metals* **2021**, *11*, 718. [CrossRef]
3. Dursun, T.; Soutis, C. Recent developments in advanced aircraft aluminium alloys. *Mater. Des.* **2014**, *56*, 862–871. [CrossRef]
4. Li, Y.; Retraint, D.; Gao, P.; Xue, H.; Gao, T.; Sun, Z. Effect of surface mechanical attrition treatment on torsional fatigue properties of a 7075 aluminum alloy. *Metals* **2022**, *12*, 785. [CrossRef]
5. Jung, D.K.; Ha, S.H.; Kim, H.K.; Shin, Y.C. Determination of plastic anisotropy of extruded 7075 aluminum alloy thick plate for simulation of post-extrusion forming. *Metals* **2021**, *11*, 641. [CrossRef]
6. Chen, F.; Qu, H.; Wu, W.; Zheng, J.H.; Qu, S.; Han, Y.; Zheng, K. A physical-based plane stress constitutive model for high strength AA7075 under hot forming conditions. *Metals* **2021**, *11*, 314. [CrossRef]
7. Su, J.X.; Zou, Y.; Chen, K.M.; Wang, Z.Y.; Guan, Q.F. Corrosion mechanism and characteristic of 7075-T6 aluminum alloy panel on airline aircraft. *J. Mech. Eng.* **2013**, *49*, 91–96. [CrossRef]
8. Zhang, R.X.; Zhao, W.D.; Zhang, H.; Yang, W.J.; Wang, G.X.; Dong, Y.L.; Ye, C. Fatigue performance rejuvenation of corroded 7075-T651 aluminum alloy through ultrasonic nanocrystal surface modification. *Int. J. Fatigue* **2021**, *153*, 106463. [CrossRef]
9. Barter, S.A.; Molent, L. Service fatigue cracking in an aircraft bulkhead exposed to a corrosive environment. *Eng. Fail. Anal.* **2013**, *34*, 181–188. [CrossRef]
10. Huang, Y.; Ye, X.; Hu, B.; Chen, L. Equivalent crack size model for pre-corrosion fatigue life prediction of aluminum alloy 7075-T6. *Int. J. Fatigue* **2016**, *88*, 217–226. [CrossRef]
11. Obert, B.; Ngo, K.; Hashemi, J.; Ekwaro-Osire, S.; Sivam, T.P. An investigation of the reduction in tensile strength and fatigue life of pre-corroded 7075-T6 aluminum alloy. *J. Mater. Eng. Perform.* **2000**, *9*, 441–448. [CrossRef]
12. Sankaran, K.K.; Perez, R.; Jata, K.V. Effects of pitting corrosion on the fatigue behavior of aluminum alloy 7075-T6: Modeling and experimental studies. *Mater. Sci. Eng. A* **2001**, *297*, 223–229. [CrossRef]
13. DuQuesnay, D.L.; Underhill, P.R.; Britt, H.J. Fatigue crack growth from corrosion damage in 7075-T6511 aluminium alloy under aircraft loading. *Int. J. Fatigue* **2003**, *25*, 371–377. [CrossRef]
14. Jones, K.; Hoeppner, D.W. Pit-to-crack transition in pre-corroded 7075-T6 aluminum alloy under cyclic loading. *Corros. Sci.* **2005**, *47*, 2185–2198. [CrossRef]
15. Genel, K. The effect of pitting on the bending fatigue performance of high-strength aluminum alloy. *Scr. Mater.* **2007**, *57*, 297–300. [CrossRef]
16. Burns, J.T.; Kim, S.; Gangloff, R.P. Effect of corrosion severity on fatigue evolution in Al-Zn-Mg-Cu. *Corros. Sci.* **2010**, *52*, 498–508. [CrossRef]
17. Burns, J.T.; Larsen, J.M.; Gangloff, R.P. Driving forces for localized corrosion-to-fatigue crack transition in Al-Zn-Mg-Cu. *Fatigue Fract. Eng. Mater. Struct.* **2011**, *34*, 745–773. [CrossRef]
18. Zupanc, U.; Grum, J. Effect of pitting corrosion on fatigue performance of shot-peened aluminium alloy 7075-T651. *J. Mater. Process. Technol.* **2010**, *210*, 1197–1202. [CrossRef]
19. Joshi, G.; Mall, S. Crack Initiation and growth from pre-corroded pits in aluminum 7075-T6 under laboratory air and salt water environments. *J. Mater. Eng. Perform.* **2017**, *26*, 2293–2304. [CrossRef]
20. Mcmurtrey, M.D.; Bae, D.; Burns, J.T. Fracture mechanics modelling of constant and variable amplitude fatigue behaviour of field corroded 7075-T6511 aluminium. *Fatigue Fract. Eng. Mater. Struct.* **2017**, *40*, 1–18. [CrossRef]

21. Song, H.P.; Liu, C.C.; Zhang, H.; Du, J.; Yang, X.D.; Leen, S.B. In-situ SEM study of fatigue micro-crack initiation and propagation behavior in pre-corroded AA7075-T7651. *Int. J. Fatigue* **2020**, *137*, 105655. [CrossRef]
22. Song, H.P.; Liu, C.C.; Zhang, H.; Yang, X.D.; Chen, Y.J.; Leen, S.B. Experimental investigation on damage evolution in pre-corroded aluminum alloy 7075-T7651 under fatigue loading. *Mater. Sci. Eng. A* **2021**, *799*, 140206. [CrossRef]
23. Zhang, S.; Zhang, T.; He, Y.T.; Du, X.; Ma, B.L.; Zhang, T.Y. Long-term atmospheric pre-corrosion fatigue properties of epoxy primer-coated 7075-T6 aluminum alloy structures. *Int. J. Fatigue* **2019**, *129*, 105225. [CrossRef]
24. Rometsch, P.A.; Zhang, Y.; Knight, S. Heat treatment of 7xxx series aluminium alloys—Some recent developments. *Trans. Nonferr. Met. Soc. China* **2014**, *24*, 2003–2017. [CrossRef]
25. Pao, P.S.; Feng, C.R.; Gil, S.J. Corrosion fatigue crack initiation in aluminum alloys 7075 and 7050. *Corrosion* **2000**, *56*, 1022–1031. [CrossRef]
26. Pao, P.S.; Gill, S.J.; Feng, C.R. On fatigue crack initiation from corrosion pits in 7075-T7351 aluminum alloy. *Scr. Mater.* **2000**, *43*, 391–396. [CrossRef]
27. Birbilis, N.; Cavanaugh, M.K.; Buchheit, R.G. Electrochemical behavior and localized corrosion associated with Al_7Cu_2Fe particles in aluminum alloy 7075-T651. *Corros. Sci.* **2006**, *48*, 4202–4215. [CrossRef]
28. Wang, Z. *Handbook of Aluminum Alloy and Its Working*; Central South University Press: Changsha, China, 2000.
29. Mcnaughtan, D.; Worsfold, M.; Robinson, M.J. Corrosion product force measurements in the study of exfoliation and stress corrosion cracking in high strength aluminium alloys. *Corros. Sci.* **2003**, *45*, 2377–2389. [CrossRef]
30. Robinson, M.J.; Jackson, N.C. The influence of grain structure and intergranular corrosion rate on exfoliation and stress corrosion cracking of high strength Al-Cu-Mg alloys. *Corros. Sci.* **1999**, *41*, 1013–1028. [CrossRef]

MDPI
St. Alban-Anlage 66
4052 Basel
Switzerland
www.mdpi.com

Metals Editorial Office
E-mail: metals@mdpi.com
www.mdpi.com/journal/metals

www.ingramcontent.com/pod-product-compliance
Lightning Source LLC
Chambersburg PA
CBHW061503210926
51850CB00008B/1806
9783036598901